In this monumental work, Fokas takes the reader through a mesmerizing journey that starts with neurophysiology, extends to the analysis of human cognitive abilities, and beautifully leads to an explanation of consciousness. Embodying interdisciplinarity, Fokas approaches knowledge and understanding under a unified, all-embracing framework that reveals the beauty of the human condition and the world around us, redefining for modern times the concept of eudemonia of the ancient Greeks. This masterpiece is a most engaging read for engineers, scientists, medical doctors, artists, and humanists alike.

Lydia E. Kavraki
Director of the Ken Kennedy Institute and
Noah Harding Professor of Computer Science
and Bioengineering, Rice University, USA

In this exceptional book, Fokas takes a deep dive into the underlying principles determining the function of the brain and uses them to explain complex human capabilities such as ideas and art. Then, remarkably, he goes further, providing an understanding of consciousness. His work covers a vast wealth of knowledge, including advanced concepts such as retinal implants. As we enter an era of more and more brain machine interfaces, whether through wearable or implantable devices, this book will undoubtedly be a must-read resource as it provides deep insight into the fields of medicine, engineering, and mathematics.

Mark Humayun
Recipient of the National Medal of Technology
and Innovation, Professor at the Keck School of Medicine
of USC and the USC Viterbi School of Engineering, USA

Fokas' truly unique, educational and thought-provoking book provides a unified treatment to sciences, medicine and the arts. This unprecedented approach is indispensable in the modern era of research convergence, where it has become clear that we cannot solve major medical, societal and human problems without combining and integrating knowledge from various fields. This book demonstrates clearly that Fokas, as has been correctly noted, is a scholar in the style of Renaissance.

Nicholas A. Peppas
Professor and Director of the Institute for Biomaterials, Drug Delivery and Regenerative Medicine, The University of Texas at Austin, USA

I am fully convinced that a work of this unprecedented depth and breadth, which manages to speak in a way that is accessible and exciting, is entirely necessary for students of all subjects, and particularly for students of the Arts and Humanities.

Charles Burdett
Professor and Director of the Institute of Languages, Cultures and Societies, School of Advanced Study, University of London, UK

An impressively beautiful book that exposes the reader to the fundamentals of brain functions as the key to understanding the interactions between neuroscience, physics, mathematics, engineering, biology, medicine and arts. For the first time, it is established that a unified knowledge of these disciplines gives rise to a deep appreciation of beauty and leads to eudaimonia. A highly recommended, uniquely brilliant book.

Marinos C. Dalakas
Professor of Neurology and Director of the Neuromuscular Division, Thomas Jefferson University School of Medicine, USA

This is a remarkable book and a true tour de force of creative thinking. It examines, with a sense of wonder, how basic physiological and anatomical principles underlying the human brain can be used to build a world of ideas, including art and consciousness. It may be impossible for us to understand and explain the connection between these two but Fokas has shown us, through his background in mathematics, engineering and medicine, the questions we need to ask ourselves if we are to get close to that elusive goal. In the process, he helps us gain a new appreciation for the beauty inside and around us.

Gislin Dagnelie
Professor of Ophthalmology and Associate Director of the Lions Vision Research and Rehabilitation Centre at the Wilmer Eye Institute,
Johns Hopkins University, USA

Ways of **Comprehending**

The Grand Illusion and the Essence of being Human

Ways of Comprehending

The Grand Illusion and the Essence of being Human

Athanassios Fokas

University of Cambridge, UK & University of Southern California, USA

World Scientific

NEW JERSEY • LONDON • SINGAPORE • BEIJING • SHANGHAI • HONG KONG • TAIPEI • CHENNAI • TOKYO

Published by

World Scientific Publishing Europe Ltd.

57 Shelton Street, Covent Garden, London WC2H 9HE

Head office: 5 Toh Tuck Link, Singapore 596224

USA office: 27 Warren Street, Suite 401-402, Hackensack, NJ 07601

Library of Congress Cataloging-in-Publication Data

Names: Fokas, A. S., 1952– author.

Title: Ways of comprehending : the grand illusion and the essence of being human / Athanassios Fokas, University of Cambridge, UK & University of Southern California, USA.

Description: New Jersey : World Scientific, [2025] | Includes bibliographical references and index.

Identifiers: LCCN 2023039948 | ISBN 9781800615137 (hardcover) | ISBN 9781800615199 (paperback) | ISBN 9781800615144 (ebook) | ISBN 9781800615151 (ebook other)

Subjects: LCSH: Consciousness. | Neurophysiology. | Comprehension. | Comprehension (Theory of knowledge) | Representation (Philosophy)

Classification: LCC QP411 .F65 2025 | DDC 612.8/233--dc23/eng/20231214

LC record available at https://lccn.loc.gov/2023039948

British Library Cataloguing-in-Publication Data

A catalogue record for this book is available from the British Library.

For any available supplementary material, please visit
https://www.worldscientific.com/worldscibooks/10.1142/Q0447#t=suppl

Desk Editors: Logeshwaran Arumugam/Shi Ying Koe

Typeset by Stallion Press
Email: enquiries@stallionpress.com

This book is dedicated to my three children,
Alexander, Anastasia, and Ioanna

ABOUT THE AUTHOR

Thanasis Fokas has a B.S. in Aeronautics from Imperial College (1975), a Ph.D. in Applied Mathematics from California Institute of Technology (1979), and an M.D. from the University of Miami (1986). In 2002 he became the first holder of the inaugural Chair of Nonlinear Mathematical Science at the Department of Applied Mathematics and Theoretical Physics in the University of Cambridge. Since 2015 he has also been an Adjunct Professor of the Viterbi School of Engineering at the University of South California. He is a Member of the Academy of Athens and of the three major European Academies. He is a Fellow of the Guggenheim Foundation, of the American Institute for Medical and Biological Engineering, and of the American Mathematical Society. He was awarded the Naylor Prize of the London Mathematical Society in 2000, the Blaise Pascal Medal of the European Academy of Sciences in 2023, and the Kruskal Award/Lecture of SIAM in 2004. He has introduced the "Fokas Method" and has published in a remarkably broad range of topics in Mathematics, Physics, Engineering, Biology, Medicine, Philosophy, and the Arts. He is the most cited mathematician of all time from the University of Cambridge.

ACKNOWLEDGEMENTS

The final version of this volume has been improved by the input of writer and editor Keith Mansfield, my wife Regina (a linguist at the Faculty of Modern and Medieval Languages and Linguistics, University of Cambridge), and my daughter Ioanna (an undergraduate at the Department of Psychology and Behavioural Sciences, Saint John's College, University of Cambridge). My sincere thanks to all three of them.

I am forever indebted to Konstantinos Tsoukalidis, the most broadly knowledgeable person I have ever met, for his extensive input to this volume and for providing a plethora of useful references. In particular, the section "The Grand Illusion" of the Prologue was decisively affected by our conversations. I am deeply grateful to biologist Nicholas Ktistakis for reading the manuscript and especially for his expert contribution regarding the sections on biology. I am truly thankful to the immunologist and poet Anastasios Germenis for many illuminating discussions. I acknowledge with sincere appreciation my former students Nicholas Protonotarios and Maria Christina van der Weele, as well as my cousin Angelos Antonatos, for their assistance and constant encouragement. I thank Mr Konstantinos Giotis, the founder of the Philosophical Forum Heraclitus, for his generous support.

I have the privilege to have benefited from the expertise of the following scholars: palaeoanthropologist Antonis Bartsiokas (Chapters 6 and 8), neurologist Siddharthan Chandran (Chapter 14), neurologist Marinos Dalakas (Chapters 2, 9, and 10), ophthalmologist Gislin Dagnelie (Chapter 12), neuroscientist Antonio Damasio (Chapter 4), radiologist

Miltiadis Krokidis (Chapters 14 and 15), neuroscientist Nikos Logothetis (Chapter 11), physician Alexios-Fotios Mentis (Prologue, Epilogue, and Chapter 17), painter Yiannis Psychopedis (Chapter 13), and leadership theorist Haridimos Tsoukas (Prologue). Also, ophthalmic surgeon Mark Humayun (regarding Argus II) and physician and neuroscientist Bradford Lowell (regarding homeostasis).

Several of my colleagues in the Academy of Athens have offered graciously their insight: mathematician Anastasios Bountis (Prologue and Chapter 18), archaeologist Michael Cosmopoulos (Prologue and Epilogue), philosopher Alexander Nehamas (the parts about Plato and several other philosophers), physician Haralabos Moutsopoulos (Prologue, Epilogue, and Chapter 17), neuroscientist George Paxinos (Chapter 3), molecular biologist Dimitris Thanos (Chapter 17), and physician and molecular biologist Filippos Tsichlis (Chapter 17).

My understanding of mental functions and mechanisms has been crucially influenced by the writings of the following leading neuroscientists: Antonio Damasio (*Self Comes to Mind: Constructing the Conscious Brain*; *Looking for Spinoza: Joy, Sorrow, and the Feeling Brain*; *The Feeling of What Happens: Body and Emotion in the Making of Consciousness*; *Descartes Error: Emotions, Reason and the Human Brain*; *The Strange Order of Things*). Stanislas Dehaene (*Consciousness and the Brain; How We Learn*) with whose ideas I resonate the most. Gerald Edelman and Giulio Tononi (*Wider than the Sky*; *Second Nature*; *Consciousness, How Matter Becomes Imagination*; *Phi, A Voyage from the Brain to the Soul*). Eric Kandel (*In Search of Memory: The Emergence of the New Science of Mind*; *The Age of Insight: The Quest to Understand the Unconscious in Art, Mind, and Brain, from Vienna 1900 to the Present*; *Reductionism in Art and Brain Science: Bringing the Two Cultures*; *The Disordered Mind*). Also, Christof Koch (*Consciousness: Confessions of a Romantic Reductionist*), Benjamin Libet (*Mind Time*), V. S. Ramachandran (*The Tell-Tale Brain*), and Dick Swaab (*We Are Our Brains*).

The following excellent books had a significant impact on writing this volume: W. Brian Arthur's *The Nature of Technology*, Roger Bartra's *Anthropology of the Brain*, Jonathan Balcombe's *What a Fish Knows*,

Daniel Bor's *The Ravenous Brain*, Daniel M. Davis' *The Beautiful Cure*, R. Douglas Fields' *The Other Brain*, Lynn Gamwell's *Mathematics and Art* and *Exploring the Invisible*, Sam Kean's *The Tale of the Dueling Neurosurgeons*, Margaret Livingstone's *Vision and Art, The Biology of Seeing*, Iain McGilchrist's *The Master and Its Emissary*, Gary Marcus' and Jeremy Freeman's (Eds.) *The Future of the Brain*, Steven Mithen's *The Singing Neanderthalis*, Arthur Miller's *Einstein and Picasso*, Alexander Nehamas' *Virtues of Authenticity*, Adrian Owen's *Into the Grey Zone*, David Quammen's *The Tangled Tree*, David Rothenberg's, *Survival of the Beautiful*, George Steiner's *The Poetry of Thought*, Max Tegmark's *Life 3.0*, Eric Topol's *Deep Medicine*, Matthew Walker's *Why We Sleep*, and James K. Wright's *Schoenberg, Wittgenstein and the Vienna Circle.*

The parts of the volume regarding painting were strongly influenced by the following outstanding books: *Abstract Expressionism*, edited by David Anfam; *Cezanne and the Modern,* Ashmolean Museum of Art and Archeology; *Rembrandt, The Late Works*, J. Bikker, G. J. M. Weber, M. E. Wieseman, and E. Hinterding; Gemma Blackshaw's *Facing the Modern: The Portrait in Vienna 1900*; Xavier Bray's *Goya: The Portraits*; Giorgio Bonsanti's *Caravaggio*; *Kandinsky, the Path to Abstraction*, edited by Hartwig Fischer and Sean Rainbird; E. H. Gombrich's *The History of Art*; *Gauguin's Portraits*; Paul Moorhouse's *Giacometti: Pure Presence*; *From Russia*, Royal Academy of Arts; *Egon Schiele the Radical Nude*, edited by Peter Vergo and Barnaby Wright.

CONTENTS

PROLOGUE

The search for understanding gives rise to deep admiration for the immense wisdom and beauty of Nature, and in particular for its greatest achievement, the human brain. Writing this book, I felt a deep sense of gratitude for the privilege of being able to enjoy a plethora of complex and multifaceted creations of Nature and humanity. I hope, and expect, that those who study this volume will experience similar feelings.

Unification and *analogical thinking* are central themes of this book. In this connection, it is worth noting that the completion of the formalism that unifies the four fundamental forces of nature, namely, the gravitational, electromagnetic, weak, and strong interactions, still stands as the holy grail of Physics. By analogy, it is natural to attempt to integrate the biological and cultural "forces" shaping life. In this volume, an effort is made to explore this unification.

By analyzing fundamental neuronal mechanisms, it will become clear that the human brain is predisposed to seek *knowledge and beauty*, without the artificial distinction between sciences and humanities. It is argued that such a grand quest requires an interdisciplinary approach. The necessity for such an integrative approach follows from the insight that *everything is related to everything else*. Perhaps no one expressed this fact more eloquently than Leonardo da Vinci, the embodiment of interdisciplinarity:

"Study the science of art. Study the art of science. Develop your senses — especially learn how to see. Realize that everything connects with everything else".

A crucial part of an interdisciplinary approach to knowledge and culture is the appreciation that life generously provides many sources of pleasure and satisfaction, beyond the utilization, efficiency, power, and beauty of technological creations. Indeed, in life there also exists that which, according to Ludwig Wittgenstein, "cannot be said". It is the part, which is not only non-verbal but also more generally non-algorithmic. It is *spiritual*, where spirituality refers to the accessibility of emotions and other creations of the *unconscious*. It is the *transcendental* part that lives in the exceedingly rich and mysterious world dominated by unconscious processes. I believe that life is incomplete without exploring this hidden world of beauty and potential *eudemonia*.

A PARADOX AND ITS RESOLUTION

Our times are characterized by an apparent paradox: on the one hand, there exists a vast amount of information and resources, and therefore the potential for broad knowledge, understanding, and personal fulfilment. On the other hand, there is a strong tendency for specialization and limited appreciation of the enormously rich scientific and cultural achievements of humanity. Moreover, the continuous bombardment of the brain with discrete, disconnected pieces of information brings confusion and a sense of alienation. The feeling of living in a world that becomes more and more difficult to comprehend. If we do not understand the world we live in, how can we evaluate our contribution?

This paradox brings to mind Erwin Schrödinger's passionate appeal to fulfil the human "longing for unified, all-embracing knowledge". This statement made in his highly influential 1944 book *What Is Life?* (Schrödinger, 1944), was followed by the Nobel Laureate's expression of regret that "[…] it has become next to impossible for a single mind fully to command more than a small specialized portion" of the existing vast amount of knowledge. According to Schrödinger, the only way out of this formidable difficulty is

"that some of us should venture to embark on a synthesis of facts and theories, albeit with second-hand and incomplete knowledge of some of them".

I fully embrace the significance of this appeal. Recognizing that this problem is indeed insurmountable, especially since general knowledge has grown immensely in the last 50 years, I will attempt a less ambitious project. My aim is to present a *framework capable of approaching knowledge and understanding in the unified, all-embracing manner* envisioned by Schrödinger. This approach has become possible as a result of the remarkable progress achieved in the last few decades regarding the *functioning of the brain*. For example, it will be shown that there now exist appropriate tools for elucidating the mechanisms responsible for the human "quest for unification", noted by Schrödinger. My attempt to delineate the above approach has been motivated by a broad education in engineering, mathematics, physics, and medicine, as well as my exposure to a wide range of areas through research activity and published work.[1]

THE GRAND ILLUSION

Answering the question "What is consciousness?", is considered one of the most important open problems in the history of sciences. Many deep scholars have written extensively on the subject, approaching it from a variety of angles including neuroscience, artificial intelligence, philosophy, mathematics, and physics. It has even motivated works of literature, for example Stoppard (2015). As discussed in the final chapter of this volume, I believe that this problem *can* be solved, provided that it is correctly defined and is placed in a proper perspective. In my opinion, this requires appreciating that *unconscious and conscious processes form a continuum*, as well as postulating *the primacy of the unconscious*. In this connection, motivated by the analysis of a plethora of neurophysiological studies, including the detailed scrutiny of visual perception presented in Chapter 4, I postulate that,

First hypothesis: *Every conscious experience is preceded by an unconscious process.*

---

[1]From differential equations and the asymptotic analysis of the Riemann zeta function to symmetries and geometry; from the general theory of relativity to particle physics; from protein folding and chronic myelogenous leukaemia to mathematical models for *C. elegans* and for the epidemiology of COVID-19; from medical imaging and "deep learning" to philosophy and the quantification of fractality in paintings of Piet Mondrian.

Let me clarify the critical importance of unconscious processes by using a simple example. My wife, Regina, enters my study in our house in Cambridge. As a result of intricate processes that are *entirely* unconscious, *my brain* perceives Regina. I will refer to the *unconscious construction* of this percept as the *unconscious structure* of Regina. As discussed in detail in this book, about a third of a second later, *my brain informs me of what it already knows.* Namely, my unconscious informs my consciousness of the presence of my wife. At this moment, the miracle of *awareness* takes place: *I* perceive Regina. I will refer to the *conscious construction* of this percept the *mental image* of Regina. As soon as unconscious processes begin to construct the unconscious structure of my wife, these processes simultaneously form a myriad of *associations* related to her. As a result of these associations, my unconscious decides that I should greet Regina with the statement "good morning my love", and about a third of a second later it instructs my consciousness to implement this decision. My wife's unconscious perceives my greeting, then it informs her consciousness, etc.

Mental images, which are three-dimensional holograms, are continuously updated, but they are always 0.3–0.5 of a second *behind* reality. Consequently, the interaction with the external environment in real time is, paradoxically, the task of the unconscious and *not* of consciousness. Indeed, it is intricate *unconscious* processes that allow us to walk, to drive without crashing, to hit or catch a ball, to perceive glimpses of colours, etc. For example, a trained athlete, following the unconscious instructions of their brain begins to race after 0.1–0.15 of a second following the sound of the starting pistol. The athlete erroneously believes that they began running as a result of becoming aware of the sound. Similarly, we are under the illusion that our *conscious self* is in charge of our communications, actions, feelings, etc. I refer to this remarkable, but largely unappreciated fact, as the "grand illusion". Following the groundbreaking work of the neurophysiologist Benjamin Libet (which will be discussed in the next volume of this tetralogy), this illusion has been discussed mainly with regard to decision making (Is there free will?) (Wegner, 2017). However, as evident from the above example, the fact that awareness *always* lags a considerable amount of time behind unconscious processes, underlines *every* mental function of *any* organism possessing consciousness. In this sense, the label "grand illusion" is, hopefully, well justified.

WHAT IS THE UNIQUE FEATURE OF HUMANS?

In contrast to our evolutionary predecessors, we have the privilege of possessing language. Undoubtedly, this enriches enormously our capacity to communicate. For example, instead of the statement "good morning my love", animals must use a variety of indirect ways to communicate their emotions. As a result of the transformative impact of language, many scholars have highlighted this great gift as the *key* difference between us and other creatures possessing consciousness. In my opinion, this is *not* entirely correct. It is argued in this volume that our qualitative advantage in comparison to other animals is that,

Second hypothesis: *We possess a predisposition to construct real versions of our mental images and our unconscious structures, or to assign to them specific symbols.* I will use the term *metarepresentations* for the emerging constructions or symbols.

In addition to language, the metarepresentations of mathematics, computations, technology, and arts, are of crucial importance for the development of our culture. Regarding the arts, as noted in Chapter 7, many great artists have explicitly stated that their creations often begin in their unconscious. This is consistent with my assertion that the origin of metarepresentations is not only mental images but also unconscious structures. The above relationships between unconscious structures, mental images, and metarepresentations are indicated in the following figure.

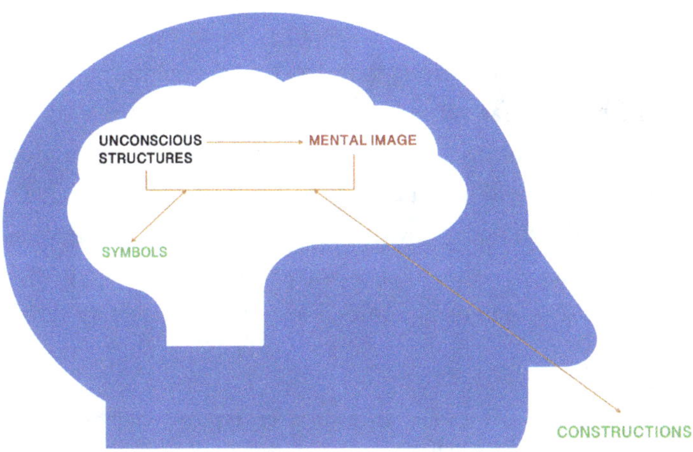

In this figure, the arrow from unconscious structures to mental images indicates that the former precedes the latter. This is consistent with the first basic hypothesis that every conscious experience is preceded by an unconscious process. Both unconscious structures and mental images give rise to symbols, a process occurring inside the brain, as well as to concrete constructions, which take place outside the brain.

As a mathematician, I have greatly benefited from the unique capacity of the human brain to create metarepresentations. Indeed, the thought process of a mathematician often compels them to *write* specific mathematical formulas; for example, $1 \times 2 = 2$. As soon as a formula is written, a dynamic interaction begins between a specific metarepresentation and related unconscious processes. This hugely generative process finally expresses itself with another formula; for example, $1 \times 2 \times 3 = 2 \times 3 = 6$. Generalizing this particular case, it becomes evident that,

the continuous and multilevel interaction between a plethora of metarepresentations and unconscious processes is the essence of the astounding human creativity.[2]

Our interaction with the environment

The first "big bang" in the biological evolution was the emergence of consciousness, or more precisely, of the defining property of consciousness, which (as Libet pointed out) is awareness. In my opinion, the second "big bang" was the emergence of metarepresentations. This is unique to humans; in fact, it constitutes the birth of humanity. As a result of the unlimited richness of metarepresentations, the interaction of humans with their environment became far more complex than that of their evolutionary predecessors. To emphasize the highly dynamic and ever-changing nature of this interaction, I will use the term *heterodynamics*. This also conveys the qualitative difference between this situation and the quasi-static state that characterizes the internal environment of living organisms, called *homeostasis*. Heterodynamics, which is decisively affected by

[2]Sentences which summarize earlier discussions or express significant positions are presented in brown colour for emphasis.

metarepresentations, dictates our cultural evolution, which has become far more important than biological evolution. In this regard, it is noted that currently the most pressing problem facing humanity is the protection of our environment. Possible approaches to this problem, whose solution is crucial for the survival of the human race, will be discussed in the fourth volume.

How do we comprehend?

The first hypothesis that unconscious processes are a necessary condition for the creation of a conscious experience, implies that *anything* we do, think, and feel, is crucially affected by unconscious processes of which we are mostly unaware. The new approach introduced in this book is based on the following triptych:

First, *the crucial importance of the "echo" created by unconscious processes.*

Second, *the impact of the fundamental notion of the metarepresentations.*

Third, the implications of the self-evident, but ignored by many scholars, position that *gaining deep insight necessitates the need to decipher fundamental biological and especially neuronal mechanisms.*

This position implies that it is necessary to identify and place special emphasis on *those notions that reflect fundamental neurophysiological processes.* It is shown in this volume that such *indispensable notions for the comprehension of our thoughts, feeling, and actions, and more generally for the search for the hidden reality*, are the following:

The notions of *continuity*, *associations*, *interconnectedness*, *analogical thinking*, *abstraction*, *generalization*, as well as the dialectic pairs of *reduction* versus *unification*, simplicity versus *complexity*, and *local* versus *global processes.*

For example, continuity was already mentioned regarding the relationship between unconscious and conscious processes. Regarding the second notion, it was noted earlier that associations dictated my unconscious decision on how to greet my wife. The relationship of the other notions with specific biological and neuronal mechanisms will be illustrated by their use in elucidating aspects of mathematics, physics, biology, neuroscience, medicine, technology, philosophy, and painting. In particular, mathematics and physics provide examples *par excellence*, respectively, of generalization and unification. Painting provides a plethora of examples of both, reduction, as in paintings by Mondrian, Malevich, and Rothko, as well as of the dialectic pair of simplicity versus complexity, as in the paintings of Kandinsky, Picasso, and Braque. Detailed analysis of some of these paintings will be presented in the third volume of this tetralogy.

Concrete outcomes of the new approach

The integrative approach to knowledge and culture based on this proposed framework has several concrete outcomes, some of which are discussed below.

It suggests a way to resolve the paradox between the abundance of information and specialization. In this regard, it is prudent to be aware of the confusion and misinformation generated by the current availability of massive amounts of data. This suggests that "ignorance" is promoted from two opposite extremes: from too few or from too many data. The relationship between information, knowledge, and wisdom was succinctly expressed by T. S. Eliot: "Where is the wisdom we have lost in knowledge? Where is the knowledge we have lost in information?" This volume presents an approach on how to acquire wide and interrelated knowledge. In particular, it provides *a framework for the interested reader to transform some of the enormous amount of available data and dispersed information into knowledge*. Using the Internet and a variety of other sources, the material presented in this volume can provide both motivation and guidelines for acquiring additional information regarding a plethora of important areas. Incidentally, I envision a synergistic relationship between this volume and the web: the Internet can immediately clarify the meaning of those concepts that I have not defined, erroneously presuming they are known. In addition, the web can be a source of supplementary information for many concrete entities and concepts presented here. For example, one may easily access the history of the paintings analyzed in this book.

The emphasis on an *all-encompassing approach to sciences and humanities* will, hopefully, help restore the arts and letters to their rightful position at the centre of what it means to be human. Our innate capacity for *computability* gives rise to the enormous computational capabilities of mathematics and the stunning achievements of artificial intelligence (and especially of "deep learning"). As a result of these developments, there is nowadays the illusion that algorithms are omnipotent. This illusion has regrettably led to many scientists questioning the importance of arts and letters. Moreover, the arrogant claim that human thought will soon be surpassed by artificial intelligence has led to a distortion of our view of the essence of humanity. However,

> mathematics reflects only a limited subset of human thought, and artificial algorithms mimic only a small part of brain processes.

Indeed, artificial intelligence not only ignores unconscious processes but also the vital role of the glial cells, the differences in the functions of the two brain's hemispheres, and the crucial fact that the mind is *embodied.* Furthermore, the important *intrinsic limitations* of both mathematics and computing are not given serious consideration. In this regard, it is noted that, in a similar manner to how Ludwig Wittgenstein exposed the limitations of language, the brilliant mathematicians, Kurt Gödel and Alan Turing, rigorously established that *neither mathematics nor computer science can reach the truth via a formal, axiomatic process* (AI is extensively discussed in the second volume). In contrast to the current apotheosis of the rational and the algorithmic, the unified, balanced approach advocated in this book suggests that,

> a basic characteristic of creative ideas and advanced artistic representations is that they are non-verbal and non-algorithmic; they are transcendental. These mental creations are generated via the interaction of unconscious processes and metarepresentations in the dynamic environment of the embodied brain, which is crucially affected by hormones and other molecules excreted by the body-proper.

Hence, I am highly sceptical as to whether such processes can be "programmed". Overall, the serious limitations of rationality, together with the pivotal role of the unconscious in arts and letters (that makes these creations even "less programmable"), imply that in the 21st century, the humanities will come to define what it means to be human *more poignantly than ever before.*

The material presented in this volume demonstrates that *a unified, integrative approach* to knowledge *is* indeed possible. For this purpose, an effort is made to refute the myth that it is supposedly impossible to be both deep and broad. Actually, breadth and depth are not antithetical; they act synergistically. Indeed,

> the more areas one is exposed to, the more extensive becomes the web of possible associations among elements and concepts of these diverse areas. Hence, the deeper the insight gained, the higher the appreciation for the value of arts and letters, and the more likely the attainment of difficult goals.

Diverse, positive experiences are accompanied by happiness, or more precisely by *eudemonia*, which is the state of elation and personal fulfilment achieved via pursuing *knowledge* and *beauty* and *attaining lofty goals.* Incidentally, the word *eudemonia* is of Greek origin; it derives from the prefix *eu*, meaning good, and *daemon*, meaning spirit. This concept was introduced by the ancient Greeks (in particular, it is discussed in Aristotle's monumental work *Nicomachean Ethics*) and was later elaborated by several scholars. This concept is much broader than *hedonia*, that derives from the Greek word *hedone*, meaning pleasure (hedonism, as envisioned by Aristippus, advocates maximizing pleasure). The joy and fulfilment that naturally accompanies the process of approaching the "essence of things" was perfectly expressed by Albert Einstein. Recalling the moment when he conceived the basic idea behind his General Theory of Relativity, Einstein wrote that "this was the happiest idea of my life". Perhaps, only the unique genius of Einstein could express so eloquently the sequel of any great achievement. He did not characterize his brilliant idea as the most profound or the most original, but simply as the happiest! Many deep thinkers have noted that understanding brings joy. For example, the philosopher Daniel Dennett, states that "I find comprehension to be one of life's greatest thrills" (Dennett, 2017). The feeling of *eudemonia* associated with knowledge is consistent with the Aristotelian understanding that "All men by nature desire knowledge".

Regarding beauty, it must be emphasized that aesthetic pleasure is not only found in arts and letters but also in many other endeavours including mathematics, science, and technology. I am often mesmerized by music, painting, and poetry; and also, by the beauty of several mathematical works. The high aesthetic value of the mathematical equations that express physical laws is perhaps a reflection of the beauty of the corresponding physical reality. Interestingly, there exist common neuronal mechanisms responsible for appreciating different aesthetic forms. For example, a study using functional MRI (magnetic resonance imaging) has shown that when mathematicians are exposed to musical or visual beauty there is activation in the *same* part of their brain as the part that is activated when they are exposed to a beautiful mathematical equation.[3] The relationship between beauty and mathematics is further discussed in Chapter 18.

The approach introduced in this volume provides the proper framework for an illuminating discussion of several important questions which, in my opinion, should concern every educated individual. They include the following: What is the origin of the distinguishing mental advantages of humans in comparison to our evolutionary predecessors? What is the relationship between innate and acquired knowledge? What does it mean to "understand" and how is insight achieved? Why is it possible for us to comprehend the universe? What is the effect of the cultural evolution on our brains? What is the neuronal origin of our emotional responses to arts and letters? Could the unbalanced emphasis on science and technology at the expense of arts and humanities "end up downgrading humans" as Yuval Noah Harari worries in his *Homo Deus* (Harari, 2016)? Can the problem of consciousness be solved? Why is beauty important in those mathematical expressions that capture basic physical phenomena? Can the impact of mathematics in biology be as essential as it has been in physics? The significance of the last question becomes evident by pointing to the claims of several leading neuroscientists that "mathematics will be crucial for solving the problem of consciousness".

Studying this volume will allow the reader to become familiar with many facts that, because of their significance, should be widely known. For example, there will be a discussion of important neuronal mechanisms

---

[3] In this study there was activation in the medial-orbitofrontal cortex. Among the equations considered beautiful were the one expressing Pythagoras' theorem, as well as Euler's equation $e^{i\pi} = -1$ (Zeki *et al.*, 2014).

related to unconscious perception and awareness. In addition, it will be shown that the transition from the unconscious to awareness is not only relatively slow but, more importantly, is accompanied by loss of information. For example, in the so-called *binocular rivalry*, analyzed in detail in Chapter 11, two *different* images are shown in the left and right visual fields. *Both* images are perceived unconsciously, but the examined individuals became aware of only *one* of them. This clearly shows that *our brains "know" much more than we do.*

It is important to note that the exposure to different areas of science and humanities is expected to train the brains of young people to *adopt a flexible, multidisciplinary way of thinking.* In my opinion, this provides the best preparation for modern life, where lines between disciplines have become blurred. Most importantly, it will make available to the reader a *new methodology* that can facilitate their analysis of a variety of phenomena. In particular, it will allow them to begin comprehending the origin of their thoughts, feelings, and actions. I hope that researchers with various areas of expertise, after studying this volume, will be motivated to revisit many disciplines including philosophy, literature, and social sciences, within the proposed methodology. For example, it will be interesting to elucidate how the extensive *network of associations* constructed by the brain motivates the creation of *social networks.* In this regard it is noted that social support, social relationships, and friendship, are not only vital components of happiness but also promote health and affect longevity. People with satisfactory social relationships improve substantially their chance of survival in comparison to those with poor ones (Holt-Lunstad *et al.*, 2010).

This volume will expose the reader to elements of biology, neuroscience, medicine, mathematics, and physics in a clear and comprehensible manner. In addition, the reader will become familiar with the stunning recent developments in brain imaging, which allow observation of specific functions of the brain in real time. This achieves a double goal: on the one hand, these elements provide *illustrations of the novel methodological approach to knowledge advocated in this volume.* At the same time, they offer *a global vista* of these important disciplines and developments. In particular, several chapters present a thorough introduction to neuroscience, from single neurons to various neuronal mechanisms crucial for perception, memory, and learning. In these chapters a number of important neurological diseases are discussed, and current treatments are noted. In Chapter 17, the impact on medicine of mathematics, computer sciences, physics, chemistry, bacteriology, pharmacology, and molecular biology is discussed. This clearly establishes the interdisciplinary nature of medicine.

The deeper one explores different realms of human endeavours by employing tools elaborated in this volume, the more relationships one discovers. For example, science, with its emphasis on the rational and religion with its reliance on metaphysics, may appear completely antithetical. However, it turns out that, at a deeper level, science and religion, paradoxically, share some common ground. Indeed, taking into consideration the limitations of the rational that were mentioned earlier, it follows that *the highest aspiration of science coincides with a fundamental goal of religion: it aims to go beyond the rational and to reach the transcendental.* In addition, it is well known that a key element of religion is the belief in a supernatural world. The creation, in the "soul" of a religious individual, of this invincible, undefined, non-material world, *has its origin in similar neurological mechanisms* that give rise, in the brain of a mature scientist, of the remote "echoes" reflecting deep, esoteric, unanswered questions of science. These mechanisms are unknown, at the moment, processes that take place in the world of the unconscious. This fundamental characteristic of science was certainly recognized by Einstein, who wrote: "The most beautiful thing we can experience is the mysterious. It is the source of all true art and science". These remarks suggest that polemical conflicts between science and religion can be avoided. Interestingly, Father and polymath Marin Mersenne (1588–1648), considered "the cause of science as the cause of God". Incidentally, Mersenne was a friend of the great scientists Galileo Galilei, René Descartes, and Étienne Pascal; he was also in contact with Pierre de Fermat and tried to find a formula that would represent all prime numbers.[4]

---

[4] Mersenne discovered the "law of a stretched string", namely, the formula expressing the frequency of the acoustic waves generated by a stretched string. After his death, his correspondence with Galileo, Fermat, Constantijn Huygens, Evangelista Torricelli, and other scientists, was found in his cell (Tononi, 2012).

It is remarkable that despite the enormous importance of unconscious processes, these processes remain largely undervalued. This is also reflected by the fact that Psychiatry, which is the area of medicine mostly concerned with unconscious mechanisms, is underestimated and underfunded. Psychiatry comes from the Greek words, *psyche* and *iatreia*, which mean *soul* and *healing*, respectively. The relation of psychiatry with the unconscious is expressed by the observation that perhaps the best interpretation of the notion of the *soul* is the *unconscious*. Considering the importance and prevalence of psychiatric illnesses, and especially of major depression, bipolar disease, and schizophrenia, this lack of appreciation of the significance of psychiatry is utterly unfortunate. Hopefully, the elucidation of several aspects of the unconscious presented in this volume will help towards elevating psychiatry to the level it deserves.

According to Aristotle, happiness is "the end to which our actions are directed". In my opinion, a crucial component of happiness is *diversity of positive experiences*. Concrete suggestions that may assist young people to achieve such diversity are presented in the next subsection.

In summary, this volume dares to suggest a unifying approach to life that is based on the elucidation of deep neuronal mechanisms, which facilitate understanding and fulfilment.

Every thought and every activity give rise to a multitude of associations and hence, if appropriately manipulated, to the potential for generating feelings of eudemonia.

ADVICE FOR THE YOUNG

The well-being of people in several parts of the world is continuously improving. Thus,

we may be approaching a point of an "eudemonic transition", namely, a crucial juncture where *eudemonia* can become a goal for the majority of society.

In this sense, young people may be characterized as "the lucky generation". However, a juncture is a *bifurcation point*, which is a point where the stability of the system may change, with possible disastrous consequences. For this reason, it is imperative that mature scholars assist young people to achieve their full potential so that ultimately they become part of something larger than themselves. This is the only way to ensure that humanity will emerge victorious. In this regard, it is noted that our huge moral obligation towards young people cannot be overestimated: our children are the result of *our* genes and the environment that *we* have created, so they are completely *our* responsibility. Hence, I find it absurd that throughout history the older generation has tended to be unjustifiably critical towards the next generation. For example, the following quote sounds contemporary:

"Young people today love luxury, they have bad manners, contempt for authority, and also, disrespect older people [...]. They no longer rise when elders enter the room, they contradict their parents, can't hold their tongues in company, gobble their food, and tyrannize their teachers".

However, it is attributed to Socrates.

While writing this book, the thought of how to help young people was constantly in my mind. Actually, one of the main reasons for starting this endeavour was the hope that the new approach to comprehending presented in this volume would contribute towards the happiness and personal fulfilment of the young people.

Although the road to growth and self-fulfilment is undoubtedly highly personal, in what follows some general principles are suggested.

Be overly optimistic: It is a moral obligation to enjoy the greatest gift of all, the gift of life. Optimism contributes to the development of resilience to setbacks. Thus, it protects against the inertia created as a result of the fear of failure. The medical community has elucidated the processes via which substance abuse causes morbidity and mortality and also has successfully promoted the understanding that healthy eating and exercising are associated with fewer diseases and longevity. However, surprisingly, the huge importance of optimism for longevity has been essentially ignored, despite the existence of an overwhelming evidence for this fact. For example, in a relevant study, patients were psychologically evaluated to determine whether they were optimists or pessimists, and subsequently underwent angioplasty. Over a 6-month period following a successful angioplasty, pessimists were 3 times more likely than optimists to have heart attacks or to require bypass surgery. Similarly, men of an average age of 61 were monitored over a period of 10 years by scientists at Harvard and Boston Universities. The most pessimistic were twice as likely to develop heart disease. In another study, 941 Dutch men and women between the ages of 65 and 85 were followed for a 9-year period. Remarkably, people who demonstrated dispositional optimism enjoyed a 45% lower risk of death (Optimism and Your Health, 2008)!

Protect your sleep: The paramount importance of sleep for a variety of mental functions including emotional stability, learning, and memory was, until very recently, also ignored by the medical community. Fortunately, this omission has now begun to be addressed. For example, the critical importance of sleep is promoted by the influential writer Arianna Huffington (Huffington, 2016) and by the leading sleep-investigator Matthew Walker (Walker, 2017). Sleep may be necessary for transferring newly learned information into long-term memories and for memories to be freely associated. During the latter process, unexpected relationships are formed, unconscious insight is gained, new solutions are dreamt, and mysterious mechanisms enhancing creativity are facilitated. Some of my best ideas have arisen during the small time-period of waking up from the rapid eye movement (REM) state, where most of dreams occur. This is also the period when I have discovered errors in my research work. This time-period provides a window to the vast information that exists in our unconscious. This window is tiny and blurry, but it still provides an almost unique possibility for accessing this information. This observation further underlines the huge importance of sleep.[5]

Be flexible and open minded: As the mathematician and French polymath Henri Poincaré had noted, "doubting everything or believing everything are the two extreme ways of dispensing with the crucial need for reflection". Flexibility is a great asset. For example, it is needed for current

[5] There is indirect evidence that the blue LED light that powers screens, if absorbed in the evening, may suppress the normal surge of melatonin. In any case the use of such devices "excites" the brain, so it delays sleep. The advice to refrain from using devices before going to sleep should be followed, or, at least, protective glasses should be worn.

approaches employed in digital technology. The modern *motto* is not to employ standard algorithms, but to optimize technology according to the specific task at hand. If, during a discussion or in some other context you encounter views that are antithetical to yours, attempt to extract from them something useful that will enrich your own ideas (instead of the typical reaction of disagreeing and criticizing). After all, there is no such thing as absolute truth. The more open to new ideas one is, the more enjoyable and fruitful will be their never-ending quest for understanding.[6]

Do not waste emotional energy: There are only a few situations, namely those that can be characterized as *irreversible* (with death being chief among them), that justify one to be dispirited. Positive affect increases both dopamine and serotonin production, which enhances happiness (Trivers, 2013: p. 131). Also, it has been shown that positive emotions enhance the vagal tone, which results in several health benefits (Kok, 2013).

Seek diversity and be mentally resilient: Failures are not only inevitable but also a measure of success. For example, the more mature and ambitious a scientist becomes, the more difficult the problems they attempt, and hence the higher the probability of failure. A key to success is the ability to cope with failures.

Be self-confident and do not succumb to peer pressure: Self-confidence fosters *authenticity*, which is an integral part of nobility and excellence of character.

Do not be afraid to ask questions and do not ignore paradoxes: There is no better path to insight than posing appropriate questions to yourself and others. Paradoxes and incongruities express a lack of understanding; thus, they provide great opportunities for further progress.

Respect your time: Do not allow time spent on idle activities, as well as on the Internet and social media, to reduce the hugely important time for

[6] There is no concrete documentation that Socrates ever made the often-quoted statement "I know one thing, namely, that I know nothing". The Neoplatonist philosopher Iamblichus, 245–325 AD, quotes Socrates stating that "I do not know anything, and I do not teach something; I only have queries". In any case, Socrates alleged, or quoted statements express a basic element of wisdom: the deeper one penetrates the essence of things, the more questions *(aporias)* are raised.

meaningful personal interactions and reflection. Invest your energy in *quality* endeavours. General criteria for quality include aesthetics, universal value, and the intriguing interplay between simplicity and complexity with emphasis on complexity.

References

Dennett, D. C. 2017. *From Bacteria to Bach and Back.* Penguin Press.

Harari, Y. N. 2016. *Homo Deus: A Brief History of Tomorrow* (Original Publication: 2015 in Hebrew). Harvill Secker.

Holt-Lunstad, J., Smith, T. B., & Layton, J. B. 2010. Social relationships and mortality risks: A meta-analytic review. *PLOS Medicine* 7, e1000316.

Huffington, A. 2016. *The Sleep Revolution: Transforming Your Life, One Night at a Time.* Virgin Digital.

Kok, B. E. 2013. How positive emotions build physical health: Perceived positive social connections account for the upward spiral between positive emotions and vagal tone. *Psychological Science* 24(7), 1123–1132.

Optimism and Your Health. 2008. Harvard Medical School. Retrieved from https://www.healt.harvard.edu.

Schrödinger, E. 1944. *What Is Life? The Physical Aspect of the Living Cell.* Cambridge University Press.

Stoppard, T. 2015. *The Hard Problem: A Play.* Grove Press.

Tononi, G. 2012. *Phi, A Voyage from the Brain to the Soul.* Pantheon Books.

Trivers, R. 2013. *Deceit and Self-Deception: Fooling Yourself the Better to Fool Others* (Original publication: 2011). Penguin Press.

Walker, M. 2017. *Why We Sleep.* Scribner.

Wegner, D. M. 2017. *The Illusion of Conscious Will.* Foreword by D. Gilbert. Introduction by T. Wheatly. MIT Press.

Zeki, S., Romaya, J. P., Benincasa, D. M., & Atiyah, M. F. 2014. The experience of mathematical beauty and its neural correlates. *Frontiers in Human Neuroscience* 8.

1 PRELIMINARIES

In an effort to promote an interdisciplinary approach to knowledge and culture, and in order to illustrate that this interdisciplinarity reflects innate neuronal mechanisms, a variety of examples from several disciplines will be discussed in detail, including those mentioned below. In mathematics: from the unexpected realization that what is called *Riemannian geometry* provides the basis for understanding our universe to the presentation of counter-arguments to the positions of Ludwig Wittgenstein regarding the essence of mathematics. In physics: from the basic laws of Newton and Maxwell to the resolution by Einstein of the fundamental question, "What is light?" In biology: from *molecular phylogeny* to *horizontal gene transfer*. In neuroscience: from the astonishing complexity of the brain's communication system to learning and memory at the molecular level. In medicine: from the importance of medications acting at the level of *neuronal synapses* to possible treatments for depression and Parkinson's disease. In technology: from *the focused ion beam transmission electron microscopy* which allows researchers to visualize every tiny cell organelle to the invention of a *retinal prosthesis* that improves the vision of certain patients. In painting: from the discussion of great works of Picasso, Egon Schiele, Goya, Édouard Manet, El Greco, Rembrandt, Alberto Giacometti, and other artists to the role of luminance and colour in the works of Claude Monet and other Impressionist painters. In philosophy: from the presentation of a summary of the main positions of Descartes, Leibniz, and Spinoza to the discussion of some of the ideas of Wittgenstein.

By reflecting on the works of many creative mathematicians, scientists, painters, composers, and writers, it will become evident that,

mathematicians and scientists, as well as artists, poets, and writers, employ similar neuronal mechanisms for elucidating deep scientific facts and creating fabulous artistic new worlds, respectively.

This fact has been appreciated by several deep thinkers. For example, Gerald Edelman (Nobel Prize in Medicine, 1972) in his inspiring book *Second Nature* (Edelman, 2006) writes:

"Science is imagination in the service of the verifiable truth [...] the brain's origins of imagination in science do not differ from those necessary in poetry, music, or the building of ethical systems [...] [thus] no divergence is necessary between sciences and the humanities".

Hopefully, this understanding will contribute towards promoting the need for an integrative approach to arts, letters, and sciences. There is an objective reason to be optimistic regarding the acceptance of this approach: it will be argued that the tendency for *unification is innate*. In this regard, it is noted that unification reaches its apogee in physics. Hence, it is not surprising that, in addition to Schrödinger, another physicist, the Indian polymath Jagadish Chandra Bose (1858–1937), also strongly expressed the importance of an integrative approach to knowledge and culture. Bose, in addition to making major contributions in physics and other sciences, was also a writer of science fiction. So, for him the concept of unification was indeed encompassing, going beyond sciences: "poets and scientists share a common goal: to find a unity in the bewildering diversity".[1]

---

[1]Bose is a stellar representative of the Bengali Renaissance, which took place in Calcutta in the period of 1840–1920. He went to London to study medicine but was allergic to formaldehyde (used for the preservation of cadavers). So, he moved to Cambridge where he studied under Lord Rayleigh (Nobel Prize in Physics, 1904). He was a close friend of the Indian poet Rabindranath Tagore (Nobel Prize in Literature, 1913), who at the age of 67 began painting, finally becoming a distinguished painter. Bose's most famous student was Satyendra Nath Bose (1894–1974), best known for the so-called Bose–Einstein statistics and for the theory of Bose–Einstein condensate. Paul Dirac introduced the terminology *bosons* to honour Satyendra Bose. Several Nobel Prizes have been awarded for research related to bosons, Bose–Einstein statistics, and Bose–Einstein condensate, but, surprisingly, Bose himself was not awarded the Nobel Prize.

In several parts of the book, a variety of different treatments for a plethora of diseases are discussed. It is important to emphasize that these treatments should *not* be taken as explicit advice for one's own medical care. The relevant discussion will hopefully be useful for understanding the underlying scientific approach to various pathological conditions. However, this insight cannot be a substitute for the physician's specific advice.

MEDICAL TERMINOLOGY, COMPLEXITY, AND NOVELTY

Regarding medical terminology, it is useful to note that the name of a disease almost always describes the relevant pathophysiology with the aid of words of Greek and occasionally Latin origin. In this regard, it is noted that the suffix "is" indicates inflammation. For example, "encephalitis" means "inflammation of the brain (encephal)". Similarly, "disseminated encephalomyelitis" means "widespread inflammation of the brain and the spinal-cord (myelo)". There are however exceptions, such as pneumonia, instead of pneumonitis.

In order to assist the reader in comprehending this volume without the need to consult other sources, I have tried to introduce every important concept from first principles. Indeed, understanding the main part of this volume does not require advanced specialized knowledge. The parts that require expert knowledge or specialized terminology are presented as footnotes. Supplementary material, sometimes of historical nature, which enriches the context but is not necessary for the understanding of the main ideas, is also provided in footnotes.

The text is at times dense. This is the inevitable consequence of the plethora, diversity, and complexity of areas covered. Also, I think that the only way for the readers to appreciate the high complexity of biological, and especially, neuronal processes, is to be introduced to some of the associated fundamental mechanisms. In particular, such detailed processes are described with respect to learning and memory.

This means that reading this volume requires a cognitive investment, which, however, I expect will be fully rewarded: the reader will be exposed to a wide vista of knowledge, and at the same time they will, hopefully, experience *eudemonia*. Moreover, the reader will be exposed to several novel concepts and will benefit from a comprehensive synthesis of key ideas from a very large number of excellent books, both in the sciences and in the humanities. More importantly, these ideas may have a serious impact on the way that the attentive reader will attempt to understand any phenomenon in the future, including their own actions, feelings, and thoughts.

Among a variety of, perhaps, novel concepts introduced in this volume are the following:

- The clearly stated notion of *continuity* between unconscious and conscious processes.
- The hypothesis that a process which starts with *homeostasis* finally gives rise to the uniquely human capacity for *metarepresentations.*
- The elaboration of the notions of *continuity* and *local versus global processes.*
- The proposition that *associations* provide the neuronal basis for the brain's capacity to reach *abstraction*, as well as to *generalize*, *reduce*, and *unify*.
- The demonstration of the importance of the concept of *simplicity versus complexity*. For example, this concept explains our ability to understand the universe, and also offers an illuminating approach to appreciating art.
- The elucidation of the dynamic interaction between cognition and culture that takes place in the ever-changing *heterodynamic* state, which humans create with respect to their environment.

I expect that, due to their fundamental importance, variations of some of these ideas have already appeared in the literature. I offer my sincere apologies to authors whose relevant works I have erroneously failed to cite. In this regard, I must emphasize that the aim of this book is *not* to claim priority, but rather to fulfil a moral obligation: to attempt to contribute even minutely towards the knowledge, culture, and happiness of people, wherever they may live "in this world, the small world, the great".[2]

[2] "Σε αυτόν τον κόσμο τον μικρό τον μέγα", from the poem *Axion Esti* (*Worth It Is*), 1959, by Greek poet Odysseas Elytis (Nobel Prize in Literature, 1979). Axion Esti (Elytis, 1974) is the name given to the icon of the Virgin Mary, now in the main church of Karyes of Mount Athos, Greece. According to tradition, the hymn Axion Estin (It Is Truly Meet) was revealed in front of this icon.

A BRIEF PHILOSOPHICAL PERSPECTIVE

The appreciation of the phenomenal ability of the brain to make *associations* and the need for a *broad, interdisciplinary approach to knowledge and culture*, are two of the basic pillars of this book. Associations are crucial for both *nature* and *nurture*. Regarding the former, it should be noted that cognition is based on the exceedingly complicated web of associations created in the brain, reflecting the brain's extremely complex topological structure. Nurture can be defined as the impact of the environment and of the prevailing culture on innate associations and on the creation of new ones. Consistent with the crucial importance of associations, the following quote by Johann Wolfgang von Goethe (1749–1832), provides an appropriate way to begin our philosophical ramifications:

"Every act of looking turns into observation, every act of observation into reflection, every act of reflection into the making of associations" (von Goethe, 1988).

As a result of his important scientific contributions, Goethe personifies the attempts to eliminate boundaries between sciences and humanities. Incidentally, Goethe was convinced that his scientific work was more important than his poetry. Among Goethe's several significant scientific contributions is his discovery in 1784 of the intermaxillary bone in the human foetal skull. This bone constitutes a vestigial remnant of the bone found in the skull of apes. This discovery demonstrated, well before Darwin's deliberations, that parts of living creatures evolved from the same stem.

Interestingly, when the English biologist and anthropologist, Thomas Henry Huxley, was asked in 1869 to write an editorial for the inaugural issue of the prestigious journal *Nature*, Huxley based his editorial on Goethe's *Nature* (1780).[3] This work contains a variety of statements regarding Nature's creations, including the following:

---

[3] T. H. Huxley (1825–1895) became famous after his 1860 debate with the English bishop in the Church of England, Samuel Wilberforce. This debate was of singular importance for the wider acceptance of the theory of evolution, earning Huxley the characterization of "Darwin's Bulldog". Among Huxley's grandchildren were the famous author Aldous Huxley and the Nobel Laureate neuroscientist Andrew Huxley, whose contributions are discussed in Chapter 2.

"Each of her works has an essence of its own; each of her phenomena a special characterization: and yet their diversity is in unity";

and also,

"Her crown is love. Through love alone dare we come near her. She separates all existences, but all tend to intermingle. She has isolated all things in order that all may approach one another".

Goethe's emphasis on *unity* is consistent with attempts in this volume to elucidate the importance of *unification*. Also, his characterization of love as 'the crowning achievement of Nature' has now been justified beyond the obvious. For example, studies reveal that acts of "social love" and "kindness" are accompanied by the production of oxytocin and dopamine. These molecules are known, respectively, as the chemicals of "love" and "happiness". In particular, the hormone oxytocin reduces inflammation and free-radicals; in turn, this protects against cardiovascular diseases and also may slow ageing (Hamilton, 2017). Moreover, meditating on "compassionate thoughts" elevates the levels of dopamine and stimulates the parasympathetic system, which in turn slows down the heart rate, reduces blood pressure, and limits inflammation (Fredrickson *et al.*, 2008). Acting in a "kind way", apparently provides an antidote to the "negativity bias", which is the tendency of organisms to be constantly on alert to face dangerous situations. Adopting such a loving way results in a more positive, happier, and healthier life (Hamilton, 2017).

Goethe was greatly influenced by his friendship with the German geographer and naturalist, Alexander von Humboldt. The latter, in his multivolume treatise *Kosmos* (from the word "cosmos" introduced by the ancient Greeks) attempted to *unify* a variety of diverse branches of scientific knowledge and culture. Goethe remarked to friends that he had never met anyone as versatile as this polymath. In 1794, Humboldt became a member of the *Weimar Classicism*. This was a German literary and cultural movement whose members aspired to establish a new humanism based on the synthesis of ideas from Romanticism, Classicism, and the Age of Enlightenment. In addition to Goethe, the German writer and physician Friedrich Schiller (whose poem *Ode to Joy*, written in 1795, was immortalized by Beethoven's monumental 9th Symphony) was also a member of this movement.[4]

Throughout his life Goethe emphasized the pursuit of *knowledge* and *beauty*, viewing this pursuit as a long and tortuous process. This concept was further illuminated by the German philosopher Martin Heidegger (1899–1976).[5] Motivated by his interpretation of the Greek notion for "truth" (*aletheia*) as "un-concealing", Heidegger formulated the hypothesis that searching for truth is a *process involving sustaining a difficult journey, which is actually part of the truth.* Heidegger's idea of considering "un-concealing" as a dynamic process is very different from the view of truth as static and an all or nothing concept.[6] The German writer and Enlightenment polymath Gotthold Ephraim Lessing (1729–1781), who like Goethe immortalized *Faust*, guided his protagonist to achieve redemption via an endless striving for knowledge. In this connection, Lessing wrote:

"It is not possession of the Truth, but rather the pursuit of Truth by which [Man] extends his powers in which his ever-growing perfectibility is to be found" (Lessing, 1979).

Philosophers and writers, as well as pioneering scientists, have appreciated that this pursuit must be imbued with the recognition of the *limits of the human understanding*. For example, the French mathematician, scientist, and theologian Blaise Pascal (1623–1662) wrote that the ultimate achievement of reason is "to recognize that there is an infinity of things which surpasses it". Indeed, in my opinion, *possession of knowledge without humility leads to*

[4] Humboldt, following his extensive travels, which included visits to Cuba, Venezuela, and Mexico, stated in 1800 and again in 1831 that humans are inducing a severe and dangerous change to the environment. In 1804, he visited USA, where he met President Thomas Jefferson (who was himself a scientist) and some of the major scientific figures of the era. They included the chemist and anatomist Caspar Wistar who was instrumental in instituting compulsory vaccination for smallpox, and the Scottish physician Benjamin Rush who was a signer of the Declaration of Independence. Humboldt's scientific approach had a significant influence on the ideology of the emerging nation.

[5] Heidegger, who is rightfully considered one of the most original and important philosophers of the 20th century, was until the end of the war a member and public supporter of the Nazi party. This raises the puzzling question of how individuals of high intellectual and aesthetic level support unthinkably atrocious ideologies.

[6] The emphasis on the "journey" as opposed to the "final goal" is immortalized in the poem *Ithaca* by the Greek-Alexandrian poet C. P. Cavafy (Cavafy, 2007).

intellectual vanity. It is imperative that "creatures" respect their "creator". In this regard, it is prudent to keep in mind that evolution had billions of years to mould and improve its creations. Thus, it is hubristic to underestimate Nature's infinite ingenuity. Unfortunately, several proclamations regarding scientific achievements, such as unrealistic speculations about the recent remarkable developments in "deep learning", betray an arrogant tendency to underestimate the immense difficulty of attempting to mimic Nature. To what extent is the scientific knowledge gained through deep learning a genuine knowledge, if we have not ourselves participated in the underlying journey? A broad cultural education is perhaps the best guard against such tendencies: the more areas we are exposed to, the more evident become the limitations of our achievements and the knowledge we have conquered.

A comprehensive, unified worldview (*Weltanschauung*) was strongly advocated by the leading German philosopher Immanuel Kant. In particular, he envisioned the unification of mathematical, scientific, and aesthetic knowledge. This grand goal is presented in his three famous books, *Critique of Pure Reason*, 1781 (on mathematics), *Critique of Practical Reason*, 1788 (on science), and *Critique of Judgment*, 1790 (on aesthetic knowledge). Kant's impact on philosophy in general, and on Vienna's modernism in particular, was far-reaching. After his retirement, Kant was replaced as professor of philosophy at Konigsberg by the influential philosopher Johann Friedrich Herbart (1776–1841), who himself influenced, among others, Hermann von Helmholtz and David Hilbert (these polymaths are mentioned extensively in this volume). Interestingly, according to Kant, "aesthetic knowledge" is based on subjective intuition, as opposed to logic and morality. These positions influenced decisively Helmholtz's studies of music.[7]

Heraclitus proposed that *truth* and *beauty* are based on the *synthesis of opposites*. This position was strongly endorsed by Friedrich Hölderlin (1770–1843). Indeed, the German poet, discussing this well-known claim of the ancient Greek philosopher, wrote that

[7] Regarding the importance of continuity, it is worth noting that Immanuel Kant (1724–1804) was strongly influenced by David Hume (1711–1776), who was an exponent of the importance of associations. Hume also influenced Albert Einstein.

"the great insight of Heraclitus could occur only to a Greek, for it is the essence of beauty and before this was discovered, there was no philosophy" (McGilchrist, 2009: p. 354).[8]

Heraclitus' union of opposites brings to mind the Apollonian and Dionysian elements, which are traditionally considered antithetical. For example, these elements are associated, respectively, with the rational and hedonistic that are supposed to form a strict antithesis. However, the connection between discoveries (for which the rational is vital) and the resulting happiness (eloquently expressed in Einstein's quote cited in the prologue), suggests that instead of an antithesis there actually exists complementarity. The antithetical relationship between the Apollonian and the Dionysian was greatly promoted by the erroneous hypothesis advocated by Friedrich Nietzsche that "reason undermines the senses and supersedes pleasure". The influential philosopher wrote in 1878:

"All our senses have in fact become somewhat dulled because we always inquire after the reason, what 'it means', and no longer for what 'it is' [...] some painters have made the eye more intellectual, and have gone far beyond what was previously called a joy in form and colour [...]. What is the consequence of this? The more the eye and ear are capable of thought, the more they reach that boundary line where they become a-sensual. Joy is transferred to the brain; the sense organs themselves become dull and weak. More and more the symbolic replaces that which exists" (Nietzsche, 1996: p. 130).

I disagree with the great philosopher: in my opinion, the more *complete* the experience, the deeper the pleasure. This completion, which is aided by logic, is achieved via associations that are crucial for science, arts, letters, and every conceivable human activity. Indeed, associations give rise to *the generation of highly complex relationships* which bring personal fulfilment to scientists, artists, intellectuals, to every thinking human being.

For example, the music director of La Scala Opera, Riccardo Chailly, noted in an interview in the *Opera News*, May 2019 that "his greatest pleasure and

[8] In literature, as it is well known, the conflict of opposites characterizes many great works from ancient Greek tragedies to the works of William Shakespeare.

source of energy" during his preparations for conducting Giuseppe Verdi's opera *Attila* (1846) was: discovering "mirror reflections and resemblances" with Verdi's opera *Macbeth* (1847), as well as noting "intimate links" with Verdi's opera *Giovanna d'Arco* (1845). Thus, essentially, it was deciphering deep relationships ("mirror reflections and resemblances", "intimate links") which brought *eudemonia* to the leading conductor.

There is another aspect of the relationship between the Apollonian and the Dionysian worth mentioning: the Dionysian is often interpreted as the "intuitive", which several German philosophers have associated with the unconscious. Heidegger has claimed that as a result of the Enlightenment, the Apollonian (rational) and the Dionysian (intuitive) have become unbalanced, in favour of the rational. This claim is consistent with the positions of Nietzsche and Arthur Schopenhauer regarding the overreliance on the rational and generally on consciousness, at the expense of the unconscious and the transcendental. In my opinion these claims are valid. As stated in the prologue, we are biased towards consciousness. This is the result of us becoming directly aware of conscious processes, whereas only the "echo" of the unconscious reaches us. This bias, which was enhanced by the important movement of the Enlightenment, is unfortunate, because, as the leading investigator of the interaction between unconscious and conscious processes Stanislas Dehaene states: the unconscious contains "an unimaginable richness to be explored" (Dehaene, 2014).

As noted in the prologue, the tendency to overestimate consciousness has been exacerbated by the tremendous progress achieved as a result of the metarepresentations of mathematics, science, technology, and artificial intelligence. The current apotheosis of technology and algorithms brings to mind the play *Prometheus Bound* by Aeschylus.[9] Prometheus, the god of technical skills and intelligence, stole fire from the gods and gave it to

[9] Aeschylus was so proud of his participation in the battles of Marathon and Salamis that he made sure that his epitaph referred to the Marathon events instead of his great successes as a playwright. This unexpected choice provides the basis of the poem *Young Men of Sidon* (400 AD) by C. P. Cavafy (Cavafy, 2007). According to Pausanias, during one of Aeschylus' dreams that occurred while sleeping in a vineyard, Dionysus, the god of wine, appeared and exhorted him to write a tragedy. This story, together with Aeschylus' perceived respect for religion and the mystic, are consistent with the importance of intuitive processes in Aeschylean tragedies. This fact perhaps explains Sophocles' statement that "Aeschylus does what is right without knowing it."

the mortals. For this crime he was chained to a rock and every day his liver would be torn out by a bird. By next day, it would grow again so that this inhumane torment could be repeated.[10] Aeschylus is profoundly compassionate towards Prometheus, but his overall stance regarding the dangers associated with the power gained via the use of tools from a fundamentally different realm, remained ambivalent.

AN IMPORTANT OMISSION

An important omission of this volume is that only Western notions, philosophies, and cultures, are elaborated. However, it is expected that the basic concepts discussed here have wider applicability.

For example, universal human values, such as the "golden rule" (one should treat others as one would like others to treat oneself), have been expressed diachronically, worldwide. Also, in the update of Bertrand Russell's comprehensive book *Western Philosophy* (Russell, 1947) recently presented by Anthony Grayling (Grayling, 2019), there is a section on non-Western philosophies where the importance of social harmony across civilizations is highlighted: from the Chinese notion of "reciprocity" (*shu*), to the idea prevalent in southern Africa called *ubuntu*, meaning that one exists only because of and through others (I am, because we are).

Similarly, as noted by Dehaene (Dehaene, 2014), the notion of the "soul" is prevalent across different cultures. This fundamental concept, which is, perhaps, a reflection of the unconscious, is often symbolized by a bird. According to metaphysical considerations, the soul takes a flight away when the body dies; hence, the bird provides an ideal metaphor. For example, there exist depictions of the Egyptian God Osiris lying on his back, while Isis hovers over his body in the form of an owl taking his sperm to engender Horus. In *Upanishads*, the collection of Hindu texts of religious and philosophical nature written between 800 and 500 BC, a dove symbolizing the soul flies away at death. In the Finnish mythology

[10] The choice of the liver is remarkable since it is the only visceral organ with the ability to regenerate. Indeed, it is known that, following chemical injury or surgical removal, even as little as a fourth of the original liver mass can generate its full size.

Sielulintu, a bird symbolizing the soul delivers the psyche to new-born babies.

The bird is not the only symbol used to represent the soul. For example, Nicholas Humphrey suggests that a rock painting in the Vilafames cave (in the Valencia region of Spain) consisting of a spiral in the head of a humanoid, made around 15,000 BC, symbolizes the soul (Humphrey, 2012: p. 158).

Interestingly, the metaphor of soul-bird is closely associated with the state of dreaming. In this state, the spirit can escape the body. For example, in one of Lascaux's Apse paintings that were made approximately 18,000 years ago, a man lies supine apparently dreaming as indicated by his erection, and there is a bird next to him. According to Michel Jouvet (Jouvet, 1999: pp. 169–171), this provides the first symbolic depiction of a dreamer.

The ancient Egyptians, the Hebrews, Aristotle, and the Stoics, placed the psyche in the heart. The Maya located it in the liver, Democritus and Galen throughout the body, Epicurus in the stomach, and Empedocles in the blood. Hippocrates and Hierophilus placed it in the brain.

Concluding, it is interesting to note that according to the Chinese cosmology of Taoism, there exists *a priori* universal energy. This abstract concept implies that the universe *always* existed. Hence, according to this philosophy, every creative act involves a *discovery*. This is consistent with Plato's ideas that learning and creating are merely acts of *recollection*.

These commonalities are consistent with the universal nature of humanity. An important aspect of this universality was first established by Darwin, who emphasized the common expression of human emotions. The great pioneer sent a set of questions to friends, other scholars, and missionaries, in Australia, New Zealand, Malaysia, Borneo, India, and Ceylon; he received 36 answers. In this way, he established that populations of distant cultures and certain aboriginal tribes displayed facial expressions and bodily postures comparable to those he was familiar with, in Britain and Europe. This study has inspired many researches. In particular, psychologist Paul Ekman collected, between the 1960s and 1980s, pictures and data from remote areas confirming Darwin's findings and delineating the role of various facial muscles in expressing emotions (Ekman, 2003). Incidentally, stimulating the volitionally controlled muscle *zygomatic major* (which draws the angle of the mouth superiorly and posteriorly) yields a fake smile. A genuine smile requires the activation of the muscle *orbicularis oculi* (the muscle around the eyes), which cannot contract on demand.[11] The universal nature of our emotions provides one more argument against the poisonous racism, which, unfortunately, is still prevalent in the world today.

References

Cavafy, C. P. 2007. *The Collected Poems*. Edited by A. Hirst. Translated by E. Sachperoglou, with an introduction by P. Mackridge. Oxford University Press. (*Ithaca* was originally published in 1911 in Greek and *Young Men of Sidon* (400 AD) was originally written in 1920 in Greek).

Dehaene, S. 2014. *Consciousness and the Brain: Deciphering How the Brain Codes Our Thoughts*. Viking.

Edelman, G. M. 2006. *Second Nature: Brain Science and Human Knowledge*. Yale University Press.

Ekman, P. 2003. *Emotions Revealed*. Henry Holt and Company.

Elytis, O. 1974. *The Axion Esti*. Translated by E. Keeley and G. Savidis. Pittsburgh: University of Pittsburgh Press. (Originally published in 1959 in Greek).

Fredrickson, B. L., Cohn, M. A., Coffey, K. A., Pek, J., & Finkel, S. M. 2008. Open hearts build lives: Positive emotions, induced through loving-kindness meditation, build consequential personal resources. *Journal of Personality and Social Psychology* 95(5), 1045–1062.

Grayling, A. C. 2019. *The History of Philosophy*. Viking.

Hamilton, D. R. 2017. *The Five Side Effects of Kindness*. Hay House.

Humphrey, N. 2012. *Soul Dust: The Magic of Consciousness*. Quercus.

Jouvet, M. 1999. *The Paradox of Sleep: The Story of Dreaming*. Translated by L. Garey. MIT Press.

[11] It is interesting that apes and even rats can laugh, but only humans shed tears. This process starts around the age of 5 months. Darwin noted that when the edge of his coat touched accidentally one of the eyes of his 2-month-old baby, this eye watered but the other remained dry. Thus, babies can produce tears much earlier, but they only begin shedding them around the age of 5 months.

Lessing, G. E. 1979. Anti-Goetze: Eine Duplik. In: Göpfert, H. (ed.) *Werke,* Vol. 8. Translated by S. Horton. (Originally published in 1778 in German).

McGilchrist, I. 2009. *The Master and His Emissary: The Divided Brain and the Making of the Western World.* Yale University Press. (The quote is from Hölderlin's *Hyperion; or, The Hermit in Greece*, which was originally published in German in two volumes in 1797 and 1799).

Nietzsche, F. 1996. *Human, All Too Human: A Book for Free Spirits.* Translated by M. Faber and S. Lehmann, with introductions by M. Faber and A. C. Danto. Lincoln: University of Nebraska Press. (Originally published in 1878 in German).

Russell, B. 1947. *History of Western Philosophy.* George Allen and Unwin Ltd. (Originally published in 1945).

von Goethe, J. W. 1988. *Theory of Color.* Edited and translated by D. Miller. Scientific Studies. (Originally published in 1810 in German).

BASIC NOTIONS

This is the central part of the volume where several novel notions are elucidated, including the concept of metarepresentations. The vital impact of this uniquely human characteristic on our cognition was captured by the highly original Caltech physicist Richard Feynman (Nobel Prize in Physics, 1965). Before dying, the charismatic Feynman, who enjoyed painting as well as playing the drums, left the following statement on his blackboard: "What I cannot built, I cannot understand". In addition, in this part of the book, the crucial role of unconscious processes in general and emotions in particular is emphasized.

In order to define certain basic notions used throughout this volume, it is necessary to first introduce elements of neurophysiology and neuroanatomy, which are presented in Chapters 2 and 3.

In Chapter 4, the problem of visual perception is considered as an *inverse problem*. The analysis of visual perception motivates the first hypothesis stated in the prologue, namely, that *every conscious experience is preceded by an unconscious process*. In addition, the anatomical difference between *local* and *global* connectivity is noted. This is of crucial importance for appreciating the role of *local versus global processes* in the neuronal mechanisms responsible for perception.

Chapter 5 discusses *homeostasis* in detail, as well as the unique ability of the human brain to construct metarepresentations. In this regard, certain anatomical differences with apes are noted, which are vital for the emergence of metarepresentations. As stated in the prologue, metarepresentations are

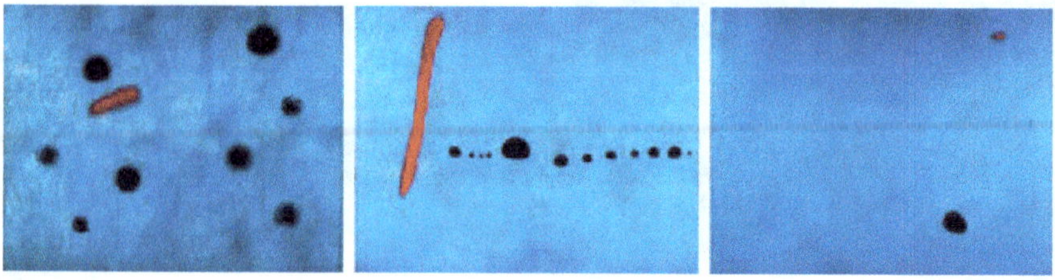

Figure PI.1. Joan Miró, *Blue Triptych* (1961).
Source: © Successió Miró/ADAGP, Paris and DACS London 2024.

creations of symbols, which is exemplified in the emergence of language and mathematics, as well as *explicit constructions of mental images and unconscious structures*, which is exemplified in technology and the arts. The terminology mental images is used for neuronal constructions associated with consciousness, whereas unconscious structures refer to the neuronal processes associated with the unconscious.

It is emphasized that metarepresentations not only reflect mental images but also unconscious structures. Highly original artists and scientists have clearly understood the unconscious origin of some of their creations. For example, the Catalan artist Joan Miró stated:

"I begin painting and as I paint, the picture begins to assure itself, or suggest itself under my brush [...] the first stage is free, unconscious [...] the second stage is calculated" (Sweeney, 1948).

Incidentally, for Miró, the colour blue had a special significance. It was the symbol of the unconscious state where his mind could flow freely without any constraints and preconditions. For him, blue was the colour of a surreal space where dreams exist in their purest form, uncensored by consciousness. In this sense, perhaps Miró's *Blue Triptych* (1961) captures the essence of this part of the volume (Figure PI.1).

Reference

Sweeney, J. 1948. Joan Miró, comment and interview. *Partisan Review*.

INTRODUCTION TO NEUROPHYSIOLOGY

Our knowledge of how the universe has evolved, and our understanding of the physical laws that dictate basic physical phenomena is truly remarkable. For example, we know that the universe was created 13,700 million years ago and that the Earth was created 4,500 million years ago. We know that *homo sapiens* appeared only approximately 200,000 years ago. Remarkably, it has been verified experimentally that physical processes, in scales ranging from the infinitesimally small to the gargantuan, are dictated by relatively simple physical laws, which, astonishingly, can be expressed in terms of beautiful mathematical equations. These and a multitude of other achievements are, to a large extent, the result of the phenomenal abilities of our brains. Actually, it is said that the human brain constitutes the most complex structure we have encountered in the universe.

Since the main tool used for the analysis of any phenomenon is the brain, it is self-evident that this analysis will be incomplete without a deep understanding of how the brain itself functions. This is where we shall begin. The first part of the book attempts to give readers the foundations for such an understanding. Unfortunately, this comes with more technical material in the opening two chapters than is usually encountered in books written for the general public. It is of course possible to convey some understanding with a presentation that avoids technical details, but that would deny readers the opportunity for genuine insight, deeper knowledge and, hopefully, *eudemonia*.

An alternative is to skip this and the next chapter and to return to them as you progress through the book.

The elementary unit of the brain is the *neuron* or nerve cell. Different types of cells possess different basic properties. For example, *cardiomyocytes*, which are the cells forming the walls of the heart, have the intrinsic property of contractility. The defining property of neurons is their ability to self-activate and spontaneously discharge. One of the most common ways to experimentally investigate neurons is to excite them by using electricity.

Substantial progress was made in neurophysiology via the technique of cutting extremely thin slides of brain tissue and then treating these slides with various chemical stains. Such *histological stains* have been of crucial importance for the discovery of the neuron and the delineation of other neuro-anatomical structures. The neuron was discovered by the Spanish neuroscientist and pathologist Santiago Ramón y Cajal (1852–1934), by employing a specific stain discovered by the Italian biologist and pathologist Camillo Golgi (1843–1926).

THE "RETICULISTS" VERSUS THE "NEURONISTS"

Looking at a collection of neurons under a microscope, it is not possible to see any gaps between them. In addition, in the late 1800s, neuroscientists thought that excited neurons form a pulsating net (reticulum). Thus, although at that time most biologists accepted that living organisms are composed of tiny discrete elements, called *cells*, neuroscientists believed in the existence of the "neural reticulum".

According to legend, in 1872, Camillo Golgi while working at the Hospital for the Chronically Sick, in Italy, spilled, by mistake, a weak solution of silver nitrate onto some slides of owl brain. Although silver was already used to stain certain tissues, Golgi thought that silver would ruin his brain samples. However, when he examined the stained slides under the microscope a few weeks later, he observed, to his surprise, that those cells that had absorbed the silver stain stood out dramatically. He could clearly

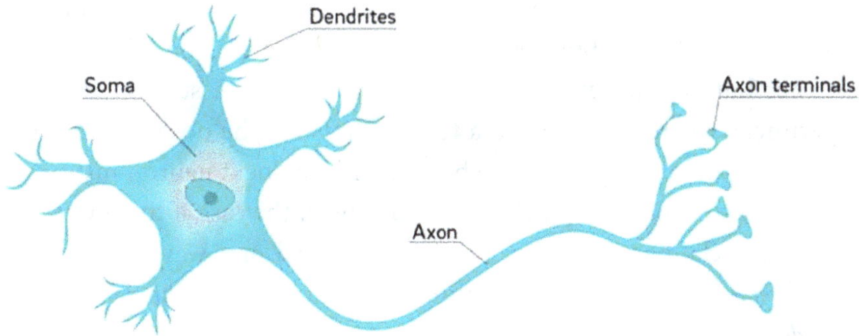

Figure 2.1. A neuron.
Source: Singh & Manure (2020).

see black silhouettes on a sherbet-yellow background. Golgi called his technique the "black reaction" (*la reazione nera*).

At the time, scientists already knew that the nervous system consists of two main types of cells, namely, neurons, which constitute the fundamental functional unit of the nervous system, and *glia* (derived from the word "glue"), which were thought to hold neurons together and provide nutrition. Golgi was the first to see these cells in detail.[1] Glia cells are rounded, and after being stained they look like black jellyfish frozen in amber. Neurons consist of a circular central body called the *soma,* a long cable-like structure called *axon*, and an intricate collection of branches, called *dendrites* spouting from the body, see Figure 2.1. Golgi could not see any space between neurons and hence he became a strong supporter of the reticulum theory.

The neuroanatomist Ramón Cajal became fascinated by the work of Golgi, proclaiming, "it renders anatomical analysis both a joy and a pleasure". Cajal's dream was to become an artist and although a local painter told his father that the 10-year-old boy had no talent, Cajal continued to harbour artistic inspirations. After turbulent beginnings that compelled his father to withdraw him from a Jesuit school and to apprentice him first to a barber and then to a cobbler, Cajal re-enrolled in school and began

[1]Golgi, in addition to using the stain noted earlier, also made several other important discoveries. In particular, in 1897, he identified the *Golgi apparatus*, which is an important organelle found in most eukaryotic cells that packages protein into membrane-bound vesicles.

concentrating on medicine and neuroscience.[2] Cajal, after examining a plethora of stained brain slides, concluded that neurons do not fuse together but are *discrete units.* His observations were supported by an experiment where he strangled a few neurons and he observed that the subsequent decay stopped at the border with the next neuron instead of killing that neuron too, which would have been the case if there were no gaps. Cajal claimed that neurons communicated in a way that allowed a *unidirectional* mode of propagating information: *from the dendrites to the cell body and then to the axon.* These findings led him to propose the *neuron doctrine.*

Cajal had a difficult time convincing other neuroscientist for the validity of his theory. He even founded his own journal to avoid rejection of his papers, but since this journal was in Spanish, this attempt had little success in promoting his ideas. A turning point towards establishing his doctrine was his self-financed participation in an international conference in Germany in 1889, where his meticulous drawings played a crucial role for the positive reception of his lecture. This led to the formation of two opposing groups, the "reticulists" and the "neuronists". Slowly, the neuronists won. Cajal and Golgi shared the Nobel Prize in Medicine in 1906.[3]

CHEMICAL VERSUS ELECTRICAL COMMUNICATION

The discovery of the neuron by Cajal led to another famous controversy in neuroscience: Do neurons communicate by sending pulses of chemicals or pulses of electricity across their gaps? By the early 1900s, biologists knew that *myocytes*, which are the basic cells forming muscle tissue, possess the intrinsic property of excitability. For example, a frog's heart removed from a frog and placed into salt water continues to beat on its own. Furthermore, it was observed that by electrically exciting suitable severed nerves leading to the heart, it was possible to either slow down or speed up the heart rate.

[2] His artistic inclination is evident in the drawings he made for an atlas of anatomy which his father, who was a Professor of Applied Anatomy in the University of Saragossa, was compiling. Regrettably, this atlas was never published.

[3] Even this great honour was not enough to establish peace between these adversaries. In his acceptance speech, Golgi accused Cajal of "deliberate omissions". Cajal regretted the "cruel irony of fate which pairs [...] scientific adversaries of such contrasting character".

This was considered evidence that communication across gaps was achieved via electrical pulses. It is interesting that, in 1921, Otto Loewi was able to establish the *opposite* conclusion using a clever modification of precisely these experiments. Loewi, already from his schooldays, showed a keen interest in humanities and throughout his life enjoyed opera, painting, and architecture. In the early years of his medical studies he was more interested in attending lectures in philosophy than in sciences. However, finally, he did concentrate on medicine. He began his clinical career in Frankfurt in 1897, but after witnessing the high mortality of countless far-advanced cases of tuberculosis and pneumonia, he decided to abandon clinical medicine and concentrate on pharmacology. Loewi, who finally became a professor of Pharmacology in Vienna, was greatly influenced by the discovery, made at Cambridge, by Walter Gaskell and John Langley, of the existence of the two basic divisions of the nervous system, the *sympathetic* and *parasympathetic.* He was also influenced by the work of T. R. Elliott, at Cambridge, on adrenaline. In 1905, Loewi established that cocaine enhances the action of epinephrine on the organs that are innervated by the sympathetic division.

In 1921, Loewi returned to the frog experiments mentioned earlier and demonstrated that the parasympathetic nerve endings are stimulated by the substance acetylcholine, whereas those of the sympathetic division use a substance closely related to adrenalin. In his Nobel Prize experiments he placed the hearts of two frogs into two separate solutions. After slowing the heart rate of one heart by stimulating appropriate severed nerve fibres (belonging to the parasympathetic division), he took saline from the container of this heart and put it in the container of the other heart. The second heart slowed down immediately. He concluded that the excited nerves caused the release of some chemical, which when transferred to the second container, caused the second heart to slow down. A similar conclusion was reached by nerve fibres of the sympathetic division which caused an increase in the heart rate.

It is fascinating that the idea for this experiment was conceived during Loewi's sleep. Apparently, he first dreamt of this experiment the night before Easter in 1920, but although he made some brief notes at night, in the morning he could not decipher his own handwriting. Fortunately, the dream recurred the following night. Loewi woke up and, rather than risking another disappointment, went straight to his laboratory!

In 1936 Loewi shared the Nobel Prize in Medicine with his lifelong friend Sir Henry Dale, who is one of 33 Nobel Laureates associated with Trinity College, Cambridge. Dale had performed painstaking investigations of the pharmacology of *ergot alkaloids*. These substances are found naturally in fungi, and their most well-known applications are the psychedelic drug LSD and the medication ergotamine which is used to treat migraines. Dale identified acetylcholine as a constituent of certain ergot extracts, and his investigation of the action of this important substance significantly extended the applicability of Loewi's discoveries. When Hitler invaded Austria in 1938, Loewi, who was Jewish, was compelled to leave his homeland, but only after he was forced to instruct the Swedish bank in Stockholm dealing with the Nobel Prize awards, to make a transfer to a Nazi-controlled bank.

Synapses

When a neuron is sufficiently excited, the effect is dramatic: electricity travels down its axon in the form of a pulse known as *action potential*, which will be further discussed in the next subsection. This is a short (about 1 millisecond) spike in voltage in the potential of the axon's membrane (from negative 70 millivolts to positive 40 millivolts).

Electricity cannot jump across the 0.02-micron gap formed between two neurons (a microns is a millionth of metre), called *synaptic cleft*. The work of Loewi suggested that, to cross this cleft, the electrical message is, somehow, translated into chemical information. This is indeed the case, and the chemical substances used to cross the synaptic cleft are called *neurotransmitters*.

The axon tip of the presynaptic neuron manufactures different types of neurotransmitters and stores them in the form of vesicles. When an action potential reaches the presynaptic end, calcium channels are activated allowing the entry of extracellular calcium. Calcium triggers the fusion of neurotransmitter-carrying synaptic vesicles with the plasma membrane,

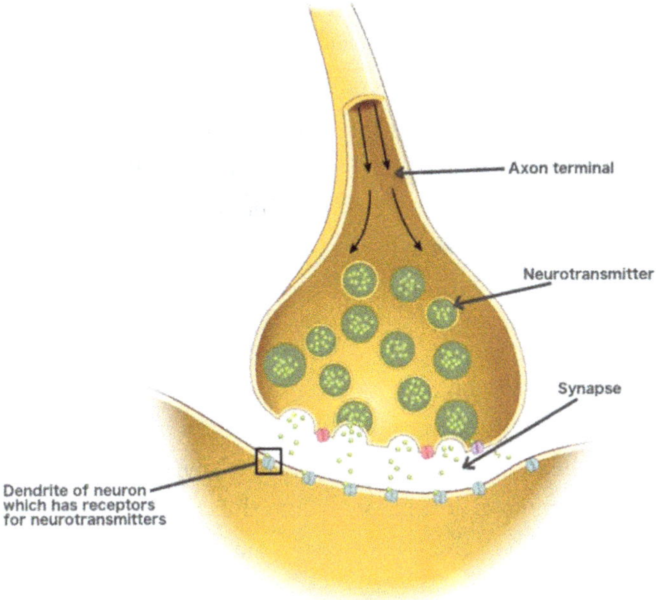

Figure 2.2. Conduction through a synapse.
Source: CNX OpenStax/Wikimedia Commons.

and this causes the release of the neurotransmitter molecules into the synaptic cleft. These molecules reach the postsynaptic neuron by the process of diffusion. There, the neurotransmitter is attached to appropriate receptors of the post-synaptic end. Depending on the specific neurotransmitter, this either excites or inhibits the postsynaptic neuron. In this way, the chemical message is translated back into an electrical one, see Figure 2.2.

The postsynaptic end belongs to a dendrite of a neighbouring neuron. In general, whereas axons of different neurons diverge, dendrites of the same neuron converge to the neuron's cell body. If the summation from all converging dendrites reaches a certain threshold of excitation, this neuron *fires*, namely it generates an action potential.

After the message is delivered, glia cells remove excess neurotransmitter molecules from the synaptic cleft, either via the release of appropriate degrading enzymes, or via a process of reabsorption (reuptake) by the presynaptic neuron.

In addition to this well-understood neurotransmitter release, which is *synchronous* with the stimulus, there also occurs a less well-understood, *asynchronous*, neurotransmitter release. This provides a less specific mode of communication between neurons. This mode is also mediated by calcium via unspecified calcium sensors (Sun *et al.*, 2007). It has been recently suggested that the age-related decline in memory is associated with age-dependent reduction of presynaptic calcium channels (Pereda *et al.*, 2019), affecting both the synchronous and the asynchronous neurotransmitter release.

The pulse-like action potential provides the universal mode of transmission along axons. This implies that to achieve *differentiation* of messages, the brain relies on the existence of *different* types of neurotransmitters. Some of them, such as glutamate, are excitatory and thus they facilitate the creation of an action potential at the postsynaptic neuron, whereas others, such as GABA and glycine, are inhibitory and prevent the creation of an action potential. Incidentally, glycine acts mainly in the brainstem and spinal cord, affecting various motor and sensory functions.

Neurons are functionally categorized as *excitatory* neurons, *inhibitory-interneurons* (which inhibit the excitatory neurons), and *neuro-modulatory* neurons. The latter are characterized according to the neurotransmitter they produce; for example, there are *dopamine*, *serotonin*, and *noradrenaline* neurons.

There are many medications that act at the level of synapses. For example, based on the understanding that the neurotransmitter serotonin has a positive influence on mood and sleep, many anti-depressant drugs, like fluoxetine (Prozac), paroxetine (Seroxat), and sertraline (Lustral), act by preventing (inhibiting) the reabsorption (reuptake) of serotonin. Thus, these medications, which are called Selective Serotonin Reuptake Inhibitors, enhance the availability of serotonin in the synapse. Ecstasy also enhances the availability of serotonin in synapses. Another example is provided by cocaine, which prevents the reuptake of dopamine, enhancing in this way the effect of this basic neurotransmitter which plays a crucial role in the anticipation of pleasure. Of clinical relevance is also the action of many psychoactive or anti-epileptic drugs that work by affecting

GABAergic or glycinergic neurotransmission, restoring the excitability/inhibitory balance.

Glia cells

The crucial importance of the glia cells was underestimated until recently. Glia cells outnumber neurons 6 to 1, but this ratio varies substantially in different parts of the brain. There are four main types of glia[4]: *Microglia* are the immunological cells of the central nervous system. *Astrocytes* support the metabolic, ionic, and neurotransmitter needs of neurons of the central nervous system. In addition, they assist microglia to eliminate unwanted substances. *Oligodendrocytes* produce myelin, namely, the outer layer of axons for the neurons of the central nervous system. Finally, *Schwann cells* myelinate the peripheral neurons. The latter cells, in contrast to the other three types of glia cells that are found in the brain and the spinal cord, do not enter the central nervous system.

Schwann cells were the first type of glia cells to be discovered. This was achieved by the German physician and physiologist Theodor Schwann (1810–1882), who also proposed that the building block of every living entity is the cell. An eukaryotic cell is a membrane-bound structure surrounding a nucleus. Cajal developed the first specific stain for astrocytes. Oligodendrocytes and microglia were discovered in the period 1919–1921 by Cajal's student Pío del Río-Hortega who developed variants of the silver stain. Glia cells will be discussed further in the second volume.

The action potential and ionic channels

The elucidation of the mechanism of an action potential has a long and illustrious history. In 1882, Sidney Ringer of the University College London established that if the solution perfusing a frog heart contains salts of sodium, potassium, and calcium in an appropriate ratio, then the heart continues to beat for a long time. This discovery is the basis of the

---

[4] Other important support cells are the *endothelial* cells that line the inside of larger blood vessels, and the *pericytes* that support smaller vessels called capillaries.

Ringer solution used today in hospitals. This solution contains sodium chloride, potassium chloride, calcium chloride, and sodium bicarbonate in the concentrations occurring naturally in bodily fluids. A variant of this solution, the Ringer–Locke solution, contains glucose and more sodium chloride than the original solution. Given intravenously, these solutions have prolonged or saved the lives of innumerable patients during surgery, after burns, and with a variety of other medical conditions.

In 1868, the neurobiologist and biophysicist Julius Bernstein, along with Hermann von Helmholtz, presented the first accurate description of an action potential. Furthermore, in 1902, Bernstein correctly proposed the *membrane theory of electrical potentials*. According to this theory, if a cell is at rest then the membrane that surrounds this cell is permeable only to potassium ions. But if the cell is excited, then the membrane becomes permeable to other ions, such as those of sodium, chloride, and calcium.

Ions carry a charge, and since any net flow of charge gives rise to an electric current, it follows that the movement of ions across the neuron's membrane generates a current. This tiny current is called *ionic* and the "gates" through which ions flow are called *ionic gates*, or *ionic channels*. An action potential is generated via an intricate mechanism based on the existence of ionic currents. This mechanism was elucidated by Kenneth Cole and Howard Curtis in the United States, and by Sir Alan Hodgkin and Sir Andrew Huxley in Great Britain.

The latter investigators succeeded in constructing a mathematical model describing the *nonlinear conductance* of potassium and sodium ionic gates. This model was consistent with experimental data that were obtained from giant squid axons via a new experimental technique, called *voltage clamp*. In earlier studies, current was applied as a stimulus and the ensuing changes in the voltage potential were measured. The voltage clamp reverses the process: voltage was applied across a membrane and then the ensuing current was measured. The availability of such data allowed Hodgkin and Huxley to derive a set of complicated mathematical equations describing characteristic features of action potentials.

Huxley and Hodgkin, who were awarded the Nobel Prize in Medicine in 1963, can be considered heirs of Lord Edgar Adrian and Sir Charles

Sherrington who shared the Nobel Prize in Medicine in 1932. Adrian was a pioneer of electrophysiological studies. In 1928, he recorded electrical discharges in single nerve fibres after applying touch, tension, and pressure on muscle cells. By placing electrodes in the neck of a rabbit and by connecting them to a loudspeaker he could hear a clicking noise every time an action potential was generated. He also studied the sense of smell, as well as brain waves recorded by an EEG (which was discovered in 1929). Sherrington, who was appointed Professor of Physiology at the University of Oxford in 1913, conducted pioneering studies in spinal reflexes during his years at Cambridge as a Fellow of Gonville and Caius College. He discovered that each of the motor neurons of the spinal cord provides signals to one or more of the body's 650 muscles. Furthermore, he identified a specific class of motor neurons, called *sensory feedback neurons*, which relay information about the relative position of the limbs and the body. This knowledge gives rise to a new sense (in addition to vision, hearing, touch, and smell) known as *proprioception* (this sense was also understood by other investigators, including Ernst Mach).

Incidentally, Sherrington's work was directly influenced by Cajal, whom Sherrington met during his visit to Spain. Hodgkin was inspired by the work of Adrian who, like Hodgkin, was also a Fellow and later Master of Trinity College, Cambridge. Additionally, Hodgkin was influenced by Kenneth Cole with whom he collaborated during a visit to the Rockefeller Institute. There, Hodgkin also met and interacted with Peyton Rous (Nobel Prize in Medicine in 1966 for his discovery that a virus can cause cancer in chickens).[5]

Although the existence of membrane channels was suggested in the 1850s, the definitive proof of their existence was finally provided in 1976 at the Max–Planck-Institute, Göttingen, by Erwin Neher and Bert Sakmann, who shared the Nobel Prize in Medicine in 1991. These neurophysiologists were able to directly measure the ionic current flowing through an ionic

[5] Through Rous, Huxley met his daughter, Marni, with whom they later married. Huxley, who in 1984 succeeded Hodgkin as the Master of Trinity College, Cambridge, was the grandson of the 19th-century scientist and writer Thomas Huxley mentioned in Chapter 1.

channel.[6] As postulated by the Huxley–Hodgkin model, depending on the polarization of the membrane, channels can become permeable to a given ion, and then *open*, allowing ions to flow through it. This tiny current was the current measured by Neher and Sakmann. This achievement was based on two technical breakthroughs: first, the development by the two Nobel Laureates and their colleagues of a method for sealing glass pipettes on the membrane of a living cell. Second, the invention by the semiconductor industry of field-effect transistors with low–voltage noise able to handle tiny currents.

A final verification of the existence of ionic channels was achieved via *crystallography*, which allowed the actual *visualization* of ionic gates.[7] Using this powerful technique, the first ion channel to be elucidated at a molecular level was the bacterial potassium channel, *KcsA*, which consists simply of a selective filter and two transmembrane helices. This work won Roderick MacKinnon of Rockefeller University a share of the 2003 Nobel Prize in Chemistry.

Several medications act at the level of ionic channels. For example, there are medications that block sodium channels located either in the central nervous system or in the peripheral nervous system. Medications blocking central neurons, which include Phenytoin, Procaine, and Benzocaine, act as anti-epileptics, namely, they protect against seizures. Medications blocking peripheral neurons, which include Lidocaine and Lamotrigine, act as local anaesthetics. Also, there is a class of medications that prevents calcium from entering the calcium channels located at the heart and blood vessel walls. This class, known as calcium channel blockers, includes the medications Diltiazem, Nifedipine, and Verapamil. These medications can reduce blood pressure and relieve chest pain (angina); in addition, they are anti-arrhythmic, i.e., they prevent irregular heart rhythm. Similarly, there exist medications which block heart-sodium channels, including

---

[6] Sakmann spent several years at the University College London investigating aspects of the neuromuscular junction. He decided to concentrate on cellular physiology after attending a lecture by Bernard Katz in Germany.

[7] The Cambridge physicist, William Henry Bragg, shared the Nobel Prize in Physics with his son Lawrence in 1915 for their work in crystallography via the employment of X-rays.

Quinidine, Procainamide, and Propafenone, which also protect the heart from developing arrhythmias.

It should be noted that basic mechanisms employed in neurons have broader validity. For example, the action potential is the universal mechanism of propagating information in the neurons of *all* living organisms. Also, ionic channels are found in the membranes of *all* cells, not only in neurons, and specifically they play a crucial role in the function of *excitable* cells. For example, they are vital for the function of muscle cells, including heart-muscle cells. Also, are important in plants; for example, they mitigate the leaf-closing response of the *Mimosa* plant. In addition, as discussed below, ionic channels are crucial for the communication between neurons and muscle cells and for the transduction occurring in sensory organs.

The synaptic mechanism allowing communication between nerves and muscles is similar to the mechanism described earlier. In this case, the presynaptic neuron is called a *motor neuron* and the synapse is called a *neuromuscular junction*. Since there is now no need for differentiating messages, only one kind of neurotransmitter is used, namely, acetylcholine. When an action potential reaches the presynaptic neuromuscular junction, the presynaptic motor nerve terminal liberates acetylcholine, which then attaches to receptors clustered on the muscle surface membrane. This binding changes the potential at the end of the muscle cell, a process called *depolarization*. This gives rise to an action potential in the muscle cell which makes the muscle fibre twitch. In the devastating auto-immune disease myasthenia gravis, which has inflicted, among others, Laurence Olivier and Aristotle Onassis, self-antibodies destroy the acetylcholine receptors at the postsynaptic part of the neuromuscular junction. Therefore, the acetylcholine molecule cannot transmit a message from a neuron to a muscle. In the healthy neuromuscular junction, unused molecules of acetylcholine are either reabsorbed by the nerve terminal or are degraded via the enzyme anticholinesterase. The medication neostigmine, by inhibiting this enzyme, has a beneficiary effect in patients with myasthenia gravis.

The elucidation of the role of ionic channels in neuromuscular junctions was achieved by Sir Bernard Katz (Nobel Prize in Medicine, 1970), in collaboration with J. C. Eccles (Nobel Prize in Medicine, 1964) and

Stephen Kuffler. Katz's Ph.D. advisor at the University College, London was A. V. Hill (Nobel Prize in Medicine, 1922) who studied Mathematics at Trinity College, Cambridge, and performed ground-breaking work in muscle cells, including the measurement of the heat produced because of muscle action.

As an example of the role of ionic channels in sensory organs, it is noted that the so-called *hair cells of the inner ear* are sensitive mechanoreceptors that are crucial for the sense of hearing. Hair cells can detect sound vibrations in the ear and also detect the direction of gravity in the vestibular apparatus. The associated ionic mechanism is similar with the one described for the neuromuscular junction. The summation of the receptor potentials of adjacent hair cells produce a micro-phonic potential at the cochlea (inner ear).

From histological stains to neurosurgery

We began this chapter by discussing the importance of Golgi's stain for the discovery of the neuron. The use of several other stains has had a significant impact on neuroscience. In particular, the *Nissl* stain, named after the psychiatrist and noted neuropathologist Franz Nissl (1860–1919), allowed the German neuroanatomist Korbinian Brodmann (1868–1918) to identify 43 distinct domains of the human cerebral cortex, which are named after him; Broadman did not map deeper brain structures. The cytoarchitectural structure of the human brain was further analyzed by the Austrian-Greek-Romanian psychiatrist and neurologist Constantin von Economo (1876–1931), and the German neuroscientist Marthe Vogt (1903–2003).[8] This information has been used extensively not only in neurology but also in neurosurgery. The latter was greatly aided from the introduction by the French psychiatrist and neurosurgeon Jean Talairach

[8] Marthe Vogt was the son of Oskar Vogt and Cécile Vogt-Mugnier, who are also known for their extensive cytoarchitectonic brain studies. In 1924, after the death of Vladimir Lenin, the Soviet authorities asked Oskar Vogt, a keen communist, to locate the seat of Lenin's genius. Vogt, faced with Lenin's shrunken brain, failed to identify any particular such region, but still declared Lenin as a "cognitive athlete" due to Vogt's claim that Lenin had a higher density of certain types of cerebral cells. These findings are now viewed with scepticism (Abbott, 2002).

of an atlas supplemented with a three-dimensional coordinate system. This atlas, first constructed in 1967, was later updated by Talairach and the neurosurgeon Pierre Tournoux, and has been extensively used in deep brain surgery. In this regard, it should be noted that there are significant anatomical variations among individuals. Karl Zilles has shown that some brain areas vary in size by a factor of 10 between different people (Zilles *et al.*, 2002).

References

Abbott, A. 2003. A new atlas of the brain. *Nature* 424, 249–250.

Pereda, D., Al-Osta, I., Okorocha, A. E., Easton, A., & Hartell, N. A. 2019. Changes in presynaptic calcium signalling accompany age-related deficits in hippocampal LTP and cognitive impairment. *Aging Cell* 18(5), e13008.

Singh, P. & Manure, A. 2020. Neural networks and deep learning with TensorFlow. In: *Learn TensorFlow 2.0.* Apress, Berkeley, CA. https://doi.org/10.1007/978-1-4842-5558-2_3.

Sun, J., Pang, Z. P., Qin, D., Fahim, A. T., Adachi, R., & Südhof, T. C. 2007. A dual-Ca 2+-sensor model for neurotransmitter release in a central synapse. *Nature* 450(7170), 676–682.

Zilles, K. *et al.*, 2002. Architectonics of the human cerebral cortex and transmitter receptor fingerprints: Reconciling functional neuroanatomy and neurochemistry. *European Neuropsychopharmacology* 12(6), 587–599.

3

INTRODUCTION TO NEUROANATOMY

I will never forget my first lecture in neuroanatomy in medical school when the lecturer said: "the most important anatomical characteristic of the brain is that everything is connected with everything else". Indeed, the complexity of the interconnections of the neurons of the human brain is astounding. However, this complexity is not uniform across the brain; it depends crucially on the specific brain areas involved. These different areas will be briefly summarized below and their basic functions, including their role in memory, will be discussed. An excellent reference is Watson *et al.* (2010).

Neurological mechanisms related to memory are discussed in detail in Chapters 9 and 10. In what follows, in order to describe the role of specific parts of the brain involved in different types of memory, a brief description of these different types is presented.

Working memory is the capacity of the brain to maintain a limited amount of information for a short time. This information can then be used to facilitate other cognitive functions, such as learning and reasoning. For example, it allows remembering, rearranging, and evaluating a small group of items, as well as placing newly acquired information within the framework of existing knowledge and plans.

Declarative memory, also called *explicit* memory, is a type of long-term memory referring to the capacity to recollect facts and events, and to be aware of the meaning of these recollections. This memory is separated into *episodic* and *semantic*. The former captures information about events; namely, what happens, where, when, and in what sequence (from the Greek word "episode" meaning "event"). A subset of this memory is the so-called *auto-biographical* memory; for example, the capacity of individuals to remember the restaurant where their 21st birthday was celebrated. Semantic memory refers to general factual knowledge that is independent of personal experience and the temporal-spatial context in which this knowledge was acquired; for example, remembering that Athens is the capital of Greece.

Non-declarative memory, also called *implicit* memory, refers to the ability to perform automatically certain procedures. This includes procedural memory, for example "remembering" how to ride a bike, or perceptual skills, such as ensuring correct grammar.

Habit memory, is a lesser-known type of memory. It involves learning a specific task after repeated training, without the employment of conscious associations between stimuli and responses. In this process, the learned task is rigidly organized and cannot be used if the task is slightly altered. For example, memorizing a part of a poem in a foreign language after mechanical repetitions.

In the early 1800s, it was realized that the brain is the continuation of the spinal cord.[1] In the 1830s, Marshall Hall (1790–1857) stated that the spinal cord consists of a chain of units that function as "reflex arcs", controlling movement. For some animals, the importance of reflexes is vital, and for such organisms many operations bypass the control of the brain. For example, the octopus' tentacles are capable of making decisions without input from its brain; actually, among the total of 500 million neurons comprising the octopus neural system, 350 million neurons are to be found along its arms (Starr, 2019).

According to Hall, a reflex is achieved via the integration of sensory and motor nerves originating from specific segments of the spinal cord.

[1] It is likely that this fact was known to Hippocrates because he had noted that a severance at the cervical level produces quadriplegia, whereas a severance at the thoracic level produces a permanent paraplegia.

As discussed in Chapter 2, spinal cord reflexes were extensively investigated by Sir Charles Sherrington. Hall's work was continued by John Hughlings Jackson (1835–1911).[2] Reflexes include blinking, coughing, being startled by a sudden noise, and pulling a limb away from a painful stimulus. Soliciting a reflex is part of a neurological examination: when nerve endings in the kneecap-tendon sense a sudden tug from a physician's mallet, they send impulses at 200 miles per hour to the spine. There, a single synapse separates each sensory neuron from each motor neuron; the latter is responsible for the subsequent knee-jerk.

According to the modern view, the spinal cord is a thick bundle of nerve fibres that conducts a steady stream of messages between the brain and the body. These messages include sensations of touch, pressure, pain, and proprioception. At its uppermost part, the spinal cord enters the skull and then grows thicker and bulbous, forming the so-called "brainstem". This part of the brain consists of three segments called "medulla", "pons", and "midbrain". The first two, control vital functions, including breathing, body temperature, and blood pressure. Thus, a haemorrhage even in a tiny part of these two segments has catastrophic effects.

THE CEREBELLUM, THE THALAMUS, AND THE BASAL GANGLIA

The *cerebellum* is located on the roof of the pons. It has an astonishingly large number of neurons, about 70 billion, which is truly remarkable, considering that the entire cerebral cortex has only about 17 billion neurons.

Two important anatomical structures surround the top part of the brainstem: these structures, which provide a link between the evolutionarily *old* and *new* parts of the brain, are the *thalamus* and the *basal ganglia*. The thalamus receives input from all senses except smell, and then relays them

[2] Jackson claimed that the evolutionary organization of the nervous system consists of three levels: the lowest level, comprising the spinal cord and the medulla, which controls automatic movements; the middle level, comprising the motor area of the cortex, which controls simple movements; and the upper level, comprising the prefrontal area of the cortex, which controls more complicated movements. Jackson's name is attached to the characteristic "march" of symptoms occurring in focal motor epilepsy, known as the "Jacksonian march".

to appropriate parts of the cortex for further processing. The basal ganglia consist of a cluster of structures including the so-called *substantia nigra* (black substance), which produces the neurotransmitter dopamine. Other structures of basal ganglia include the *subthalamic nucleus*, as well as the *striatum* which consists of the *putamen* and the *caudate nucleus*.

The putamen and the cerebellum are crucial for procedural memory. The cerebellum also coordinates movement, gait, and posture, whereas the basal ganglia control automatic movement associated with complex volitional actions. In particular, the cerebellum provides predictions about future motor and sensory states (Diedrichsen & Bastian, 2014). In the basal ganglia there exist the so-called direct pathways that facilitate movement, as well as the so-called indirect and hyper-direct pathways that suppress movement (Turner & Pasquereau, 2014). Recently, it has been established (using a technique called *virus tracing*) that there are direct connections of the cerebellum and basal ganglia with specific parts of the cortex.[3] Through these connections, these structures not only affect the control of movement but are also able to influence several aspects of cognitive behaviour, including planning, working memory, attention, and visual-spatial perception (Dum *et al.*, 2014).

The cerebral cortex and its neighbours

The top part of the brain, namely the layer closest to the skull, forms the *cerebral cortex*, which is split into two symmetrical parts, namely the right and left hemispheres. Jackson was the first to suggest that the two hemispheres deal with different mental functions. For example, the left hemisphere is crucial for logical functions and for language, whereas the right hemisphere is more important for creative functions, such as music and art. Jackson claimed that the two hemispheres inhibit each other. This suggests that damage to the left hemisphere weakens its inhibition on the right hemisphere and this may liberate the creativity capacity of the affected patients. Indeed, there have been several published reports of individuals with neurological injuries due to stroke or head trauma, describing how

[3] Namely, the motor, premotor, prefrontal, posterior parietal, and inferior temporal areas of the cortex (these parts are discussed later in this section).

these patients began practising visual art following their brain injury (Zaidel, 2014). In addition, outbursts of creativity have been observed in some patients suffering from dementia whose damage is localized in the left hemisphere (Miller & Hou, 2004).

The cerebral cortex, in addition to having approximately 17 billion neurons, it also has about 61 billion glia. In each cubic millimetre of cortical *grey matter* there are approximately 100,000 neurons, 100 million synapses, and 4 kilometres of axons (most of this axonal wire is used for *local* connections). Neuronal cell bodies have a grey colour in dead, preserved brains, and this is the origin of the term grey matter. The *white matter*, which corresponds to the *distant* wiring of the brain, contains about 9 meters of axons per cubic millimetre.

The majority of the cerebral neurons belong to a specific type called *pyramidal.* Among them, 80% possess prominent dendrites and use the excitatory molecule glutamate as their neurotransmitter; the remaining 20% do not have densely spiny dendrites and use GABA as their neurotransmitter forming inhibitory synapses.

Our cortex, in order to accommodate its extensive growth (in comparison to our evolutionary predecessors) forms a variety of topologically complicated folds. Most other mammals have a smooth cortex with at most a few folds. Incidentally, the cortex is also well developed in two other groups of mammals, the primates and the dolphins. Each hemisphere of the cerebral cortex consists of four large lobes: the *occipital*, *temporal*, *parietal*, and *frontal* lobes.

The occipital lobes are concerned with the early phase of visual processing; for this reason, each of them is called "primary visual cortex". It should be noted that all senses have their own *associated areas*, where more sophisticated information is extracted. For example, recognizing images and summoning associated memories and emotions, does not happen in the primary visual cortex, but in its associated areas; this is discussed in Chapter 12.

The temporal lobes are involved in a variety of mental functions, such as further visual processing, including face and object recognition, emotional processing associated with hearing, and processing of language. An almond-shaped structure, called the *amygdala*, is located at the front part of each temporal lobe. This structure plays an important role in emotions, including dealing with fear, phobias, and flashbacks. The upper part of the left temporal lobe contains the so-called *Wernicke's area*, which is critical for speech comprehension and the processing of semantic aspects of language. Dyslexia is associated with a defect in Wernicke's area resulting in difficulty with word comprehension.

The *hippocampus* is located within the middle temporal lobe. This structure is anatomically connected with the *memory system of the medial temporal lobe*, which consists of the *perirhinal*, *entorhinal*, and *para-hippocampal* areas. These areas, together with the hippocampus, are crucial for episodic memory. The first evidence linking this system with episodic memory was established by the neurologist Brenda Milner, who followed Henry Molaison from 1953 until his death in 2008. This famous patient, known as HM, when he was 27 years old, had a bilateral medial temporal lobectomy to alleviate epileptic seizures. Following his surgery, HM could not form new episodic memories as exemplified by the fact that he was greeting Milner every time he met her as if he was meeting her for the first time. However, his intelligence was intact, and he had normal working and procedural memories (Dittrich, 2016). Also, he could remember people he had met before his operation and he retained his gentle, polite, calm personality.

The medial temporal lobe is not needed for working memory, visual perception, or for habit memory (Christiansen & Müller, 2014). Indeed, after bilateral damage to these structures, although a patient is unable to acquire new declarative memories, memories acquired before the damage remain intact. Furthermore, the capacity for working and habit memory is also preserved.

The parietal lobes are involved in the processing of touch, pressure, and pain sensations, as well as in processing information about the state of muscles and joints. The posterior part of the parietal cortex provides a functional connection between the sensory and motor areas of the brain. By integrating sensory and motor feedback signals, neurons in this area can determine the position of the limbs at any given time and predict forthcoming motion. The parietal lobes are also involved in episodic memory (Berryhill *et al.*, 2007).

The right parietal lobe allows us to become aware of the outside world, as well as to be aware of our body's configuration and its movements. A dramatic result of extensive damage of the right parietal lobe is the phenomenon of *hemi-spatial neglect*: the patient does not perceive the left half of the visual space. For example, patients may leave the left half of the food on their plate untouched or may copy only the left half of an object. An extreme form of this phenomenon is *somatoparaphrenia*, where patients insist that their left hand belongs to someone else. The inferior part of the parietal lobe, which has expanded significantly in humans in comparison with other species, will be discussed in Chapter 6.

The frontal lobes are involved in a variety of processes, some of which are highly developed in humans, including long-term planning, complex problem solving, and humour appreciation (Shammi & Stuss, 1999). A narrow strip of each frontal lobe forms the *motor* cortex, which controls movement. The left posterior part of the frontal lobe forms *Broca's area*, which controls speech production.

The anterior part of the cortex is called the *prefrontal* cortex. The dorsolateral prefrontal cortex is vital for working memory: when a phone number is flashed on a screen and a subject is asked to remember it, there is activation in this part of the brain, and it remains activated several dozen seconds after the task is completed. The medial prefrontal cortex is activated when one thinks about one's self, as well as during interactions with other persons (but not with objects) (Eagleman, 2015).

In addition, the normal function of the prefrontal cortex is necessary for a person to behave in a moral, proper, sexually appropriate way, and to be able to control impulses. The left frontal lobe injury of a person named, Phineas Gage, and the subsequent changes in his personality have been discussed by numerous authors.[4] Patients with damage in their left

[4] It should be noted that the medical description of Henry Bigelow, a Harvard physician who examined Gage in 1849, is rather vague. His text includes statements such as, "the equilibrium [...] between his intellectual faculties and his animal propensities seems to

prefrontal cortex may withdraw from social activities, a condition called *pseudo-depression*. On the other hand, patients with damage to their right prefrontal lobe, may appear euphoric in situations where such behaviour is highly inappropriate; in rare cases, they may even present with mania.[5]

Importantly, the prefrontal cortex resolves conflicts arising from ambiguous sensory information and from uncertainties arising from lack of precise memory information.

It is worth noting that the prefrontal cortex is crucial for analogical thinking (also known as relational reasoning), namely the ability of the brain to extract analogies between disparate situations (Crone, 2014: p. 25).

Finally, the prefrontal cortex is involved in decision making, namely selecting a particular course of action among a plethora of possibilities. Apparently, the rear part is involved with immediate action selection, whereas the more frontal part is responsible for maintaining long-term tasks. Damage to the frontal cortex causes the so-called *frontal lobe syndrome*, also known as *dysexecutive syndrome*, which is characterized by a constellation of cognitive and behavioural symptoms: the former includes impairment of short-term memory, attention, speech, planning, and reasoning, whereas the latter includes difficulty to inhibit frustration, anger, aggression, excitement, and sadness. An important feature of the prefrontal cortex is that it keeps information over time; this is crucial for reflection and planning.

The limbic system

The *limbic system* is an important functional system, which is vital for emotions. This system involves different anatomical structures, and in particular, the *cingulate* cortex, which is part of the medial part of the

---

have been destroyed"; also, Bigelow wrote that friends said, "Gage was no longer Gage"; and that he was now "fitful and irreverent". A recent book by Macmillan (Macmillan, 2000) claims that many of the stories about Gage are inaccurate and suggests that later he actually recovered most of his mental functions.

[5] Mania, as discussed in Chapter 16, is usually associated with bipolar patients. However, in rare cases, mania can also appear after damage in the right ventral prefrontal area or in the right basal ganglia; this is usually caused by a stroke.

cerebral cortex and it includes the cingulate gyrus, which lies immediately above the corpus callosum. The anterior cingulate, which is evolutionary the oldest part of the cortex, is crucial for pain awareness. Also, this is the part of the brain responsible for the feeling of individuals that they have *ownership* of their self-willed actions.

Loss of the sense of self gives rise to one of the strangest disorders in neuropsychiatry called *Cotard syndrome*, where patients feel as if they are dead. V. S. Ramachandran describes such a patient in (Ramachandran, 2011: p. 279). Answering the question "What brings you to the hospital", the patient's replied, "Not much can be done: I am a corpse", and later he added, "Yes. I don't exist. You could say I am an empty shell. Sometimes I feel like a ghost that exists in another world". In general, patients with Cotard's syndrome believe that they are dead, or that they have lost their soul, or that are rotting inside, lacking a specific organs or limps. Incidentally, a multitude of antipsychotic medications have been reported to reduce the symptoms of such patients. If medications fail, electroconvulsive therapy is a useful therapeutic option.

In addition to the cingulate cortex, the limbic system involves different other anatomical structures, including the following: Several regions of the cortex and specifically the part of the frontal cortex called *orbitofrontal* cortex, which is located above the eye-sockets. This part of the cortex integrates various body-signals, giving rise to a variety of states, such as being hungry, nervous, excited, embarrassed, thirsty, and joyful (Eagleman, 2015). This domain inhibits inappropriate behaviour and encourages the delay of immediate rewards in favour of long-term advantages. Other parts of the cortex involved in the limbic system are the *olfactory bulbs* which process smell, the hippocampus, the entorhinal cortex, the *nucleus accumbens*, sub-cortical areas of the amygdala, the *hypothalamus*, and part of the thalamus.[6]

For awareness to take place it is necessary that the brain is kept at a sufficient level of arousal. This is achieved via the co-operation of hundreds of *neuronal nuclei*; a nucleus is a set of neurons with a unique neurochemical

---

[6] Part of the hypothalamus forms the *mammillary bodies* which receive signals from the hippocampus and project them to the thalamus.

profile. Nuclei necessary for arousal are found in a structure located in the midline of the upper brain stem and the hypothalamus; these nuclei form the *reticular activating system*. There exist five such nuclei that receive input from brain-stem nuclei and the cerebral cortex and have connections with every part of the cortex. The nuclei of the reticular activating system release modulatory neurotransmitters, including dopamine, serotonin, acetylcholine, and norepinephrine.

Incongruity in maturation scales

What is the earliest we remember? It is thought we are unable to make memories before the age of three. This is related with the fact that different parts of the brain mature at different rates. For example, the hippocampus matures by the age of 3 and this prevents us to have memories before that age. It is interesting that Wernicke' area, which as stated earlier is involved in language comprehension, matures by 12 months, whereas Broca's area, which is involved with speech production, matures by 18 months. This leads to babies feeling frustrated, since they can understand but cannot speak, resulting in the "terrible twos". Actually, the brain's language areas begin to function much earlier. Indeed, functional MRI studies reveal that speech activates the brain's language areas of babies as early as the age of 2 months old (playing speech backwards does not cause activation in these areas) (Dehaene, 2014: p. 238).

The *nucleus accumbens*, which produces dopamine, the pleasure neurotransmitter *par excellence*, matures much earlier than the orbitofrontal cortex. Thus, teenagers are emotionally hypersensitive. Actually, the frontal lobe, which is vital for controlling emotions, does not reach full maturity until the early 20's (Eagleman, 2015). Studies established that when subjects were shown pictures of individuals with whom they were in love, there was activation in their brain's *subcortical* structures. Taking into consideration that these structures are not under logical control, the "madness of being in love" can be justified. The state of being in love is even more irrational for young people, due to the delayed maturation of the frontal cortex. Incidentally, being in love is characterized by high levels of the hormone cortisol, which reveals high levels of stress. Perhaps, evolution's tendency for stability explains the fact that being in love is usually a transient state.

References

Berryhill, M. E., Phuong, L., Picasso, L., Cabeza, R., & Olson, I. R. 2007. Parietal lobe and episodic memory: Bilateral damage causes impaired free recall of autobiographical memory. *Journal of Neuroscience* 27(52), 1441–1442.

Christiansen, M. H. & Müller, R.-A. 2014. Cultural recycling of neural substrates during language evolution and development. In: Gazzaniga, M. S. & Mangun, G. R. (eds.) *The Cognitive Neurosciences.* MIT Press, pp. 675–682.

Crone, E. 2014. Cognitive control and affective decision making in childhood and adolescence. In: Gazzaniga, M. S. & Mangun, G. R. (eds.) *The Cognitive Neurosciences.* MIT Press.

Dehaene, S. 2014. *Consciousness and the Brain: Deciphering How the Brain Codes Our Thoughts.* Viking.

Diedrichsen, J. & Bastian, A. 2014. Cerebellar function. In: Gazzaniga, M. S. & Mangun, G. R. (eds.) *The Cognitive Neurosciences.* MIT Press, pp. 451–460.

Dittrich, L. 2016. *Patient HM: A Story of Memory, Madness and Family Secrets.* Random House.

Dum, R. P., Bostan, A. C., & Strick, P. L. 2014. Basal ganglia and cerebellar circuits with the cerebral cortex. In: Gazzaniga, M. S. & Mangun, G. R. (eds.) *The Cognitive Neurosciences.* MIT Press, pp. 419–434.

Eagleman, D. 2015. *The Brain, the Story of You.* Canongate.

Macmillan, V. 2000. *An Odd Kind of Fame Stories of Phineas Gage.* MIT Press.

Miller, B. L. & Hou, C. E. 2004. Portraits of artists, emergence of visual creativity in dementia. *Archives of Neurology* 61, 842–844.

Ramachandran, V. S. 2011. *The Tell-Tale Brain: A Neuroscientist's Quest for What Makes Us Human.* W. W. Norton & Company.

Shammi, P. & Stuss, D. T. 1999. Humour appreciation: A role of the right frontal lobe. *Brain* 122(4), 657–666.

Starr, M. 2019. *Science Alert.* ScienceAlert. June 26. https://www.sciencealert.com/here-s-how-octopus-arms-make-decisions-without-input-from-the-brain.

Turner, R. S. & Pasquereau, B. 2014. Basal Ganglia function. In: Gazzaniga, M. S. & Mangun, G. R. (eds.) *The Cognitive Neurosciences*. MIT Press, pp. 435–450.

Watson, C., Kirkcaldie, M., & Paxinos, G. 2010. *The Brain: An Introduction to Functional Neuroanatamy*. Academic Press.

Zaidel, D. W. 2014. Creativity, brain and art: Biological and neurological considerations. *Frontiers in Human Neuroscience* 8 (389).

4 PERCEPTION AS THE SOLUTION OF AN INVERSE PROBLEM

Among the most important philosophical questions posed by the ancient Greeks were the following: What are the basic characteristics of Gods? Can virtue be taught? Can truth be defined? Does innate knowledge exist? How do we comprehend reality? Various philosophical approaches to the latter question are briefly discussed in the following.

Several Greek philosophers associated knowledge with sense perception, especially visual perception. Alcmaeon, 5th Century BC, may have investigated the back of the eye using a piece of wood, in order to establish the connection of the eye with the brain. The importance of visual perception for the ancient Greeks is consistent with the observation that the most common verb in Greek for "to know" is *οἶδα*, which means "to see".

But how reliable is sensory information? Mentally disturbed people, as well as people who are dreaming, perceive their *own* worlds. How can we be sure that the *external world* is indeed the true reality, as opposed to the worlds of these people? The first relevant convincing argument was provided by Heraclitus, who was mentioned Chapter 1. This great philosopher emphasized that "the external world is the *same* for all of us", where by "us" he specified that he meant people who are mentally healthy and awake.[1] So, according to Heraclitus, the fact that our sensations and experiences to a large extent overlap implies that there exists an objective external reality. The influential ideas of Descartes are discussed in Chapter 11.

Immanuel Kant (1724–1804), under the influence of Plato, claimed that we do *not* experience "things themselves" but only their "appearances". According to him, the gap between the transcendental reality of things themselves and of what we perceive cannot be bridged: "this gap in knowledge can never be bridged, [...], it is a question which no human being can possibly answer".

However, according to Parmenides, who along with his teacher Xenophanes founded in Italy, around 480 BC, the famous philosophical school of Elea, there is no such gap: "the thing that is" (*ον*) and the "thought" (νοείν) are one and the same. Several scholars have interpreted this statement as something like, "The same thing is for being and for thinking". Parmenides' position is considered to be the endorsement of the absolute superiority of thinking and therefore of reason and logic, over senses.

The physicist and philosopher Ernst Mach also concluded that Kant's position is irrelevant but emphasized the superiority of the knowledge obtained via our senses:

"There exists no further object independent of our sensations [...]. We thus know only appearances, never the Thing-in-Itself, just the world of our own sensations. Therefore, we can never know whether there exists a Thing-in-Itself. Consequently, it makes no sense to talk about such notions".

The positions of Mach (who will be discussed further in the second volume) and Parmenides provide the two dialectically opposite answers to the basic question of whether we can comprehend the world via our senses or via our reason.

The pre-eminence of sensory information was emphasized by several Western philosophers prior to Mach. The Irish philosopher Bishop Berkeley, as well as the philosophers of the Enlightenment Denis Diderot

---

[1] The word *idiot* derives from the Greek word *ἰδιώτης*, which means "one who is in their private, own world", instead of the world shared by others.

and Jean le Rond d'Alembert, following Aristotle and Locke, claimed that knowledge can be acquired only via the systematic analysis of sensory experiences. Auguste Comte, the strongest exponent of the philosophical movement of *Positivism* since Aristotle, stated: "Nothing is intelligible which is not first seized by the senses".

The distinguished physicist and philosopher Ludwig Boltzmann (who also will be discussed in the second volume) could not decide between the prominence of senses or thinking:

"If there is something else behind perception, how can we ever find out what it is? On the other hand, if there is none, does it mean that a landscape on Mars fails to exist because no conscious being ever gazed down on it?"

The French philosopher Henri Bergson claimed that higher mental functions and especially creativity, cannot be explained by reason and science alone. According to him, humans make crucial use of their *intuition*. Bergson also noted the importance of *continuity* and *duration* at the level of cognition: what is perceived as reality is what, in retrospect, has been experienced via the sum of a continuous flow of sensations, employing unconscious processes: "Pure intellect is a contraction [...] that is exercised in the deep unconscious" (Bergson, 1911). Incidentally, Bergson in his attempts to explain the process of self-organization, introduced the notion of *vital impetus* ('élan vital').

The Scottish Enlightenment philosopher David Hume, although appreciated the importance of senses, did not think that perceptions are sufficient for understanding the laws of nature. In addition to perceptions, he emphasized the necessity of the employment of cognitive processes and in particular of associations.

The conflict between senses and the intellect is perfectly expressed in the famous dialogue of Democritus (see fragment D 125 in Schrödinger, 2014):

"Intellect: Sweet is by convention and bitter by convention, hot by convention, cold by convention, colour by convention; in truth, there are but atoms and the void. The Senses: Wretched mind, from us you are taking the evidence by which you would overthrow us? Your victory is your own fall".

The distinguished logician and mathematician Bertrand Russell (Nobel Prize in Literature, 1950) founded the branch of philosophy called *Analytical Philosophy*. This was an attempt to base philosophy on logic so that "vague or misleading statements would be transformed into logically correct propositions". The Vienna Circle of Logical Empiricism, under the influence of Ernst Mach, accepted the importance of sense-data, whereas, under the influence of Russell and his brilliant student Ludwig Wittgenstein, placed logic and language at the core of human thinking.[2] Following positions expressed in Wittgenstein's book *Tractatus Logico-Philosophicus*, the Vienna Circle claimed that there exist three types of propositions: synthetic (empirical) which can be analyzed via scientific methods, analytical (*a priori*) which are tautological (namely, they follow logically from a set of axioms), and nonsensical which cannot be refuted via empirical evidence.[3] In this sense, the analytical philosophy of Russell and the extension of this philosophy by Wittgenstein and the Vienna Circle can be characterized as follows: logic and the structure of language are the essential ingredients dictating meaningful thinking.

Wittgenstein has famously said that his goal was to understand what *can* and *cannot* be said and to say *correctly* what can be said. His statement, "[...] thus, I have solved the problem of philosophy", is often used to criticize his arrogance. However, this is only the first part of his statement; he continues:

"But this shows how little is achieved when this problem is solved. Indeed, there is what I have written, and what I have not; and the latter is much more important".

This statement reflects the intuitive understanding of Wittgenstein that fundamental philosophical questions cannot be answered without a deeper understanding not only of the structure of language but also of the nature of other basic mental functions and especially of unconscious processes (what Wittgenstein referred to as the "mystical").

---

[2]According to philosopher Victor Kraft (Kraft, 1953: pp. 24–25), the name *Logical Empiricism* was the term favoured by the Viennese philosophers Moritz Schlick and Rudolf Carnap, in order to delineate differences with the *Positivism* of Comte and Mach, and to emphasize the synthesis of logic and sense-data.

[3]Immanuel Kant had also distinguished synthetic from analytical propositions.

Taking into consideration the fact that the main tool used for the analysis of any phenomenon is the brain, it is self-evident that this analysis will be incomplete without a deep understanding of how the brain functions.

For example, several neuropsychological experiments have established the ability of the brain to go from the specific to the general. However, the members of the Vienna Circle ignored this crucial property of the brain.

Bertrand Russell, after becoming aware of important developments in the understanding of basic neuronal mechanisms of perception in the late 1930s, wrote:

"A physical process starts from a visible object, moves to the eye, there changes into another physical process in the optic nerve, and finally produces some effects in the brain simultaneously with which we see the object from which the process started, the seeing been something "mental", totally different in character from the physical processes, which preceded and accompanied it. This view is so queer that metaphysicians have invented all sorts of theories designed to replace it with something less incredible".

The last sentence clearly expresses Russell's scepticism of the ability of philosophers to solve the problem of epistemology. It will be shown later that the above statement of Russell, who was not only a great logician but also the creator of analytical philosophy that is based on precision, is *not* accurate. This is not meant as a criticism of the great man but as an endorsement of his conviction that it is impossible to answer fundamental questions regarding perception outside a rigorous scientific framework based on neuroscience.

VISUAL PERCEPTION AS THE SOLUTION OF AN INVERSE PROBLEM

Visual perception is achieved via a process of *deconstruction* and *reconstruction*. For example, for the recognition of a specific object, the brain's input is the light emitted from this object or more precisely the distribution of the constituents of the light, the *photons*, which are discussed in detail in Chapter 12. As soon as the photons arrive at the

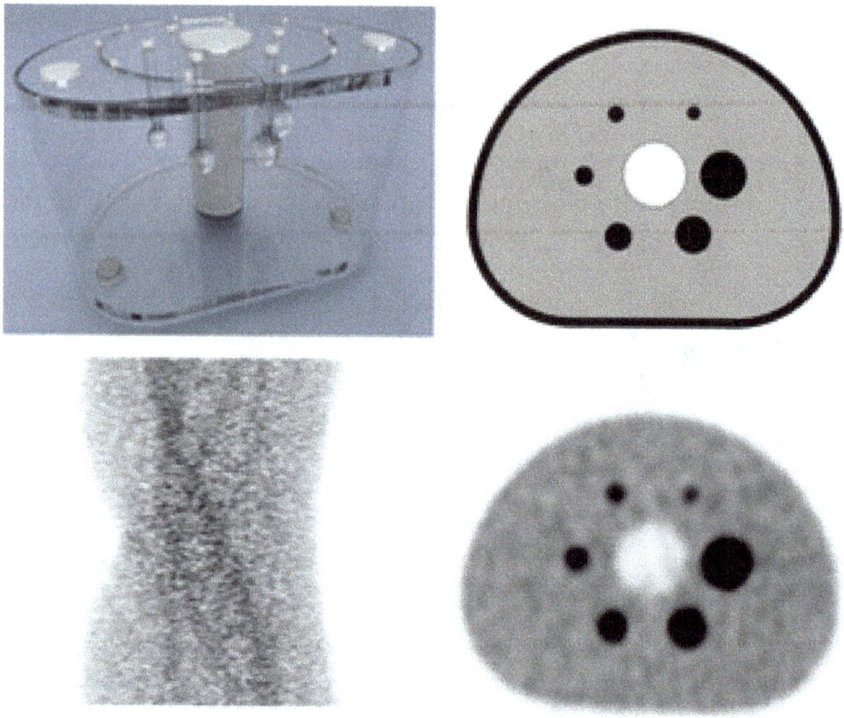

Figure 4.1. The NEMA IEC Body phantom.
Source: Reprinted from SuperTech, retrieved December 9, 2021, from https://supertechx-ray.com/QualityControlPhantoms/pro-project-pro-nm-nema-nu2.php.

retina, a complicated neural process is initiated that finally gives rise to the creation of the *mental image* of the object. This is the reason Gerald Edelman (Nobel Prize in Medicine in 1972) writes: "Every act of perception is an act of creation". Perhaps the word "creation" should be replaced by the phrase "solutions of an inverse problem".

A useful paradigm for appreciating this suggestion is provided by the functional medical imaging technique of *Single Photon Emission Computed Tomography* (SPECT). In what follows, I will discuss the employment of SPECT for the reconstruction of the object shown in the top left image of Figure 4.1. This object is a phantom consisting of a fillable large compartment, which plays the role of a background, where six fillable spheres of various diameters are placed.

A two-dimensional diagram of a horizontal section of the phantom is shown in the top right image of Figure 4.1. As is explained in detail in Chapter 14, in order to use SPECT to reconstruct this two-dimensional image, a radiotracer is placed inside the three-dimensional object. As a result of radioactive decay (discussed in the second volume), this radiotracer emits photons, which are captured by the detectors. The SPECT scanner collects these photons and stores their distribution in the form of what is called a *sinogram*, shown in the lower left image of Figure 4.1. The *inverse problem* for SPECT involves the reconstruction of the image of an object from the knowledge of its corresponding sinogram. The use of a complicated mathematical formula yields the two-dimensional image shown in the lower right part of Figure 4.1.

The above example provides an illustration of the power of mathematics: using the information specifying the distribution of the photons emitted from a given object as *input* in a mathematical formula, the two-dimensional image of this object can be reconstructed.

The distribution of the photons emitted by a given object is also the external input used by the brain in visual perception. Could it be that the brain uses analogous computations for the creation of the mental image of this object? The answer appears negative: for the reconstruction presented in the lower image of Figure 4.1, the computer follows well-defined algorithmic steps, whereas the reconstruction implemented by the brain, in addition to utilizing the distribution of the emitted photons, relies crucially on *prior information* stored in appropriate neural circuits. This information, which is used in a highly dynamic way, has been created via earlier interactions with external or internal stimuli and has been stabilized via intricate processes involving memory. The precise mechanism which allows the brain to employ in a most efficient manner this prior information remains a *mysterium tremendum*.

Hermann von Helmholtz knew that there does not exist a one-to-one correspondence between an object and its mental image. In his book, *Physiological Optics*, he uses the term "unconscious inference" to characterize how our visual apparatus unconsciously computes the *best interpretation* compatible with incoming sensory data. It is fascinating that eight centuries earlier, the Arab scientist Abu Ali al Hasan, known in the West as Alhazen (965–1040), also postulated that our brain, without our awareness, jumps to conclusions beyond its available sensory data. He called this process "unconscious inference"! He also noted that illusions are generated via this process.

The juxtaposition between the reconstruction of an object via medical imaging and its reconstruction in visual perception serves an additional useful purpose: in the same way that a physician who looks at a medical SPECT image is unaware of the complicated *mathematical computations* needed for the construction of this image, a person who becomes aware of a given external percept is completely unaware of the complicated *neural computations* responsible for the construction of the mental image of this percept. These computations are unconscious!

In summary, *visual perception involves the process of deconstruction and the solution of a complicated inverse problem. The latter solution is achieved via unconscious neural computations, which demonstrates the crucial importance of unconscious processes.*

Taking into consideration the time needed in implementing the relevant reconstruction, it follows that awareness *cannot* be instantaneous. Hence, the statement of Bertrand Russell that we see something mental *simultaneously* with the object which caused this neural image cannot be accurate.

From the above discussion, it becomes clear that in visual perception, *awareness is preceded by an unconscious process.* Actually, this time sequence is, apparently, a fundamental feature of *all* mental processes. This leads to the first hypothesis mentioned in the prologue.

The existence of an unconscious process is a *necessary*, but not a *sufficient* requirement for the generation of a conscious experience. Indeed, not *all* unconscious processes lead to awareness[4]: if a stimulus is too weak, too brief, or too noisy, then it is *subliminal*, namely, it will be perceived by the brain, but it will *not* give rise to a conscious experience (awareness).

[4] Dehaene refers to the particular subset of unconscious processes that give rise to consciousness as "preconscious".

How do we know that subliminal processes do exist? There is a substantial body of evidence, based on experiments in psychology and on studies utilizing brain imaging. As an example of a typical psychological experiment, it is mentioned that, after subjects were exposed subliminally to the word *river*, these subjects associated the double-meaning word *bank* with the word *boat* instead of the word *money*. In another experiment, volunteers looked at photos of naked bodies using the technique of *masking*, which made the photos invisible. In particular, a visual stimulus is rendered invisible if presented briefly, usually for only 30 milliseconds, and then is immediately followed by a masking visual stimulus of equal duration. Still, sensitive tests revealed that the images of naked women attracted the attention of straight men, whereas images of naked men repulsed them (Koch, 2012: p. 48).

An example of experiments using functional imaging techniques is illustrated in Figure 4.2. These images were constructed by the leading neuroscientist Stanislas Dehaene and his group using functional Magnetic Resonance Imaging (fMRI): the image on the right of Figure 4.2 depicts brain activation associated with a written word that was presented on a computer screen under masking conditions for only 32 milliseconds; this word was not perceived consciously. The same word presented for a longer period became visible; the associated brain activation is depicted on the left of Figure 4.2. It is clear from these

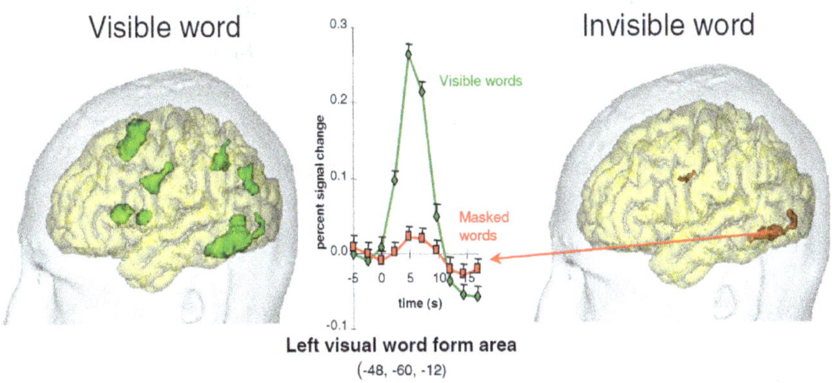

Figure 4.2. Brain activation in an invisible (right) as well as visible word (left).
Source: Dehaene & Changeux (2011).

images that this particular subliminal process gave rise to a localized activation in the brain, whereas its conscious experience gave rise to widely distributed brain activation.

In order to understand the genesis of these different types of activation, it is necessary to briefly review aspects of the brain's connectivity.

Short-, middle-, and long-range connectivity

Italian cyberneticist Valentino Braitenberg established that cortical connectivity supports the cooperation of two systems: a system distributing information *locally*, via *intra-cortical* axonal arborizations, and a system distributing information *globally*, via *cortical-cortical* and *cortical-thalamic-cortical fibres.*

Local connectivity is intimately related to the brain's *modular specialization*, namely, the fact that specific regions of the brain perform specific tasks. Modular specialization is facilitated by the existence of highly specialized neurons. For example, sensory visual neurons respond only to light, whereas sensory hearing neurons respond only to acoustic waves. Experiments in monkeys that are discussed in Chapter 12 have shown that certain sensory visual neurons respond only to vertical bars of light, whereas other respond only to horizontal bars. These *orientation* cells are organized topographically in certain cellular columns in the domain of the brain called V_1. Similarly, cells in the domain V_4 respond only to colour, and cells in the domain V_5 recognize motion. The domains V_1, V_4, and V_5 provide examples of *functional compartmentalization*, namely, examples of the existence of cells which perform a similar task and which are grouped together. Neurons in the visual area V_1 make mostly *local* connections, communicating primarily with their neighbouring areas. For example, V_1 is primarily connected to V_2, which in turn is primarily connected to V_3, and so on.

This local connectivity contrasts with the maze of remarkably rich and reciprocal connections that exist between different parts of the cortex. For example, many areas of the cortex of monkeys have direct connections with more than 10 other areas (Kötter, 2004;

Bakker *et al.*, 2012). Long-distance, global connectivity is achieved via the pyramidal neurons. These large cells, which distribute information across the brain, including information across the two hemispheres, tend to be located in layers II and III of the brain. Cajal was the first to recognize that neurons in the cortex are organized in six layers, denoted by I–VI, with layer I nearest to the skull.

The concept of continuity implies that, in addition to the short- and long-range connectivity, there must also exist a middle-range system. This is indeed the case, as shown in Braitenberg's cytoarchitectonic studies of the 1960s (Braitenberg, 1962). This followed earlier work by the German neurologist Oskar Vogt (Vogt & Vogt, 1919), who had obtained detailed diagrams of the wiring in different cortical areas (Braitenberg worked in Vogt's laboratory in the 1950s). In the last couple of decades, this type of connectivity has been further elucidated: there are long horizontal collaterals in the cortex which emit patchy ramifications over a region of a few millimetres, typically within the same functional compartment (Levitt & Lund, 2002; Voges *et al.*, 2010). A comprehensive review of studies relating cortical connectivity with data from physiological studies is presented in (Potjans & Diesmann, 2012).

THE PRIMACY OF UNCONSCIOUSNESS AND THE ESSENCE OF CONSCIOUSNESS

It is clear from the left-side image of Figure 4.2 that conscious visual perception requires enhanced activation in different parts of the brain. What is the origin of this activation? In order to answer this question, it is necessary to scrutinize the possess of *reconstruction* that takes place during perception: if a percept consists of a set of constituent elements, then the *unconscious structure* of each of these elements is constructed via the solution of a different inverse problem. For example, in visual perception, the solutions of such inverse problems reconstruct the orientation, colour, and motion of a given percept. These reconstructions are achieved by employing unconscious neural computations of specific *local neural circuits* that are located in different parts of the brain. Indeed, in early unconscious visual processing, it can be shown that short *local* loops are of vital importance for analyzing multiple fragments of a scene (Sporns *et al.*, 1991). If the presentation of the percept lasts for a sufficiently long time, then there will be a sufficient amount of incoming energy capable of exciting several of the local neural circuits *associated* with elements of this percept. This gives rise to *global activation*, which facilitates the integration of the solutions to the local inverse problems, yielding a mental image.

The importance of associations

In addition to orientation, colour, and motion, depending on familiarity, further entities associated with the percept will also be reconstructed. Suppose, for example, that a picture of a familiar person's face is presented on a computer screen. The brain, utilizing its exceedingly complex web of associations, constructs unconscious structures of the face, the voice, the body, the name, and several other characteristics of this person. In other words, many neural circuits that are involved in the construction of unconscious structures of several entities associated with this person are activated. In this connection, it is worth noting that there is a part in the temporal lobe, the *fusiform*, dedicated to face recognition. This structure is connected with the amygdala and the prefrontal cortex, which are the two key modulators of emotions. Hence, the presentation of the picture of the face excites neural circuits involved with emotional aspects of the face. Therefore, additional unconscious structures are simultaneously created, providing information about the emotional state of the person (the vital role of emotions is discussed in the following chapter in connection with homeostasis). Furthermore, Broca's and Wernicke's areas, which are responsible for naming and language in general, will also be activated in preparation for communicating with this person.

The *local* neural circuits correspond to Hebb's cell assembly that is discussed in detail in Chapter 10. The *set of different local circuits*, like those associated with the voice, friends, etc, of the given face, correspond to Hebb's *sequence of assemblies* which is also discussed in the same chapter.

The different types of connectivity discussed earlier provide the anatomical basis of the functional connectivity needed for perception. Indeed, local connectivity is ideally suited for the solution of the inverse problems constructed by local neural networks, whereas the global one is employed for the extended activation required for the emergence of consciousness. Interestingly, layers II and III of the brain, which as noted earlier house the pyramidal neurons, are thicker in the prefrontal and cingulate cortices, as well as in the associative areas of the parietal and temporal cortex. These are precisely the areas activated in a variety of conscious experiences. This shows that brain connectivity provides the basis for the optimal integration of the brain's *modular functional specialization*. It is important to note that the process of integration can also occur spontaneously, without external input. Dreaming and imagining constitute two such key examples.

The different types of connectivity used by the brain crucially affect the types of *electrical correlations* observed between different neurons. For example, electrical recordings in monkeys show that the activity of 20–50% of pairs of neurons in cortical and thalamic areas, which are highly interconnected, are broadly synchronized. On the other hand, activations of neurons in the basal ganglia, where connectivity is in the form of parallel loops, are mostly electrically uncorrelated (Bergman *et al.*, 1998; Nini *et al.*, 1995). This is consistent with the fact that *consciousness is usually associated with highly synchronized activity in broad areas of the brain.*

The anatomical basis of the global activation needed for specific conscious processes has been identified by several scientists, including neuro-anatomist Stephanie Clarke (Virgilio & Clarke, 1997). In particular, she has delineated massive connections from the fusiform gyrus to various parts of the brain, including to associative areas of both hemispheres, as well as to Broca's and Wernicke's areas.

In summary, *consciousness is the result of the creation of a mental image, which is generated following the global binding of several local neural circuits. These circuits are crucial for the construction of the mental images associated with the constituent parts of a given percept. The greater the familiarity, the larger the number of parts that are associated with the given percept, and hence, the more enhanced the activation.*

Incidentally, the process of associating different entities with a given percept is mostly unconscious, but some of these associations may become conscious. The realization that various unconscious structures obtained via the solutions of specific local inverse problems refer to a given percept, is facilitated by learning occurring over time. This fundamental mechanism, which gives rise to enriched conscious states, has been described by Gerald Edelman with the phrase that consciousness allows us to live in a "remembered present".

In certain brain abnormalities, the neural circuits responsible for the construction of the unconscious structure of a specific aspect of a percept, are disconnected from the neural circuits involved in the analysis of other elements of the same percept. Such *dissociation syndromes* include *prosopoagnosia*, namely, the inability to recognize a face, and the so-called *Capgras syndrome*, the inability to feel emotional warmth for a loved person. These fascinating syndromes are discussed in Chapter 12.

References

Bakker, R., Wachtler, T., & Diesmann, M. 2012. CoCoMac 2.0 and the future of tract-tracing databases. *Frontiers in Neuroinformatics* 6(30), 1–6.

Bergman, H. *et al.* 1998. Physiological aspects of information processing in the Basal Ganglia of normal and Parkinsonian primates. *Trends in Neurosciences* 21, 32–38.

Bergson, H. 1911. *Creative Evolution*, Henry Holt and Company, New York.

Braitenberg, V. 1962. A note on myeloarchitectonics. *The Journal of Comparative Neurology* 118, 141–156.

Dehaene, S. & Changeux, J.-P. 2011. Experimental and Theoretical Approaches to Conscious Processing. *Neuron* 70(2), 200–227.

Koch, C. 2012. *Consciousness: Confessions of a Romantic Reductionist.* MIT Press, Cambridge, MA.

Kötter, R. 2004. Online retrieval, processing, and visualization of primate connectivity data from the CoCoMac database. *Neuroinformatics* 2, 127–144.

Kraft, V., 1953. *The Vienna Circle: The Origins of Neo-Positivism.* Philosophical Library, New York.

Levitt, J. & Lund, J. 2002. Intrinsic connections in mammalian cerebral cortex. In: Schüz, A. & Miller, R. (eds.) *Cortical Areas: Unity and Diversity.* London: Taylor & Francis, pp. 133–154.

Nini, A., Feingold, A., Slovin, H., & Bergman, H. 1995. Neurons in the globus pallidus do not show correlated activity in the normal monkey, but phase-locked oscillations appear in the MPTP model of Parkinsonism. *Journal of Neurophysiology* 74, 1800–1805.

Potjans, T. C. & Diesmann, M. 2012. The cell-type specific microcircuit: Relating structure and activity in a full-scale spiking network model. *Cerebral Cortex* 24(3), 785–806.

Schrödinger, E. 2014. *Nature and the Greeks and Science and Humanism.* Cambridge University Press.

Sporns, O., Tononi, G., & Edelman, G. 1991. Modelling perceptual drooping and figure-ground segregation by means of active re-entrant connections. *Proceeding of National Academy of Sciences* 88, 129–133.

Virgilio, G. D. & Clarke, S. 1997. Direct interhemispheric visual input to human speech areas. *Human Brain Mapping* 5(5), 347–354.

Voges, N., Schüz, A., Aertsen, A., & Rotter, S. 2010. A modeler's view on the spatial structure of intrinsic horizontal connectivity in the neocortex. *Progress in Neurobiology* 92, 277–292.

Vogt, C. & Vogt, O. 1919. Results of our brain research in a broader context. *Journal für Psychologie und Neurologie* 25, 279–468.

5 FROM HOMEOSTASIS AND CONSCIOUSNESS TO METAREPRESENTATIONS: A UNIQUELY HUMAN CHARACTERISTIC

HOMEOSTASIS, EMOTIONS, AND FEELINGS

Many great philosophers and in particular Plato underestimated emotions, emphasizing instead the supremacy of reason. For example, in his dialogue, *Republic*, Plato presented a tripartition of the soul (Nehamas, 1999). The lowest part is occupied by *Appetites* (Ορμέμφυτα) which involve what today corresponds to instincts, including hunger, thirst, and, sexual desire, as well as what can be obtained with money. This part aims at immediate gratification ignoring long-term consequences. The highest part of the soul, the *Reason* which, in addition to protecting against the pain caused by overindulging in appetites, also protects against the pain caused by all passions. For example, in the *Republic*, where Plato justifies his expulsion of all poets from his ideal city accusing them of corrupting the audience by concentrating on inferior characters and vulgar subjects, it is stated:

"pleasure and pain will rule as monarchs [...] instead of the law and the rational principle which is always and by all thought to be the best".

Plato's decision to ban poetry will be further discussed in the epilogue of this volume. Incidentally, in contrast to Plato, Aristotle defended poetry, arguing that it is morally beneficial. According to Aristotle, tragedies purify emotions via the process of *katharsis*, and place them in the proper *personal* perspective.

According to Plato, reason is the *calculating* element of the soul, which seeks knowledge, and which analyzes the short-term as well as and long-term consequences of various actions. In my opinion, the emphasis on the calculating characteristic of this part of the soul, which corresponds to today's consciousness, reveals the crucial influence on Plato of the power of mathematics. It is widely accepted that, while the early dialogues of Plato reflect Socrates' ideas, for the dialogues of the middle and late periods, Socrates simply becomes the mouthpiece of Plato's own views. According to the philosopher Gregory Vlastos, this was the result of Plato being introduced to the power of mathematics, especially after meeting, on a visit to Sicily, the mathematically knowledgeable philosopher Archytas (Vlastos, 1983). In between appetites and reason, Plato placed the *Spirited* part, which has some vague similarities with today's understanding that certain emotions drive us towards honourable acts. This part is, in general, allied to reason.

Interestingly, this tripartition of the soul provides an example of *analogical thinking*. Indeed, these three parts of the soul correspond to the following three classes that are presented first in the *Republic* and then in *Timaeus*: the appetite part corresponds to the lowest class of the city, namely the *artisans*, which Plato characterizes as the "money-loving" class.[1] Reason corresponds to the *guardians*, who are responsible for the proper functioning of the city. The spirited part corresponds to the *auxiliaries*, who seek honour and are trained to assist the guardians in the governing of the ideal city.

[1]Plato associated money with greed and overindulgence. This is consistent with the following famous statement attributed to Socrates in the Platonic dialogue *Apology*: "I do nothing but go about persuading you, young and old, to have your first and greatest concern not for your body or your money, but for your soul".

To my knowledge, there is only one occasion in Plato's dialogues that irrationality played a positive role: in the dialogue *Phaedrus* (Plato, 1995), Plato revisits the concept of *eros*, which is love dominated by sexual desire. This concept was already analyzed in the *Symposium* (Plato, 1989), where Socrates, in his second speech, presents an ascending series of arguments that begin with establishing the attraction of the older lover (*erastis*) to the *beauty* of the young boy (*eromenos*). Then, follows the emphasis on the appreciation by the older lover of the importance of *knowledge*. This occurs during the process of teaching the young boy elements of virtue. Finally, there is a return to *beauty*, but now to intellectual as opposed to physical beauty. According to Socrates, this beauty is the essence of *philosophy*. In *Phaedrus*, Socrates revisits the concept of madness, which in his first speech in the *Republic* he classified as irrational and hence evil. However, now it is stated that this type of madness is more complex. Then, by employing the simile of the soul as a winged two-horse chariot driven by a charioteer, Socrates concludes, after a series of elaborate arguments, that madness is crucial for allowing the soul to pass from the beauty of the body to the beauty of the knowledge, and hence to philosophy.

In contrast to Plato's analysis, this part of the book emphasizes the indispensable role of emotions in particular, and of unconscious processes in general.

Actually, emotions are *evolutionarily older* than reason and hence are more important for our survival. In particular, emotions are crucial for *homeostasis*. In this connection, it is noted that in the same way that the formation of stars and planets in our universe requires that certain specific physical constants are in a narrow range (Rees, 2001), life in an organism is maintained only when several conditions are simultaneously met in its interior. For organisms like us these conditions include: First, in different cellular compartments it is necessary to maintain within tight permissible ranges, the amount of oxygen and CO_2, energy stores, water, sodium, temperature, and acidity (pH). Second, it is required that the secretion of various hormones and other molecules crucial for survival is tightly regulated. The process of maintaining a state necessary for survival and reproduction is called *homeostasis*. This term, which derives from the Greek words *homeo* (similar) and *stasis* (equilibrium), was introduced by the Harvard physiologist Walter Cannon (1871–1945). The predecessor of the notion of homeostasis was the concept of *milieu interieur* (internal environment) advocated by the French physiologist Claude Bernard (1813–1878), which states that the internal environment exists in a state of stability.[2] According to Bernard, the purpose of a variety of biological mechanisms is to ensure that any disturbance from equilibrium is efficiently addressed so that the organism returns to a *perfect* state.

A more dynamic and accurate notion of homeostasis arose from the ideas of Élie Metchnikoff (1845–1916) who was awarded the Nobel Prize in Medicine in 1908. By employing analogical thinking, this pioneer zoologist implemented Darwin's idea of *antagonism* inside an organism: he proposed that organisms exist in a state of *dynamic disharmony*. In this sense, Bernard's state represents only an *ideal* situation. This disharmony is the result of antagonistic efforts of different cell lines to evolve, develop, and differentiate. The process of attempting to restore harmony was named by Metchnikoff *physiological inflammation*. According to him, this curative, integrative process is carried out by particular cells, discovered in 1822 by himself, called *phagocytes*. These cells are responsible for *phagocytosis*, namely, the elimination of unwanted cells or particles via ingesting or engulfing them. The pioneering ideas of Metchnikoff regarding the role of phagocytes and generally of the immune system in homeostasis were ignored until very recently.[3] Indeed, for a long time, it was assumed that the purpose of the immune system was only to defend against pathogens. However, in a recent transformative development it has been shown that this assumption is not entirely correct: in the early 1990's it was realized that maintaining homeostasis was also a fundamental role of the immune system, returning in this way to ideas of Metchnikoff (Cohen, 1992). This crucial aspect of the immune system is characterized by its capacity to respond to signals from damaged tissues, including dead cells, and to remove them. Dead cells are produced, for example, via natural death or following a sterile injury (Matzinger, 1994).

---

[2]Bernard also introduced the use of *double-blind* experiments to avoid bias, a concept that is discussed in Chapter 17.

[3]Perhaps this is partly due to the fact that the enormous progress in *molecular biology* has, in a sense, marginalized research in the important area of *cellular biology*. It is research in the latter area that finally led to the modern understanding of immunology.

Achieving homeostasis involves reflexes as well as mostly unconscious processes, which include the following: initiation and control of breathing; regulation of the heart rate and blood pressure; regulation of endocrine glands secreting various hormones; postural adjustment of the body and limbs during walking or running; control of gastrointestinal mobility; regulation of the sleep-wake cycle and of the internal thermostat; control of the immense complexity of the immune system. Reflexes do not involve the cerebral cortex but are controlled by circuits in the spinal cord or brain stem.

It is important to emphasize that evolution adds new processes to its armament *without* necessarily eliminating earlier ones. This becomes evident in the diverse and overlapping utilization of the triptych reflexes-unconscious processes-consciousness.[4] This overlapping plays a vital role in maintaining homeostasis. For example, even single-cell organisms, as clearly shown during the growth of bacteria in culture, which migrate towards food by employing appropriate chemical signals. An even more "intelligent" behaviour to maintain homeostasis is exhibited by the worm *C. elegans*, which possesses only 302 neurons: these worms feed separately if there is plenty of food in a non-threatening environment, but in the opposite situation, such as in the case of scarcity of food or in the presence of an unpleasant odour, they feed in groups (Bargmann, 1996). The more advanced the neural system of an organism, the more effective the associated homeostatic mechanisms.

It is natural to speculate that at some stage early in the evolutionary tree, a "miracle" took place: the brain informed the organism that some homeostatic parameter was outside its optimal range.

Presumably, this first act of *the brain informing itself what already it knew*, involved pain, thirst, hunger, or temperature. Most likely, this was expressed as a vague feeling of discomfort. Antonio Damasio, in his very

[4]According to Rodolfo Llinás, the main goal of the brain is *prediction* in connection with implementing movement: "My argument is that that the sensory experience leading to active movement (motricity) through the function of prediction is the ultimate reason of the very existence of the central nervous system" (Llinás, 2002: p. 202). Prediction is, of course, crucial for motion. Llinás is the world expert in the area of movement, and perhaps he is somewhat biased towards the unitary importance of this particular function of the brain.

interesting discussion of cases of children deprived of a cerebral cortex, has presented arguments that vague feelings of pain and pleasure can arise from the brain stem regions of the *solitary nucleus* and of the *parabrachial nucleus* (Damasio, 2012: pp. 75–83). In my opinion, *this primordial form of awareness constitutes the first instance of consciousness*. This hypothesis suggests that a vital step in understanding consciousness involves elucidating the neural mechanisms that gave rise to this primordial awareness; this is further discussed in Chapter 19.

The emergence of awareness is perhaps the most advanced tool of nature's weaponry in its stochastic struggle to create organisms that have an advantage in survival and reproduction. The importance of this invention is demonstrated by its utilization in an uncountable number of other circumstances: from mechanisms of *proprioception*, namely the sense of one's bodily positions, and *kinesthesia*, namely the sense of the movement of one's bodily parts, to sensory perception, memory, learning, and imagination. In particular, by integrating parts of a variety of information, *the self becomes aware of itself*, which gives rise to *self-consciousness*. Incidentally, in this case, which is often referred to as *metacognition*, the perceived and perceiver coincide. Auguste Compte claimed that since the observed and the observing are identical, no observation can take place. Thus, he argued, it is impossible for us to think about our own mind. However, John Stuart Mill responded that there is no such contradiction because the observed and the observer are encoded within *different systems* or in *different times*.

These remarks suggest that subjectivity and self-consciousness are *not* prerequisites, but *consequences* of consciousness. This point of view is consistent with Stanislas Dehaene's position:

"becoming conscious of aspects of myself, such as my body, my feelings, or my thoughts [...] is like becoming conscious of say colour" (Dehaene, 2014: p. 24). Similarly, Rodolfo Llinás states that,

"the generated abstraction called 'self' is fundamentally no different that these secondary qualities of the senses [earlier specified as colour, sounds, etc.]" (Llinás, 2002).

Assuming that awareness is indeed equivalent to subjective experience, the hypothesis of self-consciousness advocated, among others, by Dehaene and

Llinás and adopted in this volume suggests that the so-called easy problem of consciousness, namely the identification of the neuronal circuits that support awareness, and the hard problem defined as "the elucidation of the mechanisms giving rise to personal experiences" (Chalmers, 1995) should be treated as a continuum.

Defects in the integration process giving rise to self-consciousness, result in a variety of medical syndromes. For example, in the *body integrity identity disorder*, mental images arising from the body's anatomy are inconsistent with unconscious structures generated from other parts of self. Also, in so-called *Alice in Wonderland syndrome*, which may arise as a side effect of migraines or seizures, the subjects feel that they shrink to a very tiny size, or that they grow to a gigantic size; in addition, walls may recede when approached, or the ground may feel spongy. My 20-year-old daughter, Ioanna, following the sudden onset of viral meningitis, experienced some of these events. Here is her personal description:

"I began experiencing vivid, bizarre dreams. In addition, on the third night, I experienced distortions of reality while still awake. They included: Auditory hallucinations consisting of non-harmful voices of my parents and those close to me, sometimes taking the form of conversations with them. Visual hallucinations, involving rapid growing and shrinking of my hands, movement of walls and the perception that my bed was growing as I attempted to get out of it, trapping me in it. I also felt a peculiar distortion of time which is hard to describe".

Depending on circumstances, the presence of the self may be dominant, or may vanish in the background as it happens when one is absorbed watching a movie. Brain imaging studies performed while subjects were watching a movie showed activation of areas involved in sensory, emotional, and memory processes, but relative inactivation of areas involved in introspection and planning.

The neuronal basis of homeostasis

Recent developments in neuroscience, which include studies of genetically modified mice, have allowed researchers to begin to elucidate the precise neuronal mechanisms that control basic homeostatic processes. For example, it is now possible to begin answering the question of "how do we become hungry?" This interesting question is related to the need of the organism to maintain energy balance. It is now known that if fat stores fall, the level of the hormone *leptin*, which is secreted by fat cells, also falls.[5] This activates a certain type of neurons called AgPR (agouti-related peptide). These neurons, which are located at the base of the hypothalamus, initiate a complicated process that promotes *energy restoration*. This is achieved via two basic mechanisms: first, reducing expenditure, and second, inducing the unpleasant feeling of hunger (Lowell, 2019).[6] The reduction of energy expenditure following the activation of AgPR neurons occurs via an unconscious process that causes the inhibition of the activity of the sympathetic nervous system, as well as the reduction of thermogenesis in brown adipose tissue. Regarding hunger, the precise neuronal pathways which, starting with the activation of AgPR neurons, finally give rise to *awareness* of this unpleasant feeling are currently under investigation. Apparently, these pathways involve unconscious processes in the midbrain and the hypothalamus, which are then followed by conscious processes involving the insula.[7]

Incidentally, the time scale of awareness of hunger, namely the time from the stimulation of AgPR neurons to the time of seeking food, is of the order of a minute, which is much longer than the scales occurring in perception, which as stated earlier are of the order of one-third of a second. Apparently, this is due to the involvement of many synapses, as well as to the action of several neuro-modulators whose indirect action is much slower than the usual direct action in the ionic gates of the standard neurotransmitters (Lowell *et al.*, 2021).

---

[5]Extensive training around the age of 10 decreases the levels of leptin, and this may delay puberty.

[6]After hungry mice are exposed to food, the activity of AgPR neurons falls within seconds, well before leptin levels rise. Thus, homeostasis is also regulated via a feed-forward mechanism, namely the brain also uses specific neurons that can *anticipate* the consequence on homeostasis of certain relevant actions.

[7]A particular pathway has already been delineated: it involves AgPR neurons to *melanocortin IV* receptors in the *paraventricular nucleus* of the hypothalamus, and then to neurons in the *lateral branchial nucleus* located at the boarder of pons and midbrain (Lowell *et al.*, 2021).

In the opposite situation, where leptin is raised, another type of neurons is activated. These neurons, which are also located at the base of hypothalamus, release a particular hormone, the *α-melanocyte-stimulating* hormone, which promotes weight loss.[8]

Similarly, progress has been made towards answering the question of "how we become thirsty?" The brain senses water loss by detecting blood changes of the hormone *angiotensin* II, and then restores balance via two basic mechanisms: first, reducing excretion, and second causing thirst. Reduction of excretion is achieved via unconscious changes in the activity of the *vasopressin* neurons, whereas thirst is induced via the activation of specific hypothalamic neurons, the MnPO (median preoptic nucleus) neurons (Gizowski & Bourque, 2018).

Emotions and feelings

Taking into consideration that the activation of the AgRP and MnPO neurons is a necessary and sufficient condition for hunger and thirst, respectively, it follows that the circuits formed by these neurons provide the neural basis that gives rise to the *feelings* of hunger and thirst. The unconscious process preceding a feeling was called by the leading investigator in this area, Antonio Damasio, *emotion.* This discussion suggests that

a feeling is a specific mental image expressing the conscious form of a subset of unconscious processes.

Why do I use the term "subset"? In my opinion, this term is justified by noting that, in the above examples, the full set of the relevant unconscious processes not only involves those processes that finally give rise to the conscious outcomes of hunger and thirst but also processes that *never* become conscious. The latter are the processes that yield a decrease in energy expenditure and reduction in water excretion. Thus, only the *subset* of unconscious processes related to hunger and thirst finally gives rise to consciousness, namely the feelings of hunger and thirst.

---

[8]Low levels of leptin activate AgRP neurons, whereas high levels of leptin activate the so-called *proopiomelanocortin* neurons.

Homeostatic emotions provide the response of the organism to a detected homeostatic disturbance in the structure or function of the viscera, caused by the change of a specific physiological quantity. For this reason, in my opinion, emotions should include the full set of unconscious processes preceding a feeling. For example, the unconscious processes involved in the decrease of expenditure should be part of the emotion associated with the feeling of hunger. Importantly, these remarks highlight further the fact that consciousness reveals only a small part of reality.

By generalizing the role of emotions in homeostasis, it is natural to postulate a third hypothesis (the second hypothesis, related to the genesis of metarepresentations was stated in the prologue and will be repeated in the next section):

Third hypothesis: *Emotions characterize the response of an organism to any internal change disturbing homeostasis or to any change in the exterior of the organism detected via specialized sensory probes.*

An example of an internal change is provided by the spontaneous neural oscillations giving rise to an emotionally competent memory. If the third hypothesis is indeed valid, then,

emotions capture the body's state in the reactive phase that follows internal or external changes. Such changes include the dynamic processes of creating thoughts. This suggests that every mental function is accompanied by an appropriate emotion, which depending on the circumstances, may or may not be expressed as a feeling.

Taking into consideration the fundamental role of emotions in homeostasis, it follows that emotions are of higher biological value than any other type of unconscious function. In this sense, it is not surprising that in Cotard's syndrome mentioned in Chapter 3, as well as in Capgras syndrome which is discussed in Chapter 12, emotions override mechanisms related to perception and other cognitive processes.

Summarizing, emotions are vital unconscious behavioural modulators. An example, which clearly shows that *emotional memory* has a crucial effect on behaviour, independently of consciousness, is provided by the following incident: in the 1900s the Swiss neurologist Édouard Claparède, while shaking the hand of an amnesic patient, pricked her with a pin. The next day, the patient's amnesia prevented her from recognizing Claparède, but she refused to shake hands (Dehaene, 2014: p. 53).

Feelings can take a variety of forms. In particular, pain and pleasure are the vaguest, and also the most basic of all feelings. They arise when homeostasis is, respectively, disturbed and restored. The organism employs natural analgesics, like the endogenous opioid substance *endorphin*, to attenuate emotional responses. Its action inhibits the feeling of pain, especially under highly stressful conditions, like those experienced by soldiers in battle. Several medications, including opioids, Prozac, and Valium, activate the same receptors activated by endorphin. Incidentally, the discovery of the approximately 100 known neurotransmitters has followed the same pattern: scientists, after coming across a new chemical in the brain, isolated it and then studied its effect on neurons. The exception is endorphin: the dramatic effect of opium and morphine on the brain led to the speculation that these chemicals must act on brain-receptors that are the natural target of some unknown endogenous substance. This is indeed the case, and the relevant substance, endorphin, was finally found in the 1970s after a long search involving pigs' brains.

Different feelings are associated with different parts of the brain and with different molecules. For example, social feelings, such as compassion and empathy, depend on appraising the significance of complex stimuli, including whether someone is suffering. Thus, it is not surprising that they involve the activation of part of the cortex and in particular, the ventromedial prefrontal cortex. On the other hand, fear and anger, which are more important for survival, are associated with the activation of the amygdala, which is evolutionarily older than the cortex. In addition to the hypothalamus, the insula and the ventromedial prefrontal cortex that were mentioned earlier, other parts of the brain important in the processing of feelings have been identified via functional imaging techniques. They include, the somatosensory cortex, the cingulate cortex, and various brain stem nuclei (Damasio *et al.*, 2000). In particular, the *periaqueductal area*, which is an area of grey matter in the midbrain, is important for coordinating emotional responses. This is achieved via the activation of certain motor nuclei of the *reticular formation* and nuclei of the cranial nerves, including nuclei for the *vagus* nerve.

Similar considerations are valid for other basic feelings like disgust, which presumably arose as a protection against toxic foods, and for other social feelings, like embarrassment, shame, guilt, pride, jealousy, envy, gratitude, admiration, and contempt that provides a social metaphor for disgust. The protective effect of many of these feelings is obvious. For example, we feel anger if we are offended, whereas guilt occurs in the opposite situation, which protects us from making mistakes. Similarly, the fear of the indefinite gives rise to anxiety, which motivates attempts to make the indefinite certain and to replace vagueness with precision (Frazzeto, 2013).

In an interesting experiment, subjects were asked to choose the best quality among different pairs of nylon stockings, which were actually of the same quality. There was a preference for stockings presented on the right visual field, which is controlled by the left brain (Dehaene, 2014: p. 43). What is the explanation for this unexpected result? It turns out that the left cortex is associated with more positive emotions than the right (Kawasaki *et al.*, 2001). Apparently, extraverts have the capacity to overcome some of the "negativity" of the right hemisphere. For example, when shown happy faces, extraverts exhibit more activation of the right amygdala than introverts.

METAREPRESENTATIONS

Animals certainly possess effective homeostatic mechanisms as well as consciousness. The remarkable books *Are We Smart Enough to Know How Smart Animals Are?* (de Waal, 2016) and *Beyond Words* (Safina, 2016) present clear and moving evidence of both the intelligence and the emotional richness of our evolutionary relatives. For example, chimpanzees are capable of empathy, sympathy, guilt, and shame. They also have a concept of fairness, as exemplified by an experiment where one monkey was rewarded less than another despite performing equally well, and this prompted him to break out into a frenzy of protest. Several animals even possess self-consciousness, as for example is suggested by the observation that elephants, great apes, dolphins, whales, and magpies, pass the so-called "mirror test". Namely, they recognize themselves when looking in a mirror. This becomes evident by their efforts to remove a spot placed on their forehead. Incidentally, babies older than 18 months also pass the mirror test. These considerations raise the important question of "why we differ from our evolutionary predecessors qualitatively, as opposed to quantitatively?" *I believe the answer is provided by our unique ability to construct metarepresentations.* In what follows, after arguing that Plato's world of "ideas" provides a clear illustrate of this far-reaching human characteristic, the meaning of metarepresentations will be explained.

Suppose that Hypatia, the first famous female mathematician, had decided to spend an entire day investigating geometrical properties of triangles.[9] Naturally, her brain continued to think about this problem after she stopped calculating. When she closed her eyes trying to fall asleep at night, she vividly *saw* a triangle. Plato, as a deep thinker, saw a plethora of such images. In this regard it should be noted that the brain employs different types of abstraction, which include extracting what is important, invariant, general, and representative of a given category. Thus, Plato's triangle was free of scribbles, it was *perfect*. The great philosopher could not have imagined that this *ideal* triangle was a *mental image* created by his brain. For him it was so vivid, so real, it *had* to exist *in a real space*. Thus, he *invented* a world inhabited by such ideal forms: his world of Forms (or Ideas). This act of Plato was typically human: he attempted to make mental images *real*.

Similarly, the reports in the *Iliad* of heroes that had heard the voice of gods advising or commanding them reflect Homer's efforts to make *real* the mental images generated by the contemplation of how gods would respond to the given situation. This point of view differs from the thesis of Julian Jaynes that these individuals *literally heard voices* as a result of becoming newly aware of their own intuitive thought processes (Jaynes, 2000).

According to the second hypothesis stated in the prologue,

our distinguishing characteristic with respect to other animals is our innate predisposition to construct real *versions of our mental images as well as of our*

---

[9]Hypatia was a mathematician, astronomer, and Neoplatonic philosopher who lived in Alexandria. Although she was a Pagan, she had many Christian students. She was an advisor to the Roman Prefect of Alexandria. Following a political feud of the latter with Cyril, the Bishop of Alexandria, Hypatia was accused of preventing the Prefect from reconciling with the Bishop and she was murdered, in 415 AD, by a mob of Christians.

unconscious structures, or to assign *to them specific symbols. The outcome of these constructions will be referred to as metarepresentations. Hugely important examples include the arts, language, mathematics, computations, and technology.*

Of course, there are strong interconnections between different metarepresentations. The relationship between language (words) and mathematics (numbers) was certainly clear to Aeschylus. In his play *Prometheus Bound*, the god who stole fire from heaven and gave it to mortals, says: "I invented for them Number, the chief of all devices, and how to set words in writing, Memory's handmaid, and mother of the Muses [...]".

As stated earlier, in the case of visual perception, it takes approximately a third of a second for perceived unconscious processes to give rise to awareness. In other words, this is the time needed for unconscious structures to become mental images. The "conversations" between the unconscious and consciousness are tremendously enriched by the construction of metarepresentations. For example, the formation of the mental image of a triangle allows the unconscious to scrutinize deeper the essence of this geometrical notion, and to create a plethora of associations. These associations become far more complex as a result of the drawing of this triangle. Indeed, this drawing, which provides the construction of a concrete mathematical metarepresentation, gives rise to new associations. This allows the creation of new metarepresentations. For example, the drawing of the bisectors of the angles. This procedure can be continued forever. In my opinion, *this is precisely what enabled the enormous developments achieved by humans.*

Andrey Vyshedskiy claims that the main difference between animals and humans is the ability of the latter for *mental synthesis*, which, according to him, emerged as a result of further development of the human visual system. He defines mental synthesis as the process of synthesizing a new, never seen before image, from two or more mental images. For example, generating an image of a cup on top of a keyboard, via the synthesis of the cup and the keyboard. Vyshedskiy writes: "we can invent a bicycle, because we can imagine a frame, wheels and pedals put together"

(Vyshedskiy, 2014: p. 165). In my opinion, our ability to connect mentally the various components of the bicycle is simply the result of the brain's capacity for forming *associations*. Then, the brain's ability for *unification*, which in this case takes the form of putting things together, allows the brain to construct a *mental image* of a bicycle. In contrast to Vyshedskiy I believe that the *uniquely human characteristic is the* praxis of the *construction* of the bicycle, i.e., making this mental image *real*.

Incidentally, some scholars have emphasized that the huge chasm between our achievements and those of the chimpanzees is due to the fact that human culture is *cumulative*. For example, it is stated by Steve Stewart-Williams that "as soon as we evolved the capacity for cumulative culture, as soon as we opened that Pandora's Box, culture began to evolve in its own right independently of biology" (Stewart-Williams, 2020: p. 13). In my opinion, the construction of metarepresentations, and in particular of writing as well as other forms of recording knowledge, provides, by definition, the explanation for the cumulative character of human culture.

References

Bargmann, C. 1996. Olfaction — From the nose to the brain. *Nature* 384, 512–513.

Chalmers, D. 1995. Facing up to the problem of consciousness. *Journal of Consciousness Studies* 2, 200–219.

Cohen, I. 1992. The cognitive paradigm and the immunological homunculus. *Immunology Today* 13(12), 490–494.

Damasio, A. 2012. *Self Comes to Mind.* Vintage Books.

Damasio, A., Grabowski, T., Bechara, A., Damasio, H., Ponto, L., Parvizi, J., & Hichwa, R. 2000. Subcortical and cortical brain activity during activity the feeling of self-generated emotions. *Nature Neuroscience* 3(3), 1049–1056.

de Waal, F. 2016. *Are We Smart Enough to Know How Smart Animals Are?* Granta.

Dehaene, S. 2014. *Consciousness and the Brain: Deciphering How the Brain Codes Our Thoughts.* Viking.

Frazzeto, G. 2013. *How We Feel.* Black Swan.

Gizowski, C. & Bourque, C. 2018. The neuronal basis of homeostatic and anticipatory thirst. *Nature Reviews Nephrology* 14, 11–25.

Jaynes, J. 2000. *The Origin of Consciousness in the Breakdown of the Bicameral Mind.* Mariner Books.

Kawasaki, H., Kaufman, O., Damasio, H., Damasio, A., Granner, M., Bakken, H., Hori, T., & Adolphs, R. 2001. Single-unit responses to emotional visual stimuli recorded in human ventral prefrontal cortex. *Nature Neuroscience* 4, 15–16.

Llinás, R. 2002. *I of the Vortex, from Neurons to Self.* MIT Press.

Lowell, B. B. 2019. New neuroscience of homeostasis and drives for food, water, and salt. *The New England Journal of Medicine* 380, 459–471.

Lowell, B. B., Swanson, L. W., & Horn, J. P. 2021. Chapter 41. The hypothalamus: Autonomic, hormonal, and behavioural control of survival. In: Kandel, E. R., Koester J. D., Mack, S. H., & Siegelbaum, S. A. (eds.) *Principles of Neural Science.* McGraw Hill, 6th edn.

Matzinger, P. 1994. Tolerance, danger, and the extended family. *Annual Review of Immunology* 12(1), 991–1045.

Nehamas, A. 1999. *Virtues of Authenticity: Essays on Plato and Socrates.* Princeton University Press.

Plato. 1989. *Symposium (Originally written in Ancient Greek around 385–370 BC).* (A. Nehamas & P. Woodruff, Trans.) Hackett Publishing Company.

Plato. 1995. *Phaedrus (Originally written in Ancient Greek around 370 BC).* (A. Nehamas & P. Woodruff, Trans.) Hackett Publishing Company.

Rees, M. 2001. *Just Six Numbers: The Deep Forces that Shape Our Universe.* Basic Books.

Safina, C. 2016. *Beyond Words.* Souvenir.

Stewart-Williams, S. 2020. *The Ape that Understood the Universe.* Cambridge University Press.

Vlastos, G. 1983. The Socratic elenchus. *Oxford Studies in Ancient Philosophy* 1, 27–58.

Vyshedskiy, A. 2014. *On the Origin of the Human Mind.* Greatspace Independent Publications.

THE TRANSFORMATIVE ROLE OF METAREPRESENTATIONS AND CULTURAL VERSUS BIOLOGICAL EVOLUTION

Many authors have emphasized the vital importance of language. For example, the philosopher Daniel Dennett states:

"Human consciousness is unlike all other varieties of animal consciousness in that it is a product in large part of cultural evolution, which installs a bounty of words and many other thinking tools in our brains, creating thereby a cognitive architecture unlike the 'bottom-up' minds of animals. By supplying our minds with systems of representation [....]" (Dennett, 2017).

In my opinion, the uniqueness of human consciousness is the result of the innate predisposition of humans to construct a *variety of metarepresentations*, which, in addition to language include arts, mathematics, and technology.

Chapters 6 and 7 discuss in detail the transformative metarepresentations of language, art, mathematics, computations, and technology.

In Chapter 8, it is argued that culture, as opposed to biological evolution, is now the driving force of the development of our civilization. Also, the crucial understanding regarding our evolutionary origin gained via the powerful approach of *molecular phylogenetics* is discussed. In addition,

Figure PII.1. Wassily Kandinsky, *Composition VII* (1913).
Source: The State Tretyakov Gallery.

it is noted that the concept of metarepresentations can be useful for elucidating further the dynamic interaction between cultural and biological evolutions. In particular, the novel concept of *heterodynamics* is introduced, and interactions between cognition and culture are analyzed in detail.

Which is the highest achievement of the human brain? Regarding complexity, perhaps mathematics; regarding the ability to communicate, definitely language; and regarding the revelation of the beautiful world hiding in the unconscious, arts. Of course, these amazing metarepresentations are deeply inter-related. For example, it appears that the proto language has its origin in music; mathematics, as clearly understood by many great thinkers from Pythagoras to Iannis Xenakis, is directly related to music; furthermore, there exists a deep relationship between music and painting. Interestingly, Wassily Kandinsky gave the name *Compositions* to the series of paintings that finally led to his first abstract painting. In this way, he wanted to emphasize similarities with the neuronal mechanisms that occur during the creation of symphonies. In this sense, perhaps *Composition* VII captures the essence of this part of the book (Figure PII.1).

Reference

Dennett, D. C. 2017. *From Bacteria to Bach and Back*. Penguin Press.

THE ORIGIN OF METAREPRESENTATIONS AND LANGUAGE

The concept of continuity, used extensively in this volume, does *not* preclude the occurrence of transformative changes. On the contrary, *every qualitative transformation can be traced in the cumulative effect of continuous local processes*.

The analysis of mechanisms giving rise to *breakthroughs*, which will be discussed in the second volume, provides an illustration of this fact. The only explanation for our astounding cognitive capabilities is the occurrence of a *qualitative transformation* of our mental capacity with respect to that of our evolutionary predecessors.

In physics, there are many situations when the quantitative accumulation of a particular entity leads to a qualitative change. This is reflected in the basic principles of the philosophical doctrine of *dialectic materialism* stating that "the accumulation of quantitative changes finally gives rise to a qualitative jump". For example, the process of heating water finally leads to its vaporization. This occurs when temperature reaches a critical value, which under normal conditions of pressure of one atmosphere is 100°C. The specific continuous local processes that give rise to phase transition are quite complicated. These processes finally cause transformations of the configuration or the topology of the matter under consideration.

It is natural to assume that the human *mental phase transition* is the result of the quantitative accumulation of some specific entity. Which is this entity whose accumulation led to a mental phase transition? Although at this stage it is not possible to provide a precise answer to this question, it is straightforward to give a vague answer: *biological complexity*. For example, in comparison to our ancestors, our pyramidal cortical neurons, especially in the prefrontal part of the brain, possess a far more extensive network of *dendritic spines*. These structures are membranous protrusions that connect at the synapse of a given neuron with a single axon of another neuron. Each pyramidal neuron of the prefrontal region contains approximately 15,000 spines. This highly complex network is due to a family of genes that are uniquely mutated in humans (Lai *et al.*, 2001), including the famous FOXP2 gene (Vernes *et al.*, 2011). The latter gene has two mutations specific to the *Homo* lineage. The exact DNA sequence of human FOXP2 differs from that of other species, including chimpanzees, birds, and crocodiles. Furthermore, this sequence has extremely low variability within *Homo sapiens*. This suggests that the FOXP2 gene was the target of selection in human evolution. A quantitative analysis of its variability suggests that it was fixed in its current form within the past 200,000 years (Enard *et al.*, 2002).

In 2001, the same year that the draft human genome was published, the geneticist Simon Fisher and collaborators (Lai *et al.*, 2001) showed that people with damaged versions of this gene suffer from a single-gene disorder affecting speech and language (one family, denoted as KE, has 15 affected members). Further studies have established that mutations of the FOXP2 gene cause serious impairment in articulation and speech (Christophel *et al.*, 2017). Mutant mice possessing this defective gene have reduced *plasticity* of neural circuits, namely reduced ability to modulate responsiveness to stimuli. As discussed in Chapter 10, plasticity is a key aspect of learning and memory. Interestingly, the FOXP2 gene has far-reaching implications, beyond plasticity and the modulation of our language machinery (Murakami *et al.*, 2018). Mutant mice carrying the normal FOXP2 gene grew pyramidal cells possessing a large number of dendritic spines and exhibited greater ability to learn (Squire *et al.*, 2004).

The importance of the FOXP2 gene for a variety of mental functions across different species is further supported by a study involving the fly *Drosophila*.

Flies, like humans, deliberate before making perceptual judgements. By measuring reaction times, it was established in a study reported in *Science* (Dasgupta *et al.*, 2014) that, as expected, flies take longer to choose between different smells, as the contrast between these smells decreases. Importantly, flies with a mutated FOXP gene took longer than the normal flies to make a decision.

The concept of *mental phase transition* has been used extensively by the original thinker and leading neuroscientist V. S. Ramachandran, who states that this transition occurred approximately 150,000 years ago. Ramachandran states that "This transition brought us full-fledged human language, artistic and religious sensibilities, and consciousness and self-awareness" (Ramachandran, 2011: p. 13). The inclusion of consciousness and self-awareness in this sentence is rather puzzling, since these features can be found in many animals.

ANATOMICAL DIFFERENCES WITH APES

A phase transition signifies a dramatic, transformative change. So, one may expect that it is accompanied with significant brain structural changes reflecting specific molecular alterations. This is indeed the case. Overall, it is reflected in a substantial disparity in the weight of our brain in comparison to our evolutionary predecessors: the human brain is about 1,300 grams in comparison to about 600 grams in apes. There also occurred key anatomical differences in our evolutionary tree. In this regard, it is noted that the first molecular difference between human and non-human primates traceable to an anatomical difference in the fossil records was identified in 2004. It is a mutation of the gene MYH16, which took place 2.4 million years ago. The inactivation of this gene led to a decrease in jaw-muscle size. Since altering the size of muscles can produce dramatic changes in the bones to which they attach, this had a considerable impact on the cranial morphology allowing for the size of the brain to increase (Stedman *et al.*, 2004). This may be related to the discovery of fire and cooking. Incidentally, the huge impact of the discovery of fire, and in particular its effect on cooking and on the emergence of free time for contemplation, is argued for in the excellent book *Catching Fire: How Cooking Made Us Human* (Wrangham, 2009).

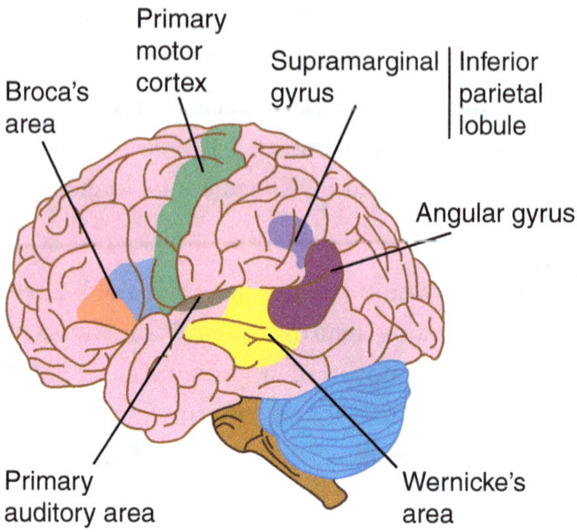

Figure 6.1. IPL and angular gyrus.
Source: Modified from https://thebrain.mcgill.ca/.

Most importantly, in the evolution from early mammals to primates, the specific region of the inferior parietal cortex called *inferior parietal lobule* (IPL) became increasingly more conspicuous, reaching a disproportionately large size in apes. It became even more pronounced in humans, where its major part split into the *angular gyrus* (Figure 6.1) and the *supra-marginal gyrus*. The IPL, as a result of its privileged position located between the occipital, parietal, and temporal lobes, receives information from the sensory modalities of vision, touch, and hearing, respectively. The IPL is also crucial for *language*. Furthermore, the angular gyrus is vital for *exact mathematics* and for *abstraction*. The latter becomes evident by noting that patients with a deficit in this part of the brain do not appreciate simple proverbs. They understand such phrases literally but not metaphorically. In addition, the supra-marginal gyrus is necessary for *dancing*, and importantly, for the effective use of *tools*. In particular, patients with a deficit in this area suffer from *apraxia*, namely inability to carry out a specific act. For example, a patient with apraxia is unable to construct a mental image of the process of combing. The patient recognizes a comb and is aware of its general use but is unable to make proper use of it.

The importance of these functions of the IPL becomes evident by the realization that the fundamental metarepresentations of *mathematics* and *language* provide the basis of the symbolic metarepresentations permeating sciences and letters. Moreover, the metarepresentation of *dancing* is an important ritual and artistic act. Finally, *tools* are the metarepresentations that provide the seeds for the development of *technology*. These considerations suggest that,

> the anatomical expansion and increased functional complexity of the IPL, together with related cultural activities, were of crucial importance for the occurrence of the mental transition characterizing the human brain.[1]

The importance of *analogical thinking* is emphasized throughout this volume. In this connection it is noted that tool-making involves the use of mental functions analogous to those used in linguistic processing. Indeed, studies using brain imaging revealed that during the process of making a specific type of an ancient stone axe, which requires different tasks to be completed in a specific order, there is activation in the region of the right hemisphere, analogous to Broca's area (Szathmáry & Számadó, 2008).

It is important to emphasize that the aforementioned arguments, which suggest that particular genetic mutations are crucial for the occurrence of the *human revolution*, must be placed in the wider context of the vital impact of culture that is discussed in Chapter 8. There is a plethora of studies analyzing these interactions. For example, lactase-persistent mutations were the *result* of dairy consumption. Similarly, mutations responsible for the higher alcohol tolerance of Europeans relative to Asians *followed* greater alcohol consumption in Europe. Finally, the invention of fire and cooking *altered* the size of human intestines (Fisher & Ridley, 2013). Hence, many genetic changes are the *consequence* rather the *cause* of cultural changes. Indeed, Leah Krubitzer writes:

"Anatomical alterations to the hand necessary for complex bimanual dexterity; to the supra-laryngeal tract necessary for speech production; and to the inner ear, which amplifies frequencies associated with human

---

[1]Painting, sculpture, and music are based on making unconscious structures real, as opposed to simply assigning symbols to mental images. Thus, their neuronal substrate is much broader.

speech, were present well before these behaviours that we attribute to modern humans were expressed within the population".

Apparently, the final development of complex language and sophisticated use of tools was the result of the "dramatic role of epigenetic mechanisms", crucially affected by "the social and cultural context in which individuals developed, rather than traditional evolutionary mechanisms" (Krubitzer, 2014: p. 190).

Strong support for the interplay of culture, genetic transformations, anatomical differences, and the resulting impact on the genesis of various metarepresentations, is provided by a variety of archaeological, anthropological, and genetic data. Relevant studies range from the remarkable works of Sally McBrearty and Alison Brooks on African archaeology of the period of 200,000–40,000 (McBrearty & Brooks, 2000) to Merlin Donald's elucidation of the extensive use of the process of *mimesis* by early humans (Donald, 1991). According to Donald, mimesis was the indispensable bridge between ape and modern human communication. In my opinion, mimesis provides the *unified predecessor* of the important metarepresentations of art (as already appreciated by Aristotle), language, music, and technology. Perhaps it would be useful if some of these studies are revisited within the framework of this new notion of metarepresentations.

LANGUAGE

In his erudite book *The Singing Neanderthals: The Origins of Music, Language, Mind and Body* (2004) Steven Mithen has presented extensive anthropological and archaeological documentation in favour of the theory that the protolanguage was *holistic*, namely, it was based on *entire messages* as opposed to words. This theory has its origin in the related hypothesis of Alison Wray (Wray, 2002), and also draws support from the computer simulations of Simon Kirby's language models (Kirby, 2002). In more detail, according to this theory hominids remained "ape-like" up to about 1.8 million years ago, both with respect to their anatomy and behaviour. At that time, a new species evolved, called *Homo ergaster*. The defining new feature of this genus, which is regarded by many as our earliest ancestor, was the emergence of a new form of *communication*, which possessed the features stated below. It was:

Holistic, for example, a specific combination of vocalizations and body gestures had the meaning of "hunt with me".

Manipulative, used for greeting, informing, threatening, requesting, appeasing, etc.

Multi-modal, combining facial expressions, vocalizations, body language, movement, etc.

Musical, involving the use of rhythm, vocalizing, etc.

Memetic, imitating a specific situation using suitable gesticulations.

Apparently, the emergence of this novel communication was greatly influenced by *bipedalism*. This hugely significant development was accompanied by a larger brain, which was necessary to control the additional sensor-motor requirements of the far more demanding movements associated with the upright position. Bipedalism had the following additional effects: First, it facilitated migration to open areas, which increased the need for communication. Second, the upright position led to the shifting of the larynx lower in the throat, which resulted in the lengthening of the vocal cords and hence the capacity to produce more diverse sounds. Third, the new posture necessitated the narrowing of the birth canal, which led to the need for smaller babies, and hence resulted in an increase of the duration and demands of parenthood. This placed more pressure on mothers to have effective ways of *communicating* with their babies. Fourth, bipedalism dramatically influenced the movement of the early humans, which in turn affected the evolution of their musical abilities. Indeed, rhythm, which is one of the most important ingredients of music, is an essential element of movement.

According to Mithen, at some later stage of evolution, the *segmentation of the holistic phrases* of the above proto language, allowed *Homo sapiens* to develop two different branches of communication: *music*, with emphasis on expressing emotions, and *language*, with emphasis on specialized transmission of information.

The tongue and the diaphragm play a crucial role in both language and music. In this regard, it is interesting to note that, in the *Homo ergaster* the sizes of the *hypoglossal canal* that carries nerves from the brain to the tongue, and of the *thoracic vertebral canal* that carries nerves from the brain to the diaphragm, are similar to those of the African apes. However, those of the *Homo neanderthalensis* are larger. Despite this fact, Mithen claims that the Neanderthals did not develop language. One of the main arguments used by Mithen, is that according to him there is no evidence in Neanderthal life of *symbolic* artefacts, which certainly constitute the most basic expression of metarepresentations:

"The absence of symbolic objects must imply the absence of symbolic thought, and hence of symbolic utterances. Without these, by definition, there was no language" (Mithen, 2006: p. 229).

However, the principle of continuity suggests that both symbolic artefacts and language must have appeared *gradually* in the evolution of humanity. Of course, historical contingencies, triggered by climatic and demographic factors, may prevent investigators to trace this continuity in fossil and other relevant domains.

The deduction that symbolic artefacts provide a *proxy* for the existence of language, requires the identification of artefacts reflecting complex-syntactical functions, such as hierarchical organization, recursion, and links between distant elements. The authors of the comprehensive review (d'Erricol *et al.*, 2009) present impressive evidence of such complex artefacts, including: employment of pigments, construction of abstract engravings, use of personal ornaments, decorations of burial places, and musical traditions. This multitude of archaeological and palaeo-anthropological evidence indicates that Neanderthals did possess cognitive abilities compatible with the development of a proper language. Thus, it appears that Neanderthals did develop a syntactical language.

More recent evidence provides further support to this conclusion. For example, in 2010, perforated and pigment-stained marine shells were found in two sites of the Neanderthal-associated Middle Palaeolithic of Iberia, which dates back to approximately 50,000 years (Zilhão *et al.*, 2010). Comparable material from Africa and the Near East is widely accepted as evidence for body ornamentation, which implies behavioural modernity. Remarkably, very recent evidence suggests that complex symbolic ability may actually predate the Neanderthals. In particular, it is shown in Hoffmann *et al.* (2018) that various complex symbolic artefacts, including ochre-painted and perforated marine shells, found in Cueva de los Aviones, southeast Spain (a site of the Neanderthal-associated Middle Paleolithic of Europe) are 115,000–120,000 years old. The authors of this study state that,

"Given our findings, it is possible that the roots of symbolic material culture may be found among the common ancestor of Neanderthals and modern humans, more than half-a-million years ago".

This possibility is consistent with genetic evidence (Krause *et al.*, 2007) showing that the critical FOXP2 gene was certainly present in the Neanderthal genome. Actually, its first appearance occurred around 300–400,000 years ago.

Regarding the importance of symbols with respect to language, it is instructive to turn to the "pump water incident", beautifully described by Helen Keller (1880–1968) who was famously blind and deaf:

"Miss Sullivan had tried to impress it upon me that 'm-u-g' is mug and that 'w-a-t-e-r' is water, but I persisted in confounding the two. We walked down the path to the well-house, [...]. Someone was drawing water and my teacher placed my hand under the spout. As the cool stream gushed over one hand, she spelled into the other the word water, first slowly, then rapidly. I stood still; my whole attention fixed upon the motions of her fingers. Suddenly I felt a misty consciousness as of something forgotten — a thrill of a returning thought; somehow the mystery of language was revealed to me. I knew then that 'w-a-t-e-r' meant the wonderful cool something that was flowing over my hand. That living word awakened my soul, gave it light, hope, joy, set it free!" (Keller, 1903).

The feeling "of something forgotten" is perhaps related to her earlier exposure to water in connection with "mug". Keller's brain was able to *associate* the combination of the specific signs that her teacher was spelling on Keller's hand, with the cool liquid she was feeling on the other hand.

In this way, *a particular symbol was assigned to the mental image* created by "the wonderful cool something that was flowing over my hand". The enormous importance of assigning symbols was evident to Keller:

"I left the well-house eager to learn. Everything had a name, and each name gave birth to a new thought. As we returned to the house every object which I touched seemed to quiver with life. That was because I saw everything with the strange, new sight that had come to me".

Incidentally, the approach used by Keller's teacher was invented by Trappists monks to allow them to communicate with each other despite their vow of silence. Keller later adopted the Braille system of raised dots, now widely used by the blind or people with low vision, which greatly expanded her capabilities and enable her to become a remarkable writer.

In the "pump water incident", assigning a symbol to a mental image was facilitated by the employment of a sequence of signs. Animals also have the ability to associate signs or signals with past, present, or future objects. In the same way that Keller associated a specific sequence of signs with water, animals can associate, for example, a specific smell with a particular flower, or a distinctive noise with an approaching predator. However, animals apparently cannot assign symbols to mental images in a hierarchical, algorithmic way as in human language and mathematics, i.e., they lack the ability to form metarepresentations.

The symbolic nature of language implies that thought *precedes* language. This fact has been succinctly expressed by Wittgenstein: "we cannot *say* what we cannot think" (Wittgenstein, 1922: p. 5.61). In this regard I believe that,

the essence of humanity is the result of continuous, dynamic interactions between neural mechanisms and metarepresentations.

In the particular case of language, these interactions are so exceedingly strong, that it is difficult to separate language from thought and the ability to reason. The ancient Greeks defined man as a "language-animal". Moreover, the Greek word *logos* equates "word", the building block of language, with "logic", which in Greek is *logiki*.

Structurally, the interaction of neural mechanisms and language expresses itself in a variety of forms, including *compositionality*. The use of symbols, coupled with the brain's ability to generalize and create associations, allow humans to *compose*, namely, to construct larger structures, like sentences, by combining smaller elements, like words.

Linguistic syntax is a complex set of rules defining how words can be combined. The importance of associations in language becomes evident with the development of a recent linguistic model which suggests that syntax is the direct consequence of word–word associations. In this model, words are associated with objects either in a referential way, such as "meat" referring to "edible organic matter", or in a non-referential way, such as "eat" being associated with "the action of eating". Since common words are less specific, they can be linked to many objects, whereas rare ones have fewer links. If the words in a text are ranked according to the number of their links, then the number of appearances of the words in this text is *inversely proportional* to their rank. This law, known as Zipf's law, is a universal feature of the architecture of such networks in all languages (Zipf, 1949). Based on this fact, Ferrer-i-Cancho and collaborators formally built a network using the simple rule that two words are linked if they share at least one object of reference. Surprisingly, the architecture of the network constructed according to the number of these links exhibits many of the fundamental features of linguistic networks (Ferrer-i-Cancho *et al.*, 2005).

The number of links affects other linguistic processes. For example, the more a verb is used, the more difficult it becomes for cultural evolution to impose unified rules, such as the past-tense marked -ed: a verb used 100 times more often than another, will be regularized 10 times more slowly. This is illustrated with the fact that the rare pair "help/holp" has been regularized, but not the common pair "go/went".

The production of language, together with the brain's fundamental property of plasticity, yields specific alterations in the brain's relevant neural circuits. For example, neuroimaging and neurophysiological studies involving literate individuals have shown that there is a specific region of the brain, in the left-hemispheric fusiform cortex, assigned to recognizing alphabetic letters (Cohen, *et al.*, 2002; Hébert & Cuddy, 2006). This region is part of the brain's domain involved with object recognition. The assignment of a specific domain for alphabet recognition is consistent with the theory that, during development, there occurs *progressive modularization*. Namely, specific areas of the brain are assigned to particular tasks (Karmiloff-Smith, 1992).

The principle of *evolutionary continuity* suggests that animals must possess rudimentary symbolic capabilities. This is consistent with the "artistic capabilities" of fish and Bowerbirds discussed in the next chapter. Indeed, animals employ organized sound patterns to communicate with each other. For example, 26 sounds have already been identified that are used by white-handed Gibbons. Also, it has finally been proven that the click sounds produced by dolphins operate like a proto language. Moreover, recent research suggests that such animal languages exhibit primitive syntactic structure. Chimpanzees provide even more direct evidence of primitive linguistic capabilities: after chimpanzees were trained to use sign language, they were able to understand and use symbols in order to ask for food and for a variety of objects. A young ape named Kanzi, after 17 months of training, was able to understand around 150 words. Furthermore, remarkably, after extensive social interaction with humans, Kanzi was able to construct phrases on electronic boards exhibiting primitive syntactic structure, and could also understand complex sentences (Savage-Rumbaugh & Lewin, 1994: pp. 278–279). These studies show that chimpanzees possess certain elements of the neuronal machinery needed for developing a primitive language. This is consistent with the existence in the brains of chimpanzees of limited versions of Broca and Wernicke's areas (in humans the connections between these two regions are much stronger) (Ardesch *et al.*, 2019). However, since such a language does not develop in their natural habitat, it follows that their relevant neuronal machinery is incomplete, requiring access to an *external* symbolic system and human environment for its function.

Humans also require linguistic interactions. This was dramatically demonstrated by the Holy Roman Emperor Frederick II. In 1211, he tried to discover the language of Adam and Eve. He thought that this language would be spoken spontaneously by children not exposed to any language. He had children raised by nuns ordered not to speak to them. These children died very young and could not speak any language.

In contrast to our evolutionary predecessors, we do possess the neuronal machinery for *ab initio* development of language. In this regard it is noted that American philosopher and educator Susanne Langer distinguishes between verbal and non-verbal modes. She also claims that language evolved in the context of non-verbal activities, including festivities, dances, music, and rituals (Langer, 1957). Such activities generate specific mental images with strong emotional context. So, it is natural for the human brain to attempt to encode them using appropriate symbols. The relationship between language and music will be presented in the third volume, which will be preceded by the analysis of the physiological basis of hearing (in the same way, that the discussion of the neuronal processes underlying vision, presented in Chapter 12, precedes the analysis of the impact of various visual mechanisms on painting, discussed in Chapter 13). Incidentally, the study of this relationship between music and language goes back to Jean-Jacques Rousseau's *Essai sur l'origine des langues* (1978) and was later studied by Otto Jespersen's *Progress in Language* (1895).

Regarding the roots of rituals, it is noted that the earliest man-made holy place was discovered in 1995 in Gobekli Tepe, Turkey, by the archaeologist Klaus Schmidt. It involves massive T-shaped slabs of limestone megaliths 3 metres high, arranged in circles and ovals. Some of them were carved images of stylized arms, whereas on others there appeared meticulously carved menageries of snakes, spiders, birds, foxes, boars, and other beasts. Radiocarbon dating and stone tool comparisons indicate that these were made 11,000 years ago, i.e., before pottery and before most signs of agriculture (Curry, 2008).

The environment not only affects language but also impacts various sensory systems, at least during certain *critical* periods. This became clear following the pioneering experiments of David Hubel and Torsten Wiesel in the 1960s. These Nobel Prize Laureates established that blocking visual input during the first few months of life in one eye of cats, had a dramatic effect on the resulting architecture of their visual circuits and could even lead to blindness of the deprived eye. For an adult cat, such deprivation did not affect the underlying architecture. It is natural to speculate that metarepresentations in general, and language in particular, are affected by the environment *more* than sensory processes. Noam Chomsky's position that there exist innate neural circuits of universal grammatical structure common to all humans is certainly important. However, his claim that language acquisition is *similar* to the growth of a physical organ of the body, seems extreme, since the growth of organs takes place without the need for external input. Chomsky states that language development is similar to the "progressive maturation of a specialized structure, thus acquisition of language does not require 'learning' but growth". In contrast to this extreme position, Jean Piaget considered language acquisition within the framework of the *epigenetic development of a symbolic system.*

The emergence of symbols necessitates the development of an appropriate *structure* for the organization of these symbols. According to the Italian jurist and scholar of the Enlightenment period, Giambattista Vico (1668–1744), humans have the innate ability to "impose structure on the chaos of the natural world".[2] In this way, humans shape the world, and at the same time humanity creates itself. Hence, structure is reflected in the human mind and it becomes a vital characteristic of the human thought. The notion of structure as applied to the particular case of language was greatly explored by the Swiss linguist Ferdinand de Saussure (1857–1913), who introduced the distinction between *langue* (language or grammar) and *parole* (speech): *langue* is the structure of language expressing its basic rules. Saussure illustrated the distinction between *langue* and parole by stating that the rules of chess constitute its *langue*, whereas a particular game of chess is an example of parole. Moreover, in the seminal text, *Course in General Linguistics* (1915), compiled by students who had taken his course, Saussure distinguishes between *diachronic* studies, which investigate the historic development of a language, and *synchronic* studies, which examine the structure of a language at a given time.

Remarkably, Saussure, echoing Poincaré, realized that meaning can be found not so much in things themselves, but in *relationships* between things. This position was further elaborated by Wittgenstein. For example, the meaning of "life" is elucidated in its relationship with "death".

[2]In 1734, Vico, motivated by the discoveries of Newton and Galileo, published the book *Principles of New Science* (*Principii di Scienza Nuova*). This volume can be considered the foundation of social, as opposed to "natural", sciences.

V. S. Ramachandran (Ramachandran, 2011: p. 169) discusses a fascinating example, which clearly shows the innate ability of humans to develop language following minimal environmental stimulation. In some areas of the world, people of different linguistic backgrounds develop a pseudo-language called *pidgin* in order to facilitate trading or working together. Although this language has limited vocabulary and rudimentary syntax, first-generation children who grew up with pidgin, spontaneously turned it into creole, which is a full-fledged language, allowing them to write novels and poetry. This further emphasizes that the remarkable features of openness, flexibility, and ambiguity, emerge naturally in human language. These characteristics are uniquely human and do not apply to the formal symbolic languages which are constructed via tools of mathematical logic and computer science (Pullum and Scholz, 2001).

It is important to note that the continuous interaction of metarepresentations with suitable neural circuits creates new metarepresentations of higher complexity. As will be discussed in the next chapter, this generative ability of humans is not only obvious in mathematics but it is also of crucial importance in language: songs, poems, and stories are the natural outcomes of these interactions, which themselves give rise to a never-ending process. An endorsement of the importance of this mechanism is expressed in the following statement of Lisa Cron (Cron, 2012): "Story, as it turns out, was crucial for our evolution — more so than opposable thumbs. Opposable thumbs let us hang on; story told us what to hang on to".

One of the crucial effects of metarepresentations is that they facilitate the processes of setting concrete future goals, as well as designing specific strategies for achieving them.

Of course, this is one of the features distinguishing humans from animals. An early form of setting or previewing goals is mentioned by Alexander Luria (Luria, 1961). According to this distinguished Russian psychologist, a two-year-old might say, "I throw bunny", just after throwing it. In this way, a concrete linguistic symbolism is assigned to this particular action. Later, the child might say "throw bunny", while throwing it, and then a short time later, this symbolism can be used to preview a *plan* of action: the child says, "I will throw this bunny", just before doing so.

Concluding this chapter, it is noted that many investigators have emphasized the symbolic ability of the human brain. In particular, Stanislas Dehaene states,

"Our brain seems to have a special knack for assigning symbols to any mental representation and for entering these symbols into entire novel combinations" (Dehaene, 2014: pp. 250–251).

Furthermore, Dehaene also appreciates that this is a uniquely human characteristic:

"In humans alone, [...there occurs] the emergence of a 'language of thought' that allows us to formulate sophisticated beliefs and to share them with others" (Dehaene, 2014: p. 253).

Numerous other scholars have connected symbolism with the crucial impact of language. For example, Mithen claims that the emergence of language changed the nature of human thought, giving rise to "cognitive fluidity", which according to him is the unique feature of modern humans.

In my opinion, in addition to symbolic ability correctly emphasized by the above and many other authors, *the tendency of humans to construct real versions of their mental images and unconscious structures, is of crucial importance to humanity.*

As discussed in the next chapter, this is perfectly expressed in the arts, where these real versions, among other creations, take the form of musical compositions or paintings.

References

Ardesch, D. J., Scholtens, L. H., Li, L., Preuss, T. M., Rilling, J. K., & van den Heuvel, M. P. 2019. Evolutionary expansion of connectivity between multimodal association areas in the human brain compared with chimpanzees. *PNAS* 116, 7101–7106.

Christophel, T., Klink, P., Spitzer, B., Roelfsema, P., & Haynes, J. 2017. The distributed nature of working memory. *Trends in Cognitive Sciences* 21(2), 111–124.

Cohen, L., Lehéricy, S., Chochon, F., Lemer, C., Rivaud, S., & Dehaene, S. 2002. Language-specific tuning of visual cortex? Functional properties of the visual word form area. *Brain* 125(5), 1054–1069.

Cron, L. 2012. *Wired for Story.* Ten Speed Press.

Curry, A. 2008. Seeking the roots of rituals. *Science* 319, 278–280.

Dasgupta, S., Ferreira, C., & Miesenböck, G. 2014. FoxP influences the speed and accuracy of a perceptual decision in Drosophila. *Science* 344(6186), 901–904.

Dehaene, S. 2014. *Consciousness and the Brain: Deciphering How the Brain Codes Our Thoughts.* Viking.

D'Erricol, F., Henshilwood, M., Maureille, G., Gambier, D., & Tillier, A. 2009. From the origin of language to the diversification of languages. In: d'Errico, F., & Hombert, J.-M. (eds.) *Becoming Eloquent: Advances in the Emergence of Language, Human Cognition, and Modern Cultures.* John Benjamins Publishing Company.

Donald, M. 1991. *Origins of the Modern Mind.* Harvard University Press.

Enard, W. *et al.* 2002. Molecular evolution of FOXP2, a gene involved in speech and language. *Nature* 418, 869–872.

Ferrer i Cancho, R., Riordan, O., & Bollobás, B. 2005. The consequences of Zipf's law for syntax and symbolic reference. *Proceedings of Royal Society London B* 272(1562), 561–565.

Fisher, S. & Ridley, M. 2013. Culture, genes, and the human revolution. *Science* 340(6135), 929–930.

Hébert, S. & Cuddy, L. 2006. Music-reading deficiencies and the brain. *Advances in Cognitive Psychology* 2(2), 199–206.

Hoffmann, D., Angelucci, D., Villaverde, V., Zapata, J., & Zilhão, J. 2018. Symbolic use of marine shells and mineral pigments by Iberian Neandertals 115,000 years ago. *Science Advances* 4(2), eaar5255.

Karmiloff-Smith, A. 1992. *Beyond Modularity.* MIT Press.

Keller, H. 1903. *The Story of My Life.* Norton.

Kirby, S. 2002. The emergence of linguistic structure: An overview of the iterated learning model. In: Cangelosi, A. & Parisi, D. (eds.) *Simulating the Evolution of Language.* Springer, pp. 121–147.

Krause, J., Lalueza-Fox, C., Orlando, L., Enard, W., Green, R.E., Burbano, H., Jean-Jacques Hublin, J.-J., Hänni, C., Fortea, J., Rasilla, M., Bertranpetit, J.,

Rosas, A. & Pääbo, S. 2007. The derived FOXP2 variant of modern humans was shared with Neandertals. *Current Biology* 17(21), 1908–1912.

Krubitzer, L. 2014. Lessons from evolution. In: Marcus, G. & Freeman, J. (eds.) *The Future of the Brain*. Princeton University Press, pp. 186–193.

Lai, C., Fisher, S., Hurst, J., Vargha-Khadem, F., & Monaco, A. 2001. A forkhead-domain gene is muted in a severe speech and language disorder. *Nature* 413, 519–523.

Langer, S. 1957. *Philosophy in a New Key: A Study in the Symbolism of Reason, Rite, and Art*. Harvard University Press.

Luria, A. R. 1961. *The Role of Private Speech in the Regulation of Normal and Abnormal Behavior.* Pergamon.

McBrearty, S. & Brooks, A. 2000. The revolution that wasn't: A new interpretation of the origin of modern human behaviour. *Journal of Human Evolution* 38, 453–563.

Mithen, S. 2006. *The Singing Neanderthals: The Origin of Music, Language, Mind and Science*. Harvard University Press.

Murakami, T. C., Mano, T., Saikawa, S., Horiguchi, S. A., Shigeta, D., Baba, K., Sekiya, H., Shimizu, Y., Tanaka, K., Kiyonari, H., Iino, M., Mochizuki, H., Tainaka, K. & Ueda, K., Ueda, H. R. 2018. A three-dimensional single-cell-resolution whole-brain atlas using CUBIC-X expansion microscopy and tissue clearing. *Nature Neuroscience* 21, 625–637.

Pullum, G. & Scholz, B. 2001. More Than Words. *Nature* 413(6854), 367.

Ramachandran, V. S. 2011. *The Tell-Tale Brain: A Neuroscientist's Quest for What Makes Us Human*. W. W. Norton & Company.

Savage-Rumbaugh, S. & Lewin, R. 1994. *Kanzi: The Ape at the Brink of the Human Mind*. Wiley.

Squire, L., Stark, C. & Clark, R. 2004. The medial temporal lobe. *Annual Review of Neuroscience* 27, 279–306.

Stedman, H., Kozyak, B., Nelson, A., Thesier, D., Su, L., Low, D., Bridges, C., Shrager, J., Minugh-Purvis, N. & Mitchell, M. 2004. Myosin gene mutation correlates with anatomical changes in the human lineage. *Nature* 428(6981), 415–418.

Szathmáry, E. & Számadó, S. 2008. Language: A social history of words. *Nature* 456(7218), 40–41.

Vernes C. S. *et al.* 2011. Foxp2 regulates gene networks implicated in neural outgrowth in the developing brain. *PLOS Genetics* 7, e1002145.

Wittgenstein, L. 1922. *Tractatus Logico-Philosophicus.* Routledge & Kegan Paul.

Wrangham, R. 2009. *Catching Fire: How Cooking Made Us Human.* Profile Books.

Wray, A. 2002. *Formulaic Language and the Lexicon.* Cambridge University Press.

Zilhão, J., Angelucci, D., Badal-García, E., d'Errico, F., Daniel, F., Dayet, L., Douka, K., Higham, T., Martínez-Sánchez, M., Montes-Bernárdez, R., Murcia-Mascarós, S. Pérez-Sirvent, C., Roldán-García, C., Vanhaeren, M., Villaverde, V., Wood, V. & Zapata, J. 2010. Symbolic use of marine shells and mineral pigments by Iberian Neandert. *Proceedings of the National Academy of Sciences* 107(3), 1023–1028.

Zipf, G. K. 1949. *Human Behaviour and the Principle of Least Effort.* Addison-Wesley.

7

THE METAREPRESENTATIONS OF ARTS, MATHEMATICS, COMPUTATIONS, AND TECHNOLOGY

THE ARTS

Several painters, sculptors, and composers have expressed their conviction that their creations *originate in their unconscious*. In particular, in 1935, Picasso, in conversation with Christian Zervos, said:

"It would be very interesting to fix photographically, not only the stage of a painting, but also its metamorphosis. Possibly one might catch a glimpse of the road by which the mind moves towards the concretization of its dream. But what is truly very curious, is to observe that the painting does not basically change, that the initial vision remains almost intact despite appearances […]. I noticed that when the work is photographed, that what I have introduced in order to correct my first vision disappeared and that, when all is said and done, the image given by the photograph corresponds to my first vision before the transformations brought about by me" (Michael & Bernadac, 1998).

It is remarkable that Picasso's characterization of the process of creating artistic metarepresentations as "the road by which the mind moves towards the concretization of its dream", is almost identical with the description of Arnold Schoenberg:

"the composer wants to find the musical laws that dictate the composition that he has conceived as in a dream" (Wright, 2007).

Praxis, i.e., *construction*, is of vital importance for the metarepresentation of art, hence, it is not surprising that Picasso and Schoenberg expressed similar points of view regarding the importance of this uniquely human activity. In 1935, Picasso, in a conversation with Christian Zervos, stated: "The important thing is to create. Nothing else matters; creation is all" (Stangos, 1981: p. 273). Similarly, when a music critic attacked Schoenberg's atonal music, he responded that "it is not important if you write tonal music or atonal music; what is important is to write music" (Wright, 2007).

Several artists have emphasized the importance of *abstraction*. For example, the Romanian sculptor Constantin Brâncuşi (1876–1957), a contemporary of Giacometti, stated:

"There are ignorant people who characterize my work as abstract; yet what they call abstraction, is most realistic. What is real, is not the appearance, but the idea, the essence of things" (Guilbert, 1957).

Presumably, "the idea" is the non-discursive reality originating in the unconscious world.

Incidentally, Brâncuşi, who studied mathematics, physics, and industrial art for four years before studying sculpture, was an admirer of Greek philosophers. In particular, in connection with his abstract sculpture *Socrates* (1922), Brâncuşi wrote: "nothing escapes the great thinker. He knows all, he sees all, he hears all". With this statue, he also expressed his respect for the French composer Erik Satie (1866–1925), whose opera *Socrates* (1918) was performed in Paris in 1920. Brâncuşi addressed Satie as Socrates and Satie addressed Brâncuşi as Plato (Satie was 10 years older than Brâncuşi).

The pioneering Cubist painter, Juan Gris, has stated "I try to make concrete that which is abstract". Perhaps, this suggests that his paintings are based on abstract unconscious structures.

Gris has revealed that the process of "concretizing the abstract" is aided by mathematics, which of course involves consciousness. Similarly, the role of consciousness is noted in the quotes by both Picasso and Schoenberg. Picasso talks about "that what I have introduced in order to correct my first vision" and Schoenberg discusses the efforts of the composer to discover *a posteriori* musical laws, which certainly affect future compositions. Clearly,

> artistic creations involve the interplay of unconscious and conscious processes, but it appears that for high-level creations the decisive starting conception belongs to the unconscious, which is revealed to consciousness much later.

This is consistent with the statement of Zervos that "Mathematics cannot replace art".

The theory of evolution implies that any human capability must appear in a rudimentary form in some earlier predecessor. Indeed, Australian Bowerbirds provide a striking example that animals do possess primitive artistic capabilities, both for "sculpture" and "dance". Actually, different Bowerbird species make completely different artistic constructions. For example, male Satin Bowerbirds, first, build a simple structure, and then decorate it elaborately with blue. A Bowerbird starts by plucking away all the leaves on bushes and low branches that block the sun from his chosen site and clears out a one-square metre area from all debris. Then, he brings hundreds of little twigs and sticks, which he uses to create a woodsy platform upon which the walls of the bower are erected. The main construction involves erecting the walls, namely, two vertical rows consisting of a foot-long twigs, which are as thick as the bird itself, leaving an empty space in the middle. Finally, the elaborate decoration takes place: the floor is decorated with grasses, the walls are painted with crushed berries, and the sunny open side is decorated with anything blue, like blossoms and parrot feathers, or even blue plastic. As soon as a female approaches the male begins to sing and dance. Some Bowerbirds even toss snails into the air during their wild dancing (Rothenberg, 2012: pp. 6–17). Bowerbirds live in tree-nests to avoid predators, so the only purpose of the entire construction and ritual is to attract females and not to create a nest. It has been shown that the females choose a partner according to the elaboration of the structure and the competence of the performance!

Apparently, the male has evolved artist capabilities for the important goal of attracting a female.

Even more remarkably, fish are capable of both "painting" and "sculpturing". For example, male Puffer fish create 2-metres-wide circular patterns on the sand at a depth of 25 metres. They achieve these creations by swimming on their side while fluttering a pectoral fin. They spend hours decorating them with snail shells that they crack in their mouths. These constructions attract females who lay their eggs in the middle of the circular construction; the furrows prevent the eggs from being carried away by currents. Regarding sculpturing, Cichlid fish, just like Bowerbirds, build bowers to improve their mating prospects. These bowers are not built as a nest; as soon as eggs are laid, the female moves them to a safer place. Some constructions take the form of volcanic-shaped sandcastles, projecting to a height of a foot or more from the bottom, with a flat platform on the top. Other fish construct even more impressive structures. In particular, male Stickleback fish produce a sticky mucus that they use as a glue to hold together pieces of algae, leaves, and grass, which they use to decorate the entrance of their bowers (Balcombe, 2016: pp. 185–186).

Concluding, it is worth noting that associations are of crucial importance not only for creating but also for appreciating art. For example, Jacques Rancière writes:

"The spectator also acts, […] she participates in the performance […] by refashioning the vital energy […] in order to make it a pure image and associate this image with a story which she has read or dreamt, experienced or invented" (Rancière, 2011: p. 13);

and also

"It is in this power of associating and dissociating that the emancipation of the spectator consists" (Rancière, 2011: p. 17).

MATHEMATICS AND COMPUTATIONS

As noted by Susanne Langer, "symbols are tools for thought". A clear illustration of this fact is provided by mathematics. This metarepresentation results from employing symbols and at the same time exploring the immense ability of the brain for *induction*, namely the ability for going from the specific to the general. Induction is facilitated by the development of mathematical logic. In this way,

> mathematics creates exceedingly complex and hierarchical relationships between abstract symbols.

Some of these relationships take the form of formulas, which express deep scientific statements in a precise, universal, and economical way. Actually,

> in terms of universality, mathematics is comparable only to music, and in terms of precision, as well as the capacity for proof or refutability, mathematics is far superior to language.

On the other hand, precisely because language can be ambiguous, it is able to generate a vast range of emotions, much like music.

Specific examples from physics that illustrate the generative power of mathematics are mentioned in Chapter 18. Arguments presented in Chapter 19 suggest that the role of mathematics in biology is qualitatively different than its role in physics. In what follows, an important related question is briefly discussed: Is it possible to "mathematize consciousness?" In this connection, it is useful to recall that mathematics cannot be completely formalized. This unexpected fact became apparent with Russell's famous barber paradox: suppose that a barber in a village is defined as the one who shaves all those, and only those, who do not shave themselves. Paradoxically, such a person *cannot* exist. Indeed, by asking the question, "does the barber shave himself?", one finds a contradiction: the barber cannot shave himself, since he only shaves those who do not shave themselves. On the other hand, if the barber does not shave himself, then he must be shaved by the barber. So, according to this definition the barber both shaves and does not shave himself, and hence this seemingly natural definition yields to a contraction.

The *coup de grace* to attempts initiated by David Hilbert of formalizing mathematics was delivered by Kurt Gödel's *incompleteness theorem*, which states that in any axiomatic system that is rich enough to include number theory, there exist infinitely many true mathematical statements which have no formal proof. Thus,

mathematics cannot provide a formal approach to reaching the truth, in the sense that it cannot develop an axiomatic system for deciding whether every statement in this system is true or false.

On the other hand, the human brain can intuitively decide the validity of many of these undecidable statements. *This implies that consciousness is much richer than mathematics.* For this reason, I am sceptical of various statements regarding the possibility that consciousness can be captured by mathematics. Such statements include the hope expressed by distinguished neuroscientists for the derivation of "a mathematical equation describing consciousness", as well as claims that consciousness has already been mathematized! For example, Dehaene claims that "only mathematical theory can explain how the mental reduces to the neural" (Dehaene, 2014: p. 139). Also, Giulio Tononi has identified consciousness with the mathematical notion of *integrated information* (Tononi, 2012). This notion is defined as the surplus information generated by a system in comparison with the information generated by its components. Tononi claims that "consciousness depends exclusively on the ability of a system to generate integrated information" (Tononi, 2008: p. 232).

According to Tononi, consciousness can emerge from any system provided that this system has sufficiently high complexity to generate integrated information.

The idea that consciousness is simply the consequence of high complexity was clearly expressed by Pierre Teilhard de Chardin (Teilhard de Chardin, 1955). He claimed that matter has an inherent compulsion to assemble into ever more complex forms, and this tends to generate consciousness. According to de Chardin, there exists "some sort of psyche in every corpuscle". Also, according to him, Cosmos evolves towards the "omega point" where the universe becomes aware of itself via maximizing complexity. Similar ideas have been expressed in the philosophical movement of *Theosophy*.

Christof Koch, who previously had argued — along with Francis Crick — that consciousness is based on the synchronous firing of neurons, has not only endorsed the above definition of consciousness, but has gone a step further. He claims that since there is a mathematical formula for computing integrated information, there now exists a mathematically precise *quantitative* measure of consciousness!

Although, in my opinion, these claims are erroneous, they do provide indirect evidence of the great power of metarepresentations: despite the fact that mathematics is only a *small subset* of human thought, its enormous computational capabilities have created the illusion of omnipotence. Thus, several neuroscientists, instead of attempting to understand the essence of mathematics by analyzing the underlying neuronal mechanisms responsible for its development, reverse this problem: they attempt to use mathematics to explain consciousness. Earlier efforts in this direction were made in Douglas Hofstadter's fascinating book, *Gödel, Escher, Bach* (1979), where it is claimed that "mathematical logic is the crux of human thinking". In other words, instead of using the brain to understand mathematics, Hofstadter attempted to use mathematical logic to understand the brain!

One of the greatest contributions of mathematics to humanity is the development of computers. The first mechanical calculator, called "Pascaline", was constructed in 1645 by Blaise Pascal (who was already mentioned in the prologue). This calculator could add, subtract, multiply, and divide any two large numbers. Gottfried Leibniz worked for 40 years on the problem of designing and constructing a mechanical calculator. He produced two such versions, in 1694 and 1706, employing the so-called "wheels of Leibniz". A machine called the arithmometer, combining Leibniz' design and Pascal's methods, was made commercially available in 1851.

The modern era of computing began with the design and construction of an *electro-mechanical programmable* computer at Bletchley Park in England, during World War II. This breakthrough, as well as parallel developments in the USA, were spearheaded, respectively, by Alan Turing and John von Neumann. Incidentally, after Turing obtained in 1938, a PhD from Princeton University, von Neumann attempted to hire him at the Institute of Advanced Studies as a post-doctoral assistant, but Turing returned to UK. Both Turing and von Neumann (who is considered by some as the most brilliant scientist that ever lived) were influenced by the mathematical brilliance of Gödel. In this regard it is noted that David Hilbert posed, in 1928, the *decision problem*, namely, whether it was always possible to give a binary answer, "valid" or "invalid", to any question arising in computing. In order to analyze this problem, Turing attempted to gain a deeper understanding of the meaning of *computational procedures*, which are also called *algorithms*. For this purpose, he devised the abstract concept of a *Turing machine*, namely a machine capable of a single, specific computation. For example, one such machine could add two numbers, another could multiply them, and yet another could decide whether a given number is an integer. If the latter machine was assigned the task of finding the largest integer, this machine would keep on computing, i.e., *it would never halt*, since there does not exist such an integer. Turing proved that answering the decision problem was equivalent to being able to tell whether such a machine would halt. In this way, he reduced a question concerning the validity of a mathematical statement to a question of machine computability.

The importance of associations is exemplified by the fact that Turing attempted to derive a result analogous to Gödel's incompleteness theorem for machine computability. For this purpose, he introduced the notion of a *universal Turing machine*, namely, a machine that could imitate the activity of any particular Turing machine. Then, by posing the brilliant question of whether such a machine could imitate itself, he was able to prove that it is *impossible* to build a machine that could always tell whether a given Turing machine will halt. In this way he proved that there is *no* solution to Hilbert's decision problem. This result is of the same importance for the foundation of computer science as Gödel's result is for the foundation of mathematics. In a sense, it shows *that*,

> the limitation of mathematics to formally reaching the truth cannot be overcome by the employment of any machine that functions algorithmically.

The necessity of facing Nazi aggression motivated Turing to transform his abstract constructions into practical solutions that resulted in the triumph of breaking the "enigma code".

It should be noted that, in the same way that the genesis of Einstein's remote associations was aided by his ability to conceive remarkable *thought experiments*, the creation of Turing's associations was aided crucially by his ability to imagine *virtual computers*.

Technology has had a tremendous impact on the development of computers, which in turn further accelerated the development of technology. In 1944, the *colossus*, which was an *electronic* computer (as opposed to Turing's electro-mechanical computer) based on 1,700 *vacuum tubes*, was delivered to Bletchley Park. It was employed to decipher the "Lorenz code" used by the German High Command.

It is worth noting that the first electronic digital computer, called ABC, was conceived by John Atanasoff, a mathematics and physics professor of Iowa State College. It was built in 1942 with the help of his student, Clifford Berry. Incidentally, the design of digital computers is based on the *analytical design* of Charles Babbage, who introduced the concept of a digital programmable computer. Babbage, who is considered the "grandfather of the computer", studied mathematics at the University of Cambridge.

The first general-purpose electronic digital computer, called ENIAC, was turned on in Philadelphia, USA, in 1946, causing lights to dim in parts of the city due to its very high electricity consumption. ENIAC had to be switched off one day per week in order to be cleaned from the bugs attracted by light and heat. This procedure, which was called "debugging", provides the origin of the term used today in connection with "software bugs". In 1952, a group of scientists at Los Alamos, led by the Greek-American physicist Nicholas Metropolis, designed and built the MANIAC I (Mathematical Analyzer Numerical Integrator and Computer Model I).[1]

The second generation of computers, developed between 1947 and 1962, used *transistors* instead of vacuum tubes. In January of 1954 engineers

[1]At Los Alamos, Metropolis also led a group of distinguished researchers, including John von Neumann and Stanisław Ulam, to develop the Monte Carlo method. This breakthrough, achieved in the 1950s, provided a powerful statistical approach, based on the use of computers, for the numerical solution of a variety of very complicated mathematical problems. A pedagogical introduction of this method can be found in (Fokas & Kaxiras, 2022).

from Bell Laboratories (with the support of the USA military) built the first computer without vacuum tubes, called TRADIC (for TRAnsistorized DIgital Computer). The size of this machine was about a cubic metre. This was an astonishing size reduction in comparison to the space of about 300 square metres occupied by ENIAC. Soon afterwards, MANIAC I was replaced by MANIAC II, also using transistors. The second generation of computers were finally replaced by the current computers that employ *integrated circuits*. Perhaps the greatest outcome of the interaction between computers and technology is artificial intelligence, which will be discussed in detail in the second volume.

TECHNOLOGY

As discussed in Chapter 6, the mental transition characterizing the exceptional cognitive capabilities of humans in comparison to other animals was affected by the anatomical expansion of specific brain areas controlling the abilities of abstraction and employing tools. These abilities are at the heart of technology. For example, the manipulation of sharp stones for a variety of human tasks, together with the ability for abstraction, led our human ancestors to hammer stone cobbles in order to create stone blades.[2] This occurred more than half a million years ago, and may be related to the last dramatic expansion of the brain size in the human lineage, about 600,000 years ago (Johnson & McBrearty, 2009). Similarly, the observation that flexible materials are potentiated when bent, i.e., when they bend they store energy, led to the use of bamboo for bow making. Combining different primitive tools led to the construction of more effective tools. Employing such tools and specially chosen materials gave rise to crafts, including pottery, weaving, boatbuilding, and metal processing. Wind and water power were harnessed for boat sailing and for windmills.

Steven Mithen considers the development of tools a prime reason for the adoption of agriculture around 10,000 years ago: "The ability to

[2]Even such primitive developments would have been most welcome by our evolutionary predecessors. This brings to mind the following fascinating story: A Tusk fish uncovered a clam buried in the sand. It carried it in its mouth 30 yards away to a rock, and then using rapid head-flicks, was able to eventually smash it open against the rock (Balcombe, 2016: p. 118).

develop tools which could be used intensively to harvest and process plant resources" (Mithen, 1998: p. 254). Some scholars consider the development of agriculture the end of human pre-history. Indeed, the investment of labour for building storage facilities and irrigation canals made moving much more difficult and facilitated the emergence of stable social structures.

The crucial importance of metarepresentations in the evolution of technology becomes clear by comparing the use of tools in humans and animals: our evolutionary ancestors can also use primitive tools,

but the unique ability of humans to interact with their metarepresentations allows them to use tools to construct tools.

This *mechanization* procedure led to the assembly of substantial technological units that consist of different elements, which themselves are technologies. At the same time, spectacular developments in science, often made possible precisely due to technological progress, led to a deeper understanding of a variety of physical phenomena. Novel technologies capturing this new understanding were used to harness thermodynamic, mechanical, gravitational, chemical, optical, electromagnetic, and quantum phenomena. In this way, more and more human needs were met, such as the need to be fed, clothed, sheltered, transported, kept in good health, and entertained. Furthermore, these advances led to the beginning of the exploration of space, but also, unfortunately, to the development of unthinkably powerful weapons of mass destruction.

According to the concept of continuity, technology evolves via the creative combination of earlier technologies and engineering practices, in the same way that science progresses via the association of different earlier scientific ideas and results. Some new technologies involve small variations, but major progress requires radical inventions. The latter are the result of scrutinizing existing assemblies of technological systems, and importantly, establishing *remote associations* among the fundamental principles of these technologies. In his thoughtful analysis of the evolution of technology, W. Brian Arthur states that invention in most cases "arrives by conscious deliberation; It arises through the process of mental association" (Arthur, 2009: p. 121).

In the process of developing and applying technology, the concept of continuity takes different forms. For example, regarding electricity, following the fundamental insight of Faraday, the brilliant inventors Edison, Westinghouse, and Tesla, figured out how to utilize electricity in applications. This was followed by the implementation of the difficult task of redesigning factories, which finally led to the huge economic impact of automation.

Some scholars consider technology an "applied science", whereas others present it as an intellectually inferior complement of science. I disagree with these assessments: there is a highly dynamic, dialectic relationship between science and technology. Indeed, the former is absolutely necessary for uncovering hidden phenomena that provide the foundation of advanced technology. On the other hand, science cannot be advanced without technological innovations that make it possible to observe and manipulate matter at scales both astonishingly small and exceedingly large, as well as to collect and store relevant data.

An example of the impact of technology on science is provided by the microscope. The transformative discovery of cells was the direct result of the invention of the microscope by the Dutch lens maker Antonie van Leeuwenhoek (1632–1723), who built microscopes capable of magnifying objects 270 times. In 1665, using such a microscope, the remarkable Oxford polymath Robert Hooke (1635–1702), widely known for his law of elasticity, discovered and named the cell.[3] The observation of animal cells that was achieved over a century later was of crucial importance for the development of cellular biology. The rich substructure of cells remained unexplored until the development, in the late 1930s, of the electron microscope with magnification of 10,000. In this way, the rich substructure of cells, including mitochondria, endoplasmic reticulum, microtubules, and ribosomes, can be observed.

In the last few years, a new technique has been developed, called *focused ion beam transmission electron microscopy*. This technique, which is a

[3]Actually, what Hooke saw was the dead cell walls of a plant cell. Still, the range and depth of Hook's contributions justify his characterization by the British historian Allan Chapman as "England's Leonardo".

remarkable generalization of electron microscopy, provides a three-dimensional depiction of the entire cell. It involves embedding a cell in a resin, bombarding it with an ion beam, and acquiring an electron microscopy image. This is repeated using many different sections of the cell in an automated procedure that lasts two days.

This new technique has revealed that the cell interior is a complicated space where organelles are not isolated entities, but traverse large sections of the cell volume making direct contact with other organelles, and existing in a state of dynamic equilibrium. This technological innovation is expected to have major ramifications for the further understanding of the function of the various components of the cell (Fermie *et al.*, 2018). Incidentally, until recently, it was very difficult to obtain three-dimensional depictions of entire cells at the level of resolution achieved by electron microscopy. The relevant procedure required laborious thin sectioning of the cell into many slices, acquiring a single image from each slice, and manually reassembling these images into a three-dimensional reconstruction. Very few laboratories in the world had the capability of obtaining such images, which required months of painstaking work.

Georges Braque emphasized the importance of depicting the *entire* space, eliminating the "interstitial space" between the percept and the background (this is discussed in detail in the third volume). No doubt, he would have been delighted by the compactness of the image shown in Figure 7.1.

Engineering constitutes the major exponent of technology. The higher the rate of development of the evolution of scientific and technological achievements, the more difficult it becomes to separate science from engineering and technology. An example of this synergy is provided by the remarkable recent progress in *organoids*. This development represents the latest innovation in the quest for developing models that mimic the function of specific human organs. In this regard, it is noted that there exist cells, called *stem cells*, that are truly *pluripotent*, meaning they can self-renew indefinitely. Furthermore, they can differentiate into any cell type. Employing such human cells, it is possible to construct an organoid *in vitro*, i.e., outside a living organism. This is a three-dimensional entity capable of simulating the architecture and functionality of a specific

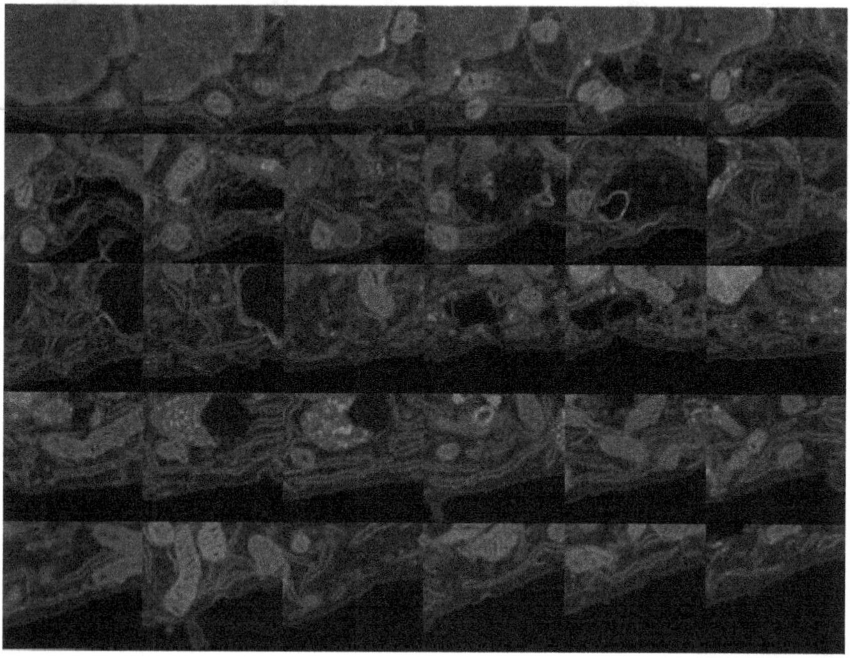

Figure 7.1. Focused Ion Beam Scanning Electron Microscopy images of a section of a human kidney cell. The sample was "sliced" into 1,600 sections along the *z* axis; a montage is shown of 1 every 50 sections. This technique allows the complete reconstruction of the volume of a biological sample in unprecedented detail.
Source: Dr. Chieko Itakura.

human organ. Such constructs may provide fundamental insights into the processes of development, homeostasis, and pathogenesis, offering new approaches to diagnosis and treatment. The first organoid, exhibiting the (crypt-villus) architecture of the intestines, was constructed in 2009 by Hans Clevers and colleagues. Their approach involved culturing stem cells in an appropriate three-dimensional extracellular matrix and treating these cells pharmacologically with appropriate growth factors, and in this way driving the formation of the desired organ (Sato *et al.*, 2009).

There now exist organoid models for many human organs, including the liver, pancreas, lungs, kidney, uterus, testis, and even the brain, inner ear, and retina (Li & Izpisua, 2019). The development of more realistic organoids, as well as their integration with other cutting-edge technologies, such as live imaging, gene therapy, and gene editing, is driven by interdisciplinary collaborations involving engineers and biologists. For example, by using a scaffold made from three-dimensional printing with bio-inks, a genuine three-dimensional renal tubule architecture has been constructed. Such "organ-on chip" systems combine organoids with microfluidics and set the stage for live imaging. This is expected to provide further insight into various biomechanical processes (Homan *et al.*, 2016).

Incidentally, a useful lesson learned from the development of organoids is that, given the proper environment and cues, developing cells have the capacity of self-organization. This provides a unique opportunity to attempt to decipher how this occurs and to discover principles of organogenesis. The success of such an effort could have an unprecedented impact in correcting a variety of birth defects, in particular regarding heart diseases.

The innate predisposition of humans to construct metarepresentations makes the progress highlighted in the above examples inevitable. Indeed,

> the interaction of metarepresentations with conscious and unconscious processes endows these processes with an even higher level of complexity. This in turn leads to the formation of more complex metarepresentations. This recursive process is the basic generative mechanism propelling technology and culture forward.

From the beginning of the industrial revolution in the early 1760s, the creative interaction of mental processes with an unlimited number of metarepresentations, including language, mathematics, computing, sciences, engineering, and technology, has led to progress involving higher and higher complexity. This has culminated in the current era of the digital revolution, where computing and communication have reached previously unimaginable levels of capabilities.

Many scholars predict that, in the same way that the 20th century was the century of physics and molecular biology resulting in remarkable applications from medicine and lasers to atomic energy and analogue circuits, the 21st century will belong to the interface of material science with biology, technology, and artificial intelligence. In particular, the potential of *neurotrophic circuits* is enormous. Ray Kurzweil, the author of *The Singularity is Near: When Humans Transcend Biology* (Kurzweil, 2005), predicted that, by the 2020s, the so-called *singularity* will take place, namely computers will have human-level intelligence. In 2017, he adjusted this date to 2029. He envisions this breakthrough as an opportunity to enhance human capabilities, predicting that there will be human brain–machine synergy in the 2030s. He mentions Parkinson's disease treatment via deep brain stimulation, which is presented in Chapter 16, as a first step in this direction.

Ray Kurzweil's prediction for the construction of a variety of suitable prostheses connected to the brain is already a reality. For example, in 2013, the Federal Drug Administration of the United States approved the Argus II system for patients with end-stage *retinitis pigmentosa* in both eyes. This system, based on the pioneering work of Mark Humayun and collaborators at the University of Southern California, consists of the following: a miniature camera mounted on a pair of glasses; an external video processing unit that is worn by the user; and specific extraocular and intraocular implants that are interconnected via a transscleral cable.[4] This device enables patients to achieve object recognition, motion detection, orientation, and mobility. Although the resolution is low at present, some patients are even able to identify letters in large print. In this connection it is noted that earlier devices, such as *cochlear implants*, which have improved the hearing of more than 200,000 people, had already demonstrated that even rudimentary technology, coupled with the brain's great ability for plasticity, can have life-changing clinical applications.

Most scholars overestimate the importance of consciousness. However, there also exist scholars on the opposite side, namely those who doubt its usefulness. Taking into consideration the crucial role of human consciousness in the construction of metarepresentations, such doubts are clearly unjustified.

[4]The camera captures visual scenes and sends the information to the video processing unit for advanced pixilation and processing. The hermetically packed extraocular electronics, along with a transceiver antenna, convert the signals they receive wirelessly from the video unit to electrical pulses. The amplitudes of these pulses correspond to the brightness of the local pixels. Stimulation pulses are delivered by the cable to an intraocular electrode array that is attached to the retina via a tack.

The interplay of the fundamental triptych of unconscious processes, experiences *(awareness), and* metarepresentations, *drives humans towards the ultimate goal of* knowing thyself, *which in turn has led to the flourishing of medicine and psychology. This in turn, has led to the remarkable prolonging of human life, which further accelerates progress.*

This famous phrase *knowing thyself* (*Γνώθι σαυτόν*), which according to Pausanias was inscribed in the pronaos of the Temple of Apollo (the God of healing and medicine) in Delphi, takes many different forms. In this connection, Iain Mc Gilchrist (McGilchrist, 2010: p. 272), quotes the classical philologist Bruno Snell: "Man must listen to an echo of himself before he may hear or know himself" (Snell, 1960: p. 200). In my opinion, this quote captures the essence of the interactions between conscious and unconscious processes as expressed in many of the Athenian tragedies: the spectator, by identifying with the protagonists, first fully experiences the instinct of vengeance and the pathos of retributions, but then is guided towards reconciliation, compassion, and above all, empathy. Through this *echo*, the fully engaged spectator discovers a range of human emotions and hence *thyself*.

Concluding, it is worth noting that, although metarepresentations have their origin in associations, metarepresentations are of higher evolutionary value than associations. For example, in an interesting study, bees were trained to learn to match the symbol *N* with two discs. Furthermore, in separate training they learned to match two discs with the symbol *N*. Subsequently, via the employment of *associations*, they could deduce several *generalizations* (Howard *et al.*, 2019). Similarly, they were trained to match an inverted *T* with three squares and in separate training they learned to match three squares with an inverted *T*. However, after learning to associate *N* with two discs, they could *not* deduce the inverse association. It is noted in the above paper that

"while bees could learn the association between a symbol and numerosity [...] bees could not spontaneously extrapolate the association to a novel, reversed task".

In my opinion, the inability of the bees to deduce the reversed task is the result of their inability to *conceive the metarepresentation* corresponding to *assigning a symbol* for the abstract entity of *two*. The authors of this study conclude that "our study therefore reveals that the basic requirement for numerical symbolic representation can be fulfilled by an insect brain [...]". In my opinion, matching is *not* equivalent to the more advanced cognitive task of *assigning symbols*. Indeed, only the latter is part of constructing metarepresentations. Assigning symbols gives rise to *robust associations* which are certainly *reversible*, while matching is not. This was clearly appreciated by Piaget who noted that children are capable of *reversible action*. For example, according to Piaget, children understand that they can turn the light on using a switch and reverse the action by flicking the switch the other way.

References

Arthur, W. B. 2009. *The Nature of Technology*. Penguin Books.

Balcombe, J. 2016. *What a Fish Knows: The Inner Lives of our Underwater Cousins*. One World.

Dehaene, S. 2014. *Consciousness and the Brain: Deciphering How the Brain Codes Our Thoughts*. Viking.

Fermie, J., Liv, N., Ten Brink, C., van Donselaar, E., Müller, W., Schieber, N., Schwab, Y., Gerritsen, H., & Klumperman, J. 2018. Single organelle dynamics linked to 3D structure by correlative live-cell imaging and 3D electron microscopy. *Wiley Traffic* 19, 354–369.

Fokas, A. & Kaxiras, E. 2022. *Modern Mathematical Methods for Computational Science and Engineering*. World Scientific.

Guilbert, C. G. 1957. Propos de Brancusi. In *Prisme des Arts 12* (pp. 5–7).

Homan, K., Kolesky, D., Skylar-Scott, M., Herrmann, J., Obuobi, H., Moisan, A., & Lewis, J. 2016. Bioprinting of 3D convoluted renal proximal tubules on perfusable chips. *Scientific Reports* 6, 34845.

Howard, S., Avarguès-Weber, A., Garcia, J., Greentree, A., & Dyer, A. 2019. Symbolic representation of numerosity by honeybees (Apis mellifera): Matching characters to small quantities. *Proceedings of the Royal Society B* 286(1904), 20190238.

Johnson, C. & McBrearty, S. 2009. 500,000-year-old blades from the Kapthurin Formation, Kenya. *Journal of Human Evolution* 58(2), 193–200.

Kurzweil, R. 2005. *The Singularity is Near: When Humans Transcend Biology.* Penguin.

Li, M. & Izpisua, J. 2019. Organoids-preclinical models of human disease. *New England Journal of Medicine* 380, 569–579.

McGilchrist, I. 2010. *The Master and the Emissary.* Yale University Press.

Michael, A. & Bernadac, M.-L. 1998. Zervos (quote) 1935. In *Picasso: Propos sur l'art.* Gallimard.

Mithen, S. 1998. *The Prehistory of the Mind.* Phoenix.

Rancière, J. 2011. *The Emancipated Spectator.* Verso.

Rothenberg, D. 2012. *Survival of the Beautiful.* Bloomsbury.

Sato, T., Vries, R., Snippert, H., van de Wetering, M., Barker, N., Stange, D., van Es, J., Abo, A., Kujala, P., Peters, O., & Clevers, H. 2009. Single Lgr5 stem cells build crypt-villus structures *in vitro* without a mesenchymal niche. *Nature* 459, 262–265.

Savage-Rumbaugh, S. & Lewin, R. 1994. *Kanzi: The Ape at the Brink of the Human Mind.* Wiley.

Snell, B. 1960. *The Discovery of the Mind.* Harper and Row.

Stangos, N. 1981. *Concepts of Modern Art.* Thames and Hudson.

Teilhard de Chardin, P. 1955. *The Phenomenon of Man.* Éditions du Seuil.

Tononi, G. 2008. Consciousness as integrated information: A provisional manifesto. *The Biological Bulletin* 215, 216–243.

Tononi, G. 2012. *Phi, A Voyage from the Brain to the Soul.* New York: Pantheon Books.

Wright, J. K. 2007. *Schoenberg, Wittgenstein and the Vienna Circle.* Peter Lang.

8 BIOLOGICAL VERSUS CULTURAL EVOLUTIONS

CULTURAL EVOLUTION

A framework for discussing the impact of culture on human evolution has been suggested by the mathematical physicist and polymath Freeman Dyson in *Biological and Cultural Evolution* (Dyson, 2019). This framework is adopted here with several modifications, which include the incorporation of the critical role of the *horizontal gene transfer*.

The generative power of culture was appreciated by writers *before* scientists. In particular, H. G. Wells, in his book *Outline of History* (Wells, 1920) clearly elevates culture as *the* driving force of history. The first part of his book describes biological evolution up to the appearance of the *Neanderthal Man* and the *Modern Man.*[1] Neanderthal Man (*Homo neanderthalensis*) lived during the so-called *Middle Palaeolithic Period*, which lasted from perhaps 250,000 years to about 45,000 years ago. This period was followed by the *Upper Palaeolithic period*, which lasted from about 45,000 years to 10,000 years ago. The Middle Palaeolithic period was preceded by the *Lower Palaeolithic period*, which started at about 2.7 million years ago, when human-like characteristics appeared in the hominid fossil record.

---

[1]Neanderthals are named after the place that the first specimen was found in 1856, namely, the German Neander Valley. Incidentally, Neander, has a Greek origin meaning new man (νεος ανδρας).

In the second part of his book, Wells concentrates on the vital impact of culture, analyzing the agricultural revolution as well as the birth of Buddhism, Judaism, Christianity, and Islam. The book ends with the catastrophe of World War One.

Following the work of Wells, the importance of cultural evolution was later discussed by several biologists, and in particular by Richard Dawkins in his famous book *The Selfish Gene* (Dawkins, 1976). In this book it is argued that, although genes are "selfish" in the sense that their only goal is to survive and replicate, in order to achieve this goal, genes promote behaviour based on cooperation, generosity, and altruism. In this way, this book provides a biological explanation of morality and ethical behaviour. Incidentally, as noted in Chapter 11, Spinoza had already advocated the biological origin of morality. Importantly, in the last chapter of his book, Dawkins introduces a novel approach that facilitates the analysis of the impact of cultural evolution. This approach provides another example of the ability of the brain for analogical thinking. Indeed, Dawkins introduced a *cultural analogue of the gene*, which he called a *meme*. Examples of memes are ideas, religious beliefs, laws, fashions in dress, and tools. It should be noted that some memes, such as tools, are important metarepresentations. He carried this analogy further by noting that, in the same way that genes propagate via *sexual* contact, memes propagate through *social* contact. According to Dawkins, genes and memes have been evolving over billions and thousands of years, respectively. This implies that the *modus operandi* of early organisms was based on *bottom-up* mechanisms, namely, it was based on biological evolution, whereas more advanced organisms, in addition to bottom-up mechanisms, also employ more and more complicated *top-bottom* mechanisms, namely mechanisms where culture is of critical importance.

Incidentally, there have been recent attempts to further elaborate these ideas using a framework that combines *evolutionary psychology* and *cultural evolutionary theory*. This approach attempts to trace fundamental aspects of behaviour to basic evolutionary processes and to the interaction of these processes with the emergence of human culture (Stewart-Williams, 2020). In my opinion, the ideas presented so far within this framework are rather simplistic. For example, it has been proposed that altruism can be explained via William Hamilton's *kin* "selection theory" and via Robert Trivers' "reciprocal altruism theory". These theories are based, respectively, on the rather elementary notions that "the closer the kinship the greater the tendency to assist" and on "the importance of reciprocity". Hamilton attempted to quantify his theory by introducing the simplistic inequality $rb>c$, where r is the degree of relatedness (for full siblings r is 0.5, for half-siblings 0.25, etc.), b is the benefit of the specific altruistic act to the recipient, and c is the cost of this act to the altruist. Taking into consideration that this area has attracted many research, in the future there may be interesting developments.

Dyson notes that, although there are indeed several analogies between genes and memes, at the same time biological and cultural evolutions have the following important *opposite* effect: genes promote diversification of species, whereas memes facilitate their unification. Indeed, although a strict interpretation of Darwin's natural selection would imply that only a few species, namely the fittest, will survive, there is in nature an abundance of species diversity. This fact can be explained via the action of a variety of *random processes* which aid diversification, including those discussed in the following.

First, *inheritance*, carried by genes, involves an element of randomness. Since Darwin was unaware of genes, he could not estimate the creation rate of new species and the extermination rate of old ones. His natural selection theory did not use a *quantitative* analysis of the statistics of inheritance. It was simply based on the general axiom of the "survival of the fittest". A proper mathematical analysis was finally presented by Motoo Kimura in his book *The Neutral Theory of Molecular Evolution* (Kimura, 1983). There, it is shown that, although the random effects caused by inheritance are unimportant for large populations, and therefore the predictions of natural selection are valid, these random effects become more important than natural selection for two types of species: for endangered and for newly emerging species. Therefore, via the random effect of inheritance, these small populations become engines of diversity.

A second source of diversity is the occurrence of *random mutations*. This effect is considerably enhanced by the remarkable fact that for many organisms, but not humans, such mutations occur at a particularly high rate in the genes controlling sexual mating (Ferris *et al.*, 1997).[2]

A third paramount source of diversity, which is not mentioned by Dyson, is the result of the fundamental biological mechanism of *horizontal gene transfer*: *genes do not only propagate vertically, from parents to off-springs but also horizontally, among elements of the same or even different species.* This hugely important mechanism, whose crucial significance has been appreciated only recently, has such serious implications that has rendered completely inadequate the standard framework, introduced by Darwin and Wallace, for understanding the origin and the development of species.

In contrast to biological evolution, which promotes diversification, cultural evolution allows people to mix and interact. Such interactions, on the negative side, contribute to the *homogenization of cultural expressions*, which in our era has the alarming effect of suppressing the development of the unique cultural manifestations arising from specific groups. On the positive side, such interactions are drivers towards the *unification of our genes*. Remarkably, *horizontal gene transfer plays a crucial role in the basic underlying biological mechanism for the implementation of this unification.*

A striking example of the process of gene unification is provided in the work of a team of palaeontologists and chemists led by Svante Pääbo (Nobel Prize in Medicine, 2022). This team has developed a technique that allows the sequencing of ancient DNA, even if it is degraded and contaminated with modern DNA. Using this technique, Pääbo's team has sequenced the DNA of the following species: Neanderthals who lived in Europe about 50,000 years ago; our ancestors who overlapped with the Neanderthals in Europe; and the "Denisovans", whose DNA was found in the Denisova cave in Siberia. Apparently, cultural evolution led to an extensive mixing among these three species: modern humans originating in Europe carry 1–3% of Neanderthals' genes. Similarly, modern humans in Papua New Guinea carry a significant percentage of Denisovan genes (Paabo, 2014).

---

[2]The only other family of genes exhibiting a very high rate of mutations is the family involved with the immune system (Ferris *et al.*, 1997). Taking into consideration the ever-diversifying world of invading microbes and allergens, the evolutionary need for such mutations is obvious. Unfortunately, mutations also increase the risk for malfunction, and perhaps this contributes to the exceedingly large number of autoimmune diseases and allergies, which are due to aberrations of the function of the immune system.

The dual role of horizontal gene transfer as a driver of both diversification and unification highlights the far-reaching implications of this fundamental biological mechanism, which is discussed further in the Epilogue.

MOLECULAR PHYLOGENETICS

The huge role of culture in shaping our evolution, should not lead to underestimating that biological evolution, which for billions of years, defined the essential mode of evolution of all living organisms. In what follows, a brief review is presented of the recent elucidation of some important steps in this evolutionary process, and in particular of the passage from prokaryote to eukaryote organisms. This magical step was of critical importance for the development of complex life.

Attempts towards a complete classification of living and inanimate matter began with Aristotle's *Scala Naturae*. Later, a comprehensive compendium was presented by Carl Linnaeus, one of Darwin's "heroes". His *Systema Natura*, published in 1735, consisted of an *Atlas* of the "three basic kingdoms of nature", namely, plants, animals, and minerals. In 1866, the highly influential zoologist and strong advocate of Darwin's theory, Ernst Haeckel, published *The Tree of Organisms*, which contained three major branches: *Plantae*, *Protista*, and *Animalia*. Moreover, in 1874, he published *The Great Oak*, presenting the "Development History of Man", which begins with monera and amoebae and ends with gorilla and man. In addition, Haeckel attempted to supplement Darwin's theory of evolution with specific "developmental rules".[3] The American biologist Robert Whittaker added *fungi* to the triptych of *Plantae*, *Protista*, and *Animalia*. This was motivated by his use of ecology, in addition to morphology: protista were characterized by morphology, whereas plants, animals, and fungi, by ecology. According to Whitaker, plants are "producers", animals are "consumers", and fungi are "absorbers". Later, Whitaker and Lynn Margulis, aware of the limitations of this classification, added to these four categories a fifth one, called *polyphyletic taxa*, consisting of elements that have more than one evolutionary origin. In emphasizing the role of morphology,

[3]The polymath Haeckel introduced the terms ecology, phylogeny, and ontology.

Whitaker was highly influenced by the botanist and microbiologist, Ferdinand Julius Cohn, who along with Pasteur played a decisive role in discrediting the theory of "spontaneous creation". Cohn, responding to the erroneous postulate of this theory, espoused by many distinguished scholars from Aristotle to Jean-Baptiste Lamarck, noted that if something "appeared from nowhere" it is not the result of spontaneous creation, but simply the result of contamination.

Scientists from several disciplines employed a variety of tools, including fossil evidence of early microbial forms, ecology, and embryology, to understand the origin and development of living organisms. An important step in this direction was taken by the microbiologists C. B. van Niel and Roger Stanier, who attempted, in 1962, to provide a rigorous definition of a bacterium. According to these leading microbiologists, all forms of *cellular* life (viruses are thereby excluded) consist of *eukaryotes*, whose defining feature is that their cells contain a nucleus, and *prokaryotes* which are single-cell organisms without a nucleus. Eukaryotes include multicellular animals, plants, fungi, and certain single-cell organisms, like amoeba. Niel and Stanier claimed that prokaryotes consist of four types of bacteria, as well as one additional group, namely, blue-green algae. Following the earlier classification scheme of Ferdinand Julius Cohn, the four types of bacteria were classified morphologically as spherical, rod shaped, filamentous, and spiral (Stanier & Van Niel, 1962). In contrast to other algae, blue-green algae were difficult to classify since they had features of both plants and bacteria. In their paper, Stanier and van Niel emphasized the following two distinguishing characteristics of prokaryotes: first, cell division occurs by simple fusion instead of mitosis involving chromosomal pairing; second, the wall of prokaryotes is strengthened by a molecule called *peptidoglycan.*

The most powerful techniques for revealing the evolution history of organisms are based on the analysis of information hidden in DNA and RNA sequences. The origin of this powerful approach, known as *molecular phylogenetics*, can be traced back to a talk given by Francis Crick in 1958. In this talk, later published in (Watson and Crick, 1958), the Nobel Laureate suggested that evolutionary history could be deduced from evidence provided by proteins:

"Biologists should realize that before that long we shall have a subject which might be called 'protein taxonomy' — the study of amino acid sequences of the proteins of an organism and the comparison of them between species".

In 1964, Emile Zuckerkandl and Linus Pauling also proposed that the study of certain molecules could reveal the evolutionary history of organisms; they named this approach *chemical paleogenetics* (Zuckerkandl & Pauling, 1965). These scientists suggested that minor changes in molecular variants reflect the time elapsed between eons. So, these changes could provide "a molecular evolutionary clock".[4]

The key scientist to explore the enormous potential of molecular phylogeny was Carl Woese, who spent most of his academic life at the University of Illinois, Urbana. In addition to discovering a completely new category of life, called *archaea*, this American biophysicist and microbiologist, also played a major role in the rigorous proof of the remarkable fact that both mitochondria and chloroplasts have bacterial origin. Woese's first important step towards these transformative discoveries was to provide an answer to the following question raised by the suggestions of Crick as well as of Zuckerkandl and Pauling: Which molecule carries useful information relevant to evolutionary history?

He decided to concentrate on a short RNA molecule which is located on one of the two units forming ribosomes. For bacteria, this molecule is called 16S ribosomal RNA (16S rRNA), whereas, for eukaryotes cells it is called 18S rRNA. RNA is part of the "translational apparatus", which is crucial for translating the instructions of DNA needed for constructing proteins. This apparatus consists of messenger RNA, the small subunits of ribosomes, and some raw material.

In addition to its participation in this apparatus, RNA also has a structural function. For example, each ribosome unit consists of structural RNA and proteins. Taking into consideration that ribosomes constitute an important

[4]Zuckerkandl was raised in Vienna. In 1938 his family relocated to Paris and later to Algiers to escape the poisonous atmosphere against the Jewish people created by the Nazi ideology (to my knowledge, he is not related to the anatomist Emil Zuckerkandl, who taught Gustav Klimt anatomy). In 1958, he met Pauling, who invited Zuckerkandl to become his postdoctoral assistant at Caltech. Zuckerkandl is considered one of the founders of evolutionary biology.

ingredient of every living cell, as well as noting the multiple functions of RNA, Woese decided to concentrate on this short molecule which forms part of the large subunit of ribosomes. His approach was similar to that used for sequencing DNA, but now applied to RNA, namely: extract ribosomal RNA, cut it into variously sized fragments using appropriate enzymes, and separate these segments via electrophoresis.

After analyzing various bacteria, as well as some eukaryotes, he decided to study a specific microbe of the so-called methanogen species (which at the time was thought to be a bacterium). This was a difficult task because this species can only grow in an environment lacking oxygen. By 1976, his team had studied several methanogenic microbes. The surprising result was that the sequences obtained from these organisms were *very different* from those of any sequences encountered before. In 1977, Woese and his colleagues speculated that methanogens comprise a completely *new category of life*. These new microbes were first called archaebacteria and later *archaea* (Fox *et al.*, 1977).

This hypothesis was strengthened by the investigations of the German botanist and microbiologist Otto Kandler, who had earlier shown that the wall of methanogens does *not* contain peptidoglycans, establishing in this way that these organisms were different than prokaryotes. Soon afterwards, new members were added to archaea, which included, *halophiles* which are microbes that live in high concentrations of salt, and *thermoacidophiles* which are organisms that live in extremely hot and very acidic environment. The definitive proof of the novelty of this new category was provided in 1996: four months after the publication of the complete genome of *Haemophilus influenza*, which was the first complete genome of a free-living organism, the genome of a member of the archaea was obtained. Half of its genes were indeed completely new. Woese appears in this publication as the penultimate author next to the American biotechnologist Craig Venter (Bult *et al.*, 1996).

Woese was initially sceptical about the idea of *endosymbiosis*, namely, the idea that bacteria somehow entered other single-cell creatures where they achieved lasting compatibility yielding a new kind of cell. However, ironically, his techniques provided the main tools for elucidating this remarkable phenomenon, which is at the heart of the origin of complex life. Endosymbiosis was first suggested, in the early 1900s, by Konstantin Sergeevich Mereschkowski under the name of *symbiogenesis*, and by the American biologist Ivan Wallin in the 1920s under the name of *symbionticism* (Wallin, 1927). Actually, Mereschkowski claimed that a plant cell is simply an animal cell where a bacterium has been added, allowing this cell to perform photosynthesis. Somehow, these ideas were ignored until Hans Ris and Walter Plaut published a related paper in 1962 (Ris & Plaut, 1962). Endosymbiosis began to be studied seriously only after the publications of the biologist and evolutionary theorist Lynn Margulis, who was a student of Ris (Margulis, 1970). Definitive results were obtained using the techniques of Woese. In particular, in a series of papers, using Woese's techniques, the evolutionary biologists Linda Bonen and Ford Doolittle elucidated the origin of different algae, proving that the blue-green algae is actually a bacterium, which was named *Cyanobacterium* (Bonen & Doolittle, 1978). Furthermore, the same investigators established that this particular bacterium provides the origin of *chloroplasts*. These organelles are crucial for photosynthesis, the fundamental process, used by green plants and certain other organisms, for converting light energy to the ingredients necessary for their survival (Bonen & Doolittle, 1976).

The question of determining the specific bacterium that gave rise to mitochondria became a long-standing open problem, which was finally settled by Woese and colleagues in 1985, using rRNA genes. It is a particular bacterium that belongs to a subgroup of the so-called purple bacteria (Yang *et al.*, 1985). It is now widely accepted that the capture of this purple bacterium by another bacterium, an event that apparently took place between 1.6 and 2.1 billion years ago, had profound consequences: the captured bacterium finally became mitochondria delivering vast amounts of energy, which allowed the recipient cell to increase its size and complexity. Presumably this gave rise to the need for containment and protection of the *essence of the cell*, namely, its genetic material. Finally, this led to the formation of the nucleus, and in this way, prokaryotes evolved into eukaryotes.

The aforementioned discussion suggests that the crucial notion of interconnectedness not only characterizes the function of the brain but also provides the basic features of our evolutionary origin. Indeed, despite the efforts of van Niel and Stanier, the occurrence of *endosymbiosis* made it clear that it is *not* possible to define a "pure" classification scheme. Such a task is further complicated by the occurrence of *horizontal gene transfer*, which, as noted in the epilogue, is indeed ubiquitous.

HETERODYNAMICS AS OPPOSED TO HOMEOSTASIS

In my opinion, the new concept of metarepresentations can aid in the further elucidation of the dynamic relationship between biological and cultural evolutions. In this regard, it is noted that, several scholars have argued that, in the same way that the brain imposes a state of homeostasis in the interior of an organism (as discussed in Chapter 5), the brain also achieves an equilibrium with the sociocultural state of the environment; see for example (Damasio, 2012: p. 176). In my opinion, the attainment of such an equilibrium with the environment may be true for our evolutionary ancestors, but not for us. It will be argued below that,

as a result of our unique ability to construct metarepresentations, our relationship with the environment is actually heterodynamic, as opposed to homeostatic.

I shall explain. In 1973, the Hungarian endocrinologist Hans Selye introduced the term heterostasis to describe the *systemic adaptation* of an organism in the presence of an exogenous substance, namely, of a substance originating *outside* the organism (Selye, 1973). In heterostasis, the word *homeo*, appearing in homeostasis, is replaced by the Greek word *hetero* (other). In this way, heterostasis denotes the equilibrium achieved in the presence of an exogenous substance. For example, in alcoholism, a new homeostatic equilibrium is achieved that depends on the presence of alcohol.

If one generalizes the notion of heterostasis to include adaptation in an ever-changing external environment, then one arrives at the notion of heterodynamics.

Incidentally, as stated in Chapter 5, according to Metchnikoff, homeostasis is also a highly dynamic process. Indeed, this process is not static, but it is time dependent. For example, the values of vital signs, like blood pressure, as well as the values of various hormones vary throughout the diurnal cycle. In particular, cortisol, the hormone that helps us cope with stress, peaks in the morning, then decreases throughout the day, dipping to its lowest at 11 pm, which is the time that melatonin levels are high. Also, about 30 minutes after falling asleep, growth hormone begins to be excreted from the pituitary gland, stimulating cellular recovery.[5] However, the organism is in a *quasi-static state*, namely, time variations are small, cyclic, and predictable. This is to be contrasted with the time-varying human dependency on metarepresentations, especially in the modern era. *The term heterodynamics expresses this ever-changing dependency.*

Our *neurophysiological dependence* on a variety of metarepresentations is irrefutable. For example, reading is associated with the activation of a particular part of the brain that is located in a small region of the left ventral visual pathway. The neurophysiological assignment of this region as a specific functional module which is clearly vitally important for the development of culture is the result of the relatively recent development of the important symbolic metarepresentation of writing. The underlying dynamic element of heterodynamics has been enhanced enormously in recent years due to the exponential rate of technological developments. For example, new electronic discoveries compel humans to continuously acquire new and more sophisticated gadgets. This has led to *neuropsychological dependency*, as exemplified by the current addiction to mobile phones, tablets, and the Internet.

The impact of culture on cognition

Complicated cognitive processes generate metarepresentations which transform our culture and this in turn further affects our cognitive development.

The transformative impact of culture on our cognitive development was first addressed by social and cultural scientists and by philosophers, before being addressed by scientists.

[5]In this sense, perhaps the term homeodynamics, used by my colleague and friend at USC Vasilis Marmarelis, may be better than homeostasis.

Sir Francis Galton, a polymath, half-cousin of Darwin, and a child prodigy, in his famous book *Hereditary Genius* (1869) argued that greatness and genius are determined genetically. Galton was the first scholar to use statistical methods to study various human characteristics, including the impact of genetics on intelligence. In particular, he made extensive use of the mathematical law of "regression toward the mean". In 1888, he published a paper reporting that the average head size of the best students at the University of Cambridge was slightly larger than that of the worst students. MRI results have confirmed this correlation: on average, people with larger brains have slightly higher IQs, and this correlation is slightly stronger with the size of the frontal and parietal lobes. Galton also coined the terms *nurture versus nature* and *eugenics*. Unfortunately, he was a strong exponent of the latter.

Galton's contemporary, the French-Swiss botanist Alphonse Pyramus de Candolle, in *Histoire des Sciences et des Savants depuis deux Siècles* (1873), argued in favour of the environment. Later, the cultural anthropologist Alfred Kroeber emphasized culture. More recently, the cognitive philosophers Andy Clark and David Chalmers advocated the notion of "active externalism", the position that the environment plays an active role in driving cognitive processes (Clark, 2008). The cognitive neuroscientist Elkhonon Goldberg claims that there is a shift from the right to the left brain due to cognitive templates that are stored externally through cultural means (Goldberg, 2001). The cognitive philosopher Robert Wilson notes that the brain is embedded in the environment, where "consciousness is supported by cultural scaffolding" (Wilson, 2004). The ability of humans to develop culture has been addressed by many other scholars. For example, the developmental psychologist Michael Tomasello states that humans "possess biologically inherited capacity to live culturally" (Tomasello, 1999: p. 53).

Perhaps the most radical analysis of the impact of culture on human cognitive function is presented by the leading Latin American intellectual and anthropologist Roger Bartra. He has advocated that consciousness *cannot* be separated from the technological and cultural environment:

"What we have is a complex technological and cultural system that is a prosthesis that substitutes functions that we cannot perform, or that we carry out slowly and inadequately. This system together with the brain is the foundation of consciousness" (Bartra, 2014).

I agree with the above statement, provided that the word "consciousness" is replaced with the word "humanity". As clearly exemplified by animals, a "complex technological and cultural system" is not a prerequisite for consciousness. Indeed, animals do not possess such a system but certainly do possess consciousness. Bartra claims that humans have the "Genetically inherited incapacity to live as other animals do-naturally, biologically", in the sense that they possess neural circuits that are *incomplete* and require the environment for their completion (Bartra, 2014: p. 13). My position is different.

Fourth hypothesis: *Humans are predisposed to construct metarepresentations, which generate tremendous cultural and technological advances that have metamorphosed our civilization. As a result of the creation of powerful prostheses caused by the dynamic interaction of appropriate neural mechanisms with specific metarepresentations, human capabilities have been enhanced enormously.*

For example, consider the process employed by a geometer to derive new mathematical properties of triangles via the use of paper and pencil: the innate predisposition of humans to make mental images *real*, impels the geometer to *draw* a triangle. The interaction of unconscious and conscious processes with this figure, which is a specific metarepresentation, may motivate the drawing of the bisector of one of the angles of the triangle. This creates new interactions that, in turn, motivate further constructions. Clearly,

> the process of building mathematical relations involves the continuous and highly dynamic interplay between conscious and unconscious processes on one hand, and the resulting metarepresentations on the other. Certainly, paper, pencil, and the computer are indispensable tools for this interplay, but these external objects are facilitators, as opposed to the internal neural mechanisms which are creators.

The complex, hierarchical structures of metarepresentations create a highly complicated external network. This network has such an enormous impact on human behaviour that it is tempting to consider it part of consciousness, as Bartra does. However, for the reasons discussed earlier, it seems that Bartra statement "consciousness [is] understood as a process that links neuronal activity with the symbolic exocerebral network" (Bartra, 2014: p. 115), should be modified to the following position:

consciousness facilitates the interaction between neural activity and the infinitely complex network of metarepresentations.[6]

Despite the above caveat, Bartra is right that the impact on our cognitive development and on our civilization of what he calls the "exocerebral" network and I call metarepresentations, is indeed immense. For example, starting with the instinct of survival, and combining our innate predisposition for constructing metarepresentations with the phenomenal ability of the human brain to generalize, unify, and invent, we have excelled in constructing houses, making cloths, preparing elaborate foods, etc. These endeavours cannot be justified only by their usefulness. They represent remarkable "cultural prostheses", which along with kinship become essential parts of our existence. As Bartra notes, this fact has been appreciated both by scientists and writers. For example, William James states:

"[...] a man's Self is the sum total of all that he *can* call his, not only his body and his psychic powers, but his cloths and his house, his wife and his children, his ancestors and friends[...]" (James, 1890).

William James' brother, the famous writer Henry James, writes:

"There is no such thing as an isolated man or a woman; we are each of us made up of some cluster of appurtenances. What shall we call our 'self'? Where does it begin? Where does it end? [...] I know a large part of myself is in the cloths I chose to wear. I have a great respect for *things*! One's self-for other people-is one's expression of one's self; and one's house, one's garments, the books one reads, the company one keeps-these things are very expressive" (James, 1881).

[6]With the word "connects", Bartra apparently means "combines". Perhaps for this reason he criticizes Antonio Damasio, Clark and Chalmers, and other authors, for claiming that consciousness is something that happens *internally*.

The use of *symbols* to represent people, objects, locations, situations, intentions, actions, rituals, relationships, artistic creations, etc., in conjunction with gigantic databases that were earlier found in libraries and are now deposited on the Internet, allow humans to have access to an unprecedented abundance of information. In this way, the limited human memory is supplemented with an artificial memory whose capacity is boundless.

Bartra claims that "many neuroscientists tend to be allergic to using discoveries made in the social and cultural sciences" (Bartra, 2014: p. 20). This is a harsh statement. Nevertheless, it is true that unfortunately, there exists a divide between social and cultural sciences and the so-called "hard" sciences. For example, neuroscientists have mostly ignored the interesting analysis presented by Bartra in his book *Anthropology of the Brain* (Bartra, 2014). On the other hand, Bartra, while correctly emphasizing the importance of cultural prostheses, it seems that he understates the immense role of mathematics, computers, sciences, and technology, in the development of culture.

References

Bartra, R. 2014. *Anthropology of the Brain.* Cambridge University Press.

Bergey, D. 1932. *Manual of Determinative Baceriology.*

Bonen, L. & Doolittle, W. 1976. Partial sequence of 16S rRNA and the phylogeny of blue-green algae and chloroplasts. *Nature* 261.

Bonen, L. & Doolittle, W. 1978. Ribosomal RNA Homologies and the Evolution of the Filamentous Blue-Green Bacteri. *Journal of Molecular Evolution* 10.

Bult, C., White, O., Olsen, G., Zhou, L., Fleischmann, R., Sutton, G., Blake, J., FitzGerald, L., Clayton, R., Gocayne, J., Kerlavage, , Dougherty, B., Tomb, J., Adams, M., Reich, C., Overbeek, R., Kirkness, E., Weinstock, K., Merrick, J., Glodek, A ., Scott, J., Geoghagen, N., & Venter, J. A. 1996. Complete genome sequence of the methanogenic archaeon, methanococcus jannaschii. *Science* 273(5278), 1058–1073.

Clark, A. 2008. *Supersizing the Mind: Embodiment, Action, and the Cognitive Extension.* Oxford University Press.

Damasio, A. 2012. *Self Comes to Mind.* Vintage Books.

Dawkins, R. 1976. *The Selfish Gene.* Oxford University Press.

Dyson, F. 2019, February 2. Biological and Cultural Evolution (An EDGE Original Essay). *Edge.*

Ferris, P., Pavlovic, C., Fabry, S., & Goodenough, U. 1997. Rapid evolution of sex-related genes in Chlamydomonas. *Proceedings of the National Academy of Sciences* 94(16), 8634–8639.

Fox, G., Magrum, L., Balch, W., Wolfe, R., & Woese, C. 1977. Classification of methanogenic bacteria by 16S ribosomal RNA characterization. *Proceedings of the National Academy of Sciences* 74(10), 4537–4541.

Goldberg, E. 2001. *The Executive Brain: Frontal Lobes and the Civilized Brain.* Oxford University Press.

James, H. 1881. *The Portrait of a Lady.* London: Macmillan.

James, W. 1890. *The Principles of Psychology.* New York: Holt.

Kimura, M. 1983. *The Neutral Theory of Molecular Evolution.* Cambridge University Press.

Margulis, L. 1970. *Origin of Eukaryote Cells.* Yale University Press.

Paabo, S. 2014. *Neanderthal Man: In Search of Lost Genome.* Basic Books.

Ris, H. & Plaut, W. 1962. Ultrastructure of DNA-containing areas in the chloroplast of Chlamydomonas. *The Journal of Cell Biology* 13(3), 383–391.

Selye, H. 1973. Homeostasis and heterostasis. *Perspectives in Biology and Medicine* 16, 441–445.

Stanier, R. & Van Niel, C. 1962. The concept of a bacterium. *Archiv für Mikrobiologie* 42(1), 17–35.

Stewart-Williams, S. 2020. *The Ape that Understood the Universe.* Cambridge University Press.

Tomasello, M. 1999. *The Cultural Origin of Human Cognition.* Harvard University Press.

Wallin, I. 1927. *Symbionticism and the Origin of Species.* Wewverly Press.

Watson, J. & Crick, F. 1958. On protein synthesis. *The Symposia of the Society for Experimental Biology* 12, 138–163.

Wells, H. 1920. *Outline of History.* George Newnes.

Wilson, R. A. 2004. *Boundaries of the Mind: The Individual in the Fragile Sciences.* Cambridge University Press.

Yang, D., Oyaizu, Y., Oyaizu, H., Olsen, G., & Woese, C. 1985. Mitochondrial origins. *Proceedings of the National Academy of Sciences* 82(13), 4443–4447.

Zuckerkandl, E. & Pauling, L. 1965. Molecules as documents of evolutionary history. *Journal of Theoretical Biology* 8(2), 357–366.

THE INTERPLAY BETWEEN UNCONSCIOUS AND CONSCIOUS PROCESSES

The basic characteristics of ancient Greek thinking were the quest for knowledge and the admiration of beauty. Therefore, it is not surprising that ancient Greek philosophers were the first to attempt to analyze the process of *knowing*, and start reasoning about *thinking*. These contributions constitute the beginning of the philosophical foundations of neuroscience.

The ancient Greeks understood that the search for knowing requires developing appropriate methodologies for reaching the *hidden*. This was first understood by Parmenides, and clearly stated by Anaxagoras, a distinguished pre-Socratic philosopher who after moving from the Ionian Asia Minor to Athens became a friend and protégé of Pericles. Anaxagoras wrote: "phenomena are the sight of the hidden".

The first attempt to answer the fundamental question "What is knowledge?" was made by Plato in *Theatetus* (Nehamas, 1994). In this, and several other dialogues, Plato developed a theory regarding "knowledge and how it is attained". An important part of this theory is the notion of "*a priori* knowledge". This is an innate form of knowledge that "every soul possesses" (Nehamas, 1999).

Aristotle proposed a different definition of knowledge. In his *Posterior Analytics* he stated: "We have knowledge of a fact when we know: first, the cause from which this fact results, and second, that this fact cannot be otherwise than it is".

The ancient Greek philosophers understood that if we had no knowledge other than *demonstrative knowledge* then we would end up in an *infinite regression*, namely: to know something, one must be able to prove it is based on something else that one knows, which in turn one must be able to prove in terms of something else, etc. Aristotle attempted to overcome this difficulty by introducing the idea that there exist certain *principles* that humans can comprehend *intuitively*.

The different points of view expressed by Plato and Aristotle regarding the essence of knowledge are closely related to their dialectically opposite understanding of the role of innate knowledge. Following a brief summary of these positions, Chapter 9 addresses the interaction of innate versus acquired knowledge from the point of view of modern neuroscience. This includes the presentation of the Nobel Prize-winning work of Eric Kandel regarding *learning at the molecular level.*

The postulates of Donald Hebb, as well as detailed neurophysiological mechanisms crucial for learning and memory, are discussed in Chapter 10. Medical implications of the elucidation of these fiendishly complicated mechanisms are also discussed.

The modern understanding of learning is intimately related to the interaction of conscious and unconscious processes. Apparently, the first attempt to elucidate the relationship between these processes was the works of physician and physiologist Carl Gustav Carus (1789–1869). In 1846, he wrote:

"The key to the knowledge of the nature of the soul's conscious life, lies in the unconscious. This explains the difficulty, if not the impossibility of getting a real comprehension of the soul's secret" (Carus, 1846).

If the word "soul" is replaced with "brain" then I believe that this phrase captures the essence of the unconscious–conscious relationship. Carus was a friend of the German romantic artist and writer, Caspar David Friedrich

Figure PIII.1. Caspar David Friedrich, *Wanderer above a Sea of Fog* (1818).

(1774–1840).[1] The influence of Carus' ideas on Friedrich is evident in the iconic painting *Wanderer above a Sea of Fog* (1818), where the striking remoteness convey a feeling of intense introspection (Figure PIII.1).

In the 20th century, science and technology reached the necessary level of sophistication, so that the precise interaction between conscious and unconscious processes could be rigorously investigated. Some of these developments are presented in this part of the volume. In particular, Chapter 11 discusses several aspects of the interplay between conscious and unconscious processes, including the phenomena of *confabulation*, exhibited by individuals whose left and right hemispheres have been surgically separated, and *binocular rivalry*, where the left and right

[1]In 1821, while Carus and Goethe were both in Weimar they became close friends. This is consistent with their common ideology.

visual fields receive two different images. These works further highlight the primacy of the unconscious: they show that the transition from the unconscious to awareness involves *loss of information*. In addition, the recent discovery of the *concept neurons* is also discussed in this chapter.

Chapter 12 presents details of visual perception: from intricate neurophysiological mechanisms crucial for vision, to several abnormalities including blindsight, Anton's syndrome, prosopagnosia, and Capgras' syndrome. Specific neuronal mechanisms for face recognition are also discussed.

Chapter 13 begins with a discussion of the nature of light. Then, it focuses on the intuitive employment by artists of some of the neural mechanisms elucidated in Chapter 12. This includes the usage by all great portraitists of the fact that *the highest resolution is achieved at the centre of the gaze*. In this regard, various portraits by Picasso, El Greco, Manet, and Goya are analyzed. In addition, the extensive use by many artists, including Picasso and Goya, *of the aesthetic power of caricatures*, is discussed. The neural basis of the strong effect of caricatures is due to the fact that, as discussed in Chapter 12, the visual system concentrates on the analysis of *how a particular face differs from an average face*, which is precisely the essence of caricatures. Chapter 13 also discusses the *importance of associations in painting*. On the one hand, the *creative impact* of associations is illustrated by analyzing one of the late portraits of Picasso. On the other hand, it is noted that the *continuous generation of associations can hinder an artist's attempt to perceive reality*. No one understood this potential difficulty deeper than Alberto Giacometti. Finally, the importance of luminance and colour for the Impressionists is analyzed.

References

Carus, C. G. 1846. *Psyche: On the History of the Development of the Soul*. Spring Publications.

Nehamas, A. 1994. Meno's paradox and Socrates as a teacher. In *Plato's Meno In Focus*. Routledge.

Nehamas, A. 1999. *Virtues of Authenticity: Essays on Plato and Socrates*. Princeton University Press.

INNATE KNOWLEDGE AND LEARNING

INNATE VERSUS ACQUIRED KNOWLEDGE

Does there exist innate knowledge? As noted in the introduction of Part III, according to Plato, the answer is yes. Indeed, Plato and later Gottfried Leibniz and Immanuel Kant emphasized the existence of *a priori knowledge.* Based on the Platonic dialogue *Phaedon*, the Neo-Platonic philosopher Syrianus, who belonged to the philosophical school of Athens established by Iamblichus, suggested that there exist three categories of souls: at the first level, there are the souls of gods and their fellow travellers. At the second level, the *pure* (achrantes) souls which are sent to Earth to save the third category of souls, namely those that are *corrupt*. For example, Socrates was sent to Earth on a *soteriological* mission, especially to save the souls of the young. The pure souls contain in themselves knowledge, and this is the object of the study of *Dialectics*. Even the souls that have fallen from grace and are corrupted were pure at some point earlier, hence they contain in themselves *a priori knowledge*. For this reason, the process of learning and discovering is merely a process of *recollection*. The teacher stimulates this process by asking appropriate questions. Within this framework, the *maieutic method* of Socrates, based on penetrating and repeated questions, acquires an indispensable character. This Platonic position becomes clear in the reaction of Socrates after he led with his questions a slave to discover certain geometrical truths; he exclaimed: "the slave had always this knowledge within his soul". Importantly, Plato emphasized that knowledge is *not* acquired via the senses but via *theorizing.*[1]

Plato's first attempt to define knowledge was made in his dialogue *Theatetus*, where Socrates proposes identifying knowledge with "true belief (*doxa*), accompanied by logos". However, with striking honesty Plato has the dialogue end in failure, because, despite several attempts, the interlocutors do not succeed in defining *logos*.

Regarding this definition it should be noted that according to the early Platonic dialogues, a belief is true, at least until proven otherwise, if it *cannot* be overthrown via the method of the *elenchus*. It is stated explicitly by Socrates in the Platonic dialogue *Meno* that a belief "is worth little until it is tied down by reasoning" (Nehamas, 1994). The method of *elenchus* is closely related to the concept of *logos*, which constitutes a collection of logical arguments (*logismos*) employed in the process of verifying that a specific belief is indeed true.

What is the *origin* of "true beliefs"? A crucial contribution of Plato's theory of knowledge is that it provided a partial answer to this important question: in addition to the knowledge acquired in the past, there also exists *a priori* knowledge. As stated earlier, this is innate knowledge that can be accessed via the process of "recollection". Plato's emphasis on the "earlier knowledge" provides a partial resolution to that which Peter Geach called "Socrates' error", namely, the fact that supposing Socrates assumed that it is not possible to analyze a given category without first providing its precise definition. For example, after Socrates persuaded Meno to admit that he does not know the definition of virtue, the statement is made that "one cannot search for what one does not know for one does not even know what to search for" (Nehamas, 1994). From this statement it may be inferred that it is impossible to acquire any knowledge. However, as a result of the existence of earlier knowledge and especially of *a priori* knowledge, one *does* possess preliminary knowledge, which provides a starting point for further contemplation. For example, regarding attempts to define virtue, one can recognize that certain acts are virtuous.

In my opinion, the inability of Plato to define *logos* is the result of Plato's deep understanding of *interconnections*. Indeed, by employing *logos*,

[1]In Greek, this concept is expressed by the word θεωρεῖν, which means the simultaneous process of observing and contemplating.

the dialectic method of *elenchus* attempts to transform an existing belief to knowledge via the following process:

it delineates the logical interrelations that the properties of the given concept satisfy between themselves, as well as with respect to other entities.

However, this task is impossible due to the vast number (perhaps an infinity) of such interrelations.[2]

Aristotle responded to the positions of his teacher with an undeniably *empiricist* reply: in *Posterior Analytics*, he grants full scientific status to several disciplines, including astronomy, acoustics, harmonics, and optics. Each of these sciences has its own *principles* that cannot be proven. These principles become known to humans at the end of a cognitive process which involves perception, memory, experience, and the *mind* (νους). For Aristotle, *mind is the cognitive apparatus that comprehends principles.* Hence, the combination of the mind and demonstrative knowledge, which is based on *logical syllogisms*, leads to the truth.

In his analysis, Aristotle emphasizes that learning is impossible without *associating ideas*. Following Aristotle, the British empiricist John Locke proposed that the human mind is a *tabula rasa*, i.e., a blank slate, which is inscribed as a result of experience.

Who was right? Plato or Aristotle? Hopefully, it will become clear in this chapter that, in a sense, they were both correct. For the first time ever, there now exists a proper scientific framework for the comprehensive study of such questions. This framework, whose basic elements are introduced in the next section, is based on neuroscience.

FROM BEHAVIOURISM TO NEUROSCIENCE

It is well known that Ivan Pavlov (Nobel Prize in Medicine, 1904), as well as Edward Thorndike, provided experimental proof that there exists a type of learning that is based on *associating information* from different stimuli.

---

[2]In the two dialogues that follow *Theatetus*, namely, the *Sophist* and the *Politicus*, Plato attempts to define sophistry and the characteristics of a statesman. In these dialogues, where Socrates is no more the key person, the importance of interrelations is clearly noted.

Pavlov, guided by his advisor Ivan Sechenov, the "father of Russian physiology", who was obsessed with reflexes, studied the dogs' salivation reflex. He was trying to measure the amount of saliva produced in response to a given amount of food. Pavlov was surprised to notice that the dog started salivating as soon as it heard the food being prepared. In order to investigate this unexpected "psychic reflex", Pavlov designed, in 1902, his classical experiments: the dog was led to expect the food whenever it heard a bell. Then, the dog soon began salivating to the sound of the bell alone.[3] After establishing his famous "conditional learning", he was able to show that a dog with no cerebral cortex could still reflexively salivate when fed, but not when alerted by the bell. Hence, a dog's associative abilities reside in its cerebral cortex. The classical experiments of Pavlov and Thorndike led to the emergence of the empirical school of *behaviourism.* It was clearly argued by John B. Watson and Burrhus F. Skinner that behaviour can be studied experimentally as rigorously as phenomena in the physical sciences. The behaviourists, following Sigmund Freud, deliberately avoided any speculation about the relation of behaviour with brain activity, and focused exclusively on observable behaviour.

The rigorous experiments of the behaviourists led to a deeper understanding of the process of acquiring knowledge. For example, in 1969, Leon Kamin demonstrated that animals do not simply learn that a neutral stimulus, such as sound, *precedes* a reward, such as food, but rather that the stimulus *predicts* the reward. This suggests that *associative learning* is based on the ability of the brain to couple events that occur together regularly, so that it can predict the occurrence of an outcome. This basic form of learning underlines the huge importance of *associations.* This concept, which was already emphasized in Chapter 4, provides one of the guiding principles of this volume.

Interestingly, recent results suggest that active learning is enhanced when there is some *discrepancy* between the expected outcome and what actually happens. In this sense, *surprise* is a stronger facilitator of learning than *predictability.* For example, the prediction of a reward causes the release

[3]Since Pavlov was strongly influenced by his advisor, this seminal discovery illustrates the importance of continuity but also the impact of serendipity.

of the neurotransmitter dopamine. If the reward equals expectation, then dopamine continues to be released at the same rate, whereas if it exceeds expectations, then the rate of dopamine release increases. Similarly, dopamine release decreases if the reward is below expectations (Schultz, 1998). If the cellular predictions are correct and the reward arrives on time then the organism "rewards itself for being right" with a surge of dopamine. It was shown in Zaghloul *et al.* (2009) that neurons in substantia nigra, a region in the brain producing dopamine, exhibited high-level of activation after unexpected gains (Zaghloul *et al.*, 2009). This explains the addictive behaviour, especially gambling, of some patients suffering from Parkinson's disease who are treated with a medication containing dopamine.

The evolutionary origin of associations becomes evident by noting that primitive organisms make extensive use of this basic mechanism. For example, the nematode worm "remembers" the smells and temperature levels that are previously associated with food and employs this information to choose an optimal path for finding new food.

The existence of learnt associations between stimulus and award is consistent with Aristotle's position. However, several other observations and experiments support the validity of *tabula inscripta*, which means, as suggested by Plato, that the brain is inscribed. For example, regarding animals, it is noted that dolphins are born swimming, giraffes learn how to stand within hours of their birth, and zebras can run within 45 minutes after they are born. Clearly, these animals are born genetically hardwired to perform these tasks, vital for their survival (Eagleman, 2015). Experimental support for this fact was provided in a series of studies performed by Nikolaas Tinbergen (1948), who was awarded the Nobel Prize in Medicine in 1973. It turns out that a gull chick seeks food by pecking on a prominent red spot on its mother's yellow beak. The mother responds by feeding the chick. When Tinbergen showed the chick an isolated beak unattached from a mother's body, the chick pecked at the red spot just as vigorously as it had on its mother's beak. Moreover, remarkably, the chick behaved the same way when Tinbergen showed the chick a yellow stick with a red spot on it. These experiments provide direct evidence for the existence of *elemental innate knowledge*. Indeed, in order to survive the chick's brain is wired to be attracted to a red spot in a yellow background.

Of course, the amount of innate knowledge is limited. For example, wild monkeys fear snakes, but monkeys born in captivity do not, which implies that this fear is not innate. Clinical psychologist Susan Mineka has proven that monkeys acquire a fear of snakes by observing the reaction of other monkeys. Indeed, she took six baby monkeys born in captivity to wild-born mothers. When she exposed them to snakes, they were not afraid. However, the mother's terrified reaction to snakes was immediately picked up by the baby monkeys who subsequently became fearful of them. Furthermore, she was able to prove that there was nothing special about snakes. By splicing different parts of tapes, Mineka prepared a tape showing mothers not afraid of snakes but being terrified by a flower. After the baby monkeys watched this tape, they continued to be fearless of snakes but became fearful of this particular flower (Ridley, 2003: p. 193).

Innate knowledge has also been demonstrated in humans. This knowledge is in addition to the existence of several reflexes, like sneezing, gagging, and yawning, which are controlled by circuits in the lower brain and are vital for survival.

For example, it has been established that a few months old babies understand that objects are discrete entities occupying space; that objects cannot be in two different places at the same time; and that they cannot disappear spontaneously (Baillargeon & DeVos, 1991). Also, even at this very early age, babies can perform simple arithmetic. For, example, in a specific experiment, babies observed one object being hidden behind a screen, which was followed by a second object; then the screen was removed. If two objects were revealed, the babies looked at them only briefly. However, if only one object appeared, the babies expressed *cognitive surprise* by looking at the scene for a prolonged period (Koechlin *et al.*, 1997). Remarkably, according to Dehaene (2020: p. 61), babies possess the concept of probability! For example, 1 green and 3 red balls were placed in a transparent box and were moving randomly; at some point the box was occluded and then a ball came out at the bottom. If the ball was red, babies looked at it only briefly, but if it was green, they expressed their surprise by looking at it for much longer. Similarly, by observing that the majority of the balls emerging from an occluded box were red, as opposed to green, babies could predict that the majority of the balls in the box were red. Incidentally, one of the most important results in the theory of mathematical probability is a theorem by Reverend Thomas Bayes, which formalizes this important innate concept. Namely, it allows the inference of rigorous conclusions from the distribution of observations.

It is well known that, regardless of the linguistic wealth of their environment, children at a very young age manage to acquire their native language with great ease. Within the first four years of their life they learn about 12 words a day, and at the same time they are able to produce grammatically correct sentences of complex structure, many of which they have never heard before. For instance, in a well-known experiment designed by Steven Pinker, 4-year-old children were instructed to ask questions to the doll Jabba (from Star Wars). The statement "it is raining" corresponds to the question "Is it raining?", i.e., the auxiliary verb is moved to the beginning of the clause. The more complex sentence, "The horse that is in the field is eating a flower", is associated with the question, "Is the horse that is in the field eating the flower?" Remarkably, 90% of the children moved the auxiliary verb correctly. Furthermore, it appears that many grammatical structures are valid universally. For example, in the prepositional phrase "from the song" there is a *structural relationship* between the preposition, the determiner, and the noun (object). This word order differs across languages, but the structural relationship is the same. For example, in French, English, and Greek, the prepositions are heading the phrase, while in Japanese the preposition is placed at the end. The speed and accuracy of learning the vocabulary of a specific language, as well as the existence of universal grammatical structures, led the distinguished linguist and philosopher Noam Chomsky to the conclusion that there exists an *innate universal grammar.*

Several studies suggest that in addition to language, basic geometrical concepts, as well as the ability to *approximate* arithmetic are also innate. The latter is consistent with the studies in babies noted earlier. For example, a study published in *Science* in 2006 (Dehaene *et al.*, 2006) compared the *geometrical intuition* of 14 children of the indigenous group Munduruku, 30 adults of the same group, and 26 children from United States. The Munduruku, who live in an isolated area of the Amazon, do not have words for basic geometrical concepts such as parallel or symmetrical, and they do not possess maps or other relevant tools that facilitate the development of geometrical intuition. Moreover, they have words only for the numbers 1–5. In one of the relevant experiments, children were presented with six diagrams, one of which was slightly less symmetric than the other five, and were asked to point out the "strange" or "ugly" shape among these six diagrams. Remarkably, all three of the above groups gave similar answers.

Regarding the ability for arithmetic, in another publication in *Science* (Pica *et al.*, 2004), Munduruku children were shown 1–25 dots on a computer screen and were asked to find their number. From 1 to 3 dots the answer was exact. But from 4 to 25 the answer was approximate. For example, even though there exists a word for 5, among the answers for 5 dots were: 3, 4, 5, and "about as many as the fingers of one hand". For 13 dots, among the answers was "two hands and something more". Other similar experiments demonstrated the ability of the children for *approximate* addition and subtraction. These studies suggest that humans are born with the ability to appreciate certain geometrical features including symmetry, as well as with the ability for exact arithmetic for numbers up to 3, and for approximate arithmetic for larger numbers. The impact of numbers on human culture is discussed in Everett (2019).

It seems that face perception is also innate. Pascalis and collaborators (Pascalis *et al.*, 2002) have shown that 6-month-old babies can differentiate between human faces as well as monkey ones: they look at a familiar human face on average for 3.6 seconds and at a novel human one for 4.6 seconds. Similarly, they look at a familiar monkey face for 2.5 seconds and at a new monkey one for 4 seconds. However, 9-month-old babies can only differentiate between human faces: they look at a familiar human face for 3.6 seconds and at a new human one for 4.5 seconds. They look both at a familiar monkey face and a new monkey one for 3.8 seconds. For adults the analogous times are 1.6–2.8 and 2.4, respectively. In analogy with these experiments, it is noted that babies between 4 and 6 months old not only can discriminate between phonetic variations in their mother tongue but also in other languages. However, babies between 10 and 12 months old can only discriminate phonetic variations in their mother language.

In summary: *both animals and humans are born with innate universal algorithms that allow them to extract information from the physical world. These algorithms are modified as a result of interactions with the environment.*

Is it possible to decipher the neuronal mechanisms responsible for the interplay between innate and acquired knowledge? The answer is yes. For the sake of brevity, only two relevant works are described below.

By recording the electrical activity of brain cells of mice in the hippocampus (the region of the brain important for memory), the neuroscientist John O'Keefe (Nobel Prize in Medicine, 2014) showed that this part of the brain creates a *representation* of the external space. This is accomplished by using certain cells which O'Keefe named *place cells*. As the mouse moves in a cage, specific cells of its brain are activated only when the mouse is in a specific position. Apparently, the brain subdivides the external space into many overlapping areas, and each one of them is represented in the brain by a specific cell. The general ability of the brain to create representations of the external space is innate, consistent with the notion of *tabula inscripta*, whereas a specific map is created as the result of experience, consistent with Aristotle's position.

Eric Kandel (Nobel Prize in Medicine, 2000) has shown that the interplay between innate and acquired knowledge can be studied at the molecular level. His remarkable results were obtained by studying the giant marine snail *aplysia*. To breathe, *aplysia* uses an external organ, called *gill*, which lies in a protected cavity. Weak tactile stimuli at the tail of *aplysia*, called *siphon*, cause the gill to withdraw into the cavity. After a repetition of weak touches, the snail becomes *habituated*, namely, adapts to the presence of repeated *harmless* stimuli, and therefore the withdrawal reflex diminishes. However, when the weak touch is paired with a *shock* to the tail, the snail is *sensitized*, and henceforth even a weak touch leads to a strong gill withdrawal reflex. How does *aplysia* "learn" to do this? The weak stimulus

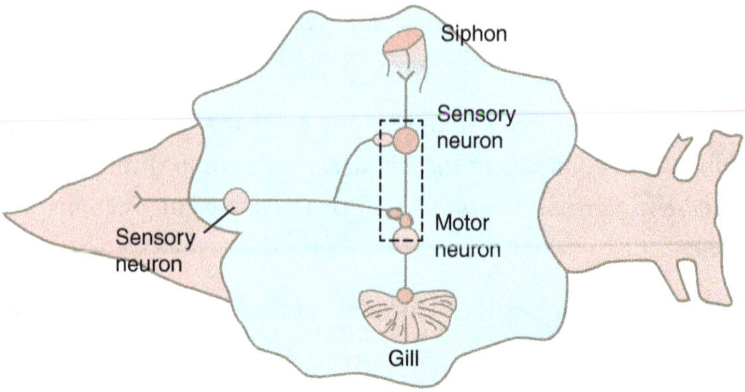

Figure 9.1. A diagram of the neuronal mechanisms employed by aplysia for learning.
Source: Modified from Kandel (2006).

activates a sensory neuron, and the action potential of this neuron causes the release of the excitatory neurotransmitter glutamate, which in turn stimulates the motor neuron responsible for the withdrawal reflex (see Figure 9.1).

Repeated touches cause glutamate depletion and hence the withdrawal reflex diminishes. The shock at the tail stimulates another sensory neuron, and the action potential of this neuron causes the release of serotonin. In turn, this induces the release of a larger amount of glutamate and this yields a stronger withdrawal reflex. This behaviour lasts only for a few minutes.

If instead of one shock, five shocks are delivered, the changes at the neuronal level are truly dramatic. The associated higher amount of serotonin not only causes the release of more glutamate but it also activates a specific gene which initiates the process of creating *new* synapses. Hence, an anatomical change takes place, and the behaviour of a stronger withdrawal reflex lasts for weeks instead of minutes.

It should be noted that these fundamental mechanisms have broader validity. For example, the mouse uses learning mechanisms similar to those of *aplysia*. Indeed, during the process of memorizing a specific place, it is possible to observe quantitative changes of the relevant synapses in the hippocampus of a mouse. Furthermore, long-term memory requires the creation of new synapses. Instead of the neurotransmitter serotonin used by *aplysia*, the mouse uses the neurotransmitter dopamine, but it is interesting that the creation of new synapses involves the activation of the same gene as in *aplysia*, namely of the gene called CREB.

Clearly, the underlying neuronal architecture provides the basis for both the existence of the elemental innate knowledge possessed by aplysia, namely that it "knows" to withdraw following a touch at the siphon, and its ability to learn from experience. The latter, a form of acquired knowledge, is achieved via qualitative and quantitative changes at the synaptic level.

In summary, complicated neuronal circuits, together with the astounding dynamic behaviour of their synapses, provide the substrate for the existence of elemental innate knowledge, as well as for a predetermined predisposition for learning. The brain converts this predisposition into knowledge by employing intricate neural mechanisms that are crucially affected by the continuous bombardment of external and internal stimuli.

Returning to the beginning of this chapter, it can be stated that,

Plato's position reflects the existence of neuronal circuits and the innate information genetically encoded in their architecture, whereas Aristotle's position reflects the changes occurring in the synapses.

Aristotle's understanding of the importance of associations, which was popularized by John Lock and other philosophers, led to the psychological theory of *associationism*. This has been elaborated upon by many psychologists and neuroscientists. For example, Sebastian Seung, who has emphasized that "we are the connections formed between our neurons", writes: "Ideas are represented by neurons, associations of ideas by connections between neurons, and a memory by a cell assembly or synaptic chain" (Seung, 2012: p. 73).

THE VITAL IMPORTANCE OF THE UNCONSCIOUS

A plethora of experiments support the basic postulate stated in Chapter 4: *any experience is preceded by an unconscious process*. According to this hypothesis, before we are aware of any object, our brain *perceives* this object, before we become aware of a particular decision, our brain "has made" this decision. In general, before we become aware of any thought and emotion, our brain "knows" and "feels" them.

Naturally, since unconscious processes do not reveal themselves in a direct way, the acceptance of the crucial importance of the unconscious in *every mental process* has been met with strong resistance. Starting with the pioneering experiments of Benjamin Libet regarding the genesis of our decisions and the role of free will, which will be discussed in detail in the second volume, many other experiments attempted to establish the pre-eminence of the unconscious in specific situations. These works were met either with astonishment or with scepticism.

Such works include several investigators' attempts to establish that, before a subject becomes aware of the *meaning* of a number, a word, an object, or a situation, there is an unconscious understanding of this meaning. In particular, in experiments performed in 1983, psychologist Anthony Marcel used the technique of masking, mentioned in Chapter 4, to investigate particular instances of the unconscious processing of the *meaning of words*. In a key experiment carried out in the Medical Research Council Applied Psychology Unit in Cambridge, he flashed the word *blue* on a screen. The use of masking prevented the participants of the study from becoming aware of this word. Nevertheless, when they were subsequently shown several patches of different colours and asked to choose a patch of a particular colour, there were faster, by 1/20th of a second, when they were asked to choose the *blue* patch. Marcel concluded that the meaning of the word blue was understood unconsciously, and this knowledge *primed* the participants when choosing the blue patch. Marcel also claimed that, for words which have two different meanings, *both* of these meanings are known at the unconscious level. For such words, their conscious meaning becomes clear in the context of a given sentence. For example, the word "bank" appearing in a sentence involving the word "river", has a completely different conscious meaning from the word bank in a sentence involving the word "money". Marcel claimed that he was able to experimentally establish that the word bank primes both money and river, thus proving that the word bank retains both its meanings at the unconscious level. However, the above experiments were later criticized for certain technical aspects. Similarly, although the Seattle psychologist, Anthony Greenwald, published, in 1996, an article in the prestigious journal *Science* claiming that the *emotional meaning of words* is first understood unconsciously, he later raised doubts about the validity of his own work.

The definitive proof that the meaning of words is indeed first processed unconsciously was presented in 1998 by Stanislas Dehaene and Lionel Naccache. These investigators flashed a subliminal number, say 9, which was followed by a visible number. The participants were asked whether the visible number was larger than 5. If 9 was *congruent* with the visible number, namely if both were larger than 5, then the participants answered faster than if the two numbers were *incongruent*, namely one was smaller and the other larger than 5. For example, in the case of the invisible number 9, the response was faster for the visible numbers 6 and 8, but slower for the visible numbers 4 and 1. Similar results were obtained when the investigators used as a subliminal stimulus the word *nine* instead of the number 9. Additional support was provided by simultaneously imaging via fMRI of a particular region of the left and right parietal lobes, which is activated by numbers. If both the invisible and visible numbers were 9, the activity in this region *decreased* in comparison to the usual activity generated by the number 9. This phenomenon, known as *repetition suppression*, is a consequence of the fact that the relevant neurons were able to recognize that this number is shown twice, even if the first time the number 9 was invisible. The associated suppression is due to the neuronal process of *habituation*, discussed earlier in connection with the *aplysia* experiments.

Additional experiments have elucidated the importance of the unconscious with respect to the *meaning of sentences*. In this connection, it should be noted that it has been well established by several groups of investigators that if the final word in a sentence has an *unexpected* meaning, then the EEG exhibits a particular signal known as N400. This is a negative brainwave recorded on the top of the scalp, 400 milliseconds after the occurrence of the unexpected word. For example, such a signal occurs with the sentence "in the morning I have coffee with sugar and socks", due to the occurrence of the word "socks" instead of the expected word "cream". Interestingly, the size of the signal reflects the level of absurdity: slightly inappropriate words generate small signals, whereas completely unexpected words generate larger signals. EEG recordings show a similar signal even if the meaning of a visible word was unexpected due only to the presence of an *invisible* word (van Gaal *et al.*, 2014). For example, an N400 wave was generated if the visible word "war" was preceded by the subliminal stimulus "happy". Furthermore, the signal was larger if the subliminal stimulus was "very happy". Remarkably, the relevant signal was of exactly the *same* size whether very happy was subliminal or visible. Commenting on this surprising fact, Dehaene states:

"It also means that unconscious stimuli do not always generate minuscule events in the brain. Brain activity can be intense even though the stimulus that causes it remains invisible" (Dehaene, 2014).

These findings are consistent with the discussion in Chapter 4 where it was noted that during the solution of a *local* inverse problem, the unconscious neuronal operations are of high level of complexity. In general, the difference between an unconscious process and a conscious experience is that, in the former case, the activation is localized in the particular domain of the brain specializing in the solution of the particular inverse problem. On the other hand, in a conscious experience, different local circuits are integrated, and thus the activation in the brain is extensive. In the case of both the conscious and the unconscious processing of "very happy", activation is limited in a small area of the temporal cortex, which is the domain where semantics is analyzed.[4]

Another study that provides direct evidence for the occurrence of extensive brain activation during unconscious processes involved a patient suffering from progressive volume loss in the cerebral, midbrain, and cerebellar domains (Schubert *et al.*, 2020). This patient, identified as R.F.S., has lost the ability to consciously perceive the form of the Arabic numerals 2–9. So, he cannot describe or copy these numerals. When R.F.S. was asked

[4]Regarding the difference between conscious and unconscious processes, Dehaene states: "[...] unconscious activity is confined to a narrow and specialized brain circuit. During unconscious processing, brain activity remains within the boundaries of the left temporal lobe, which is the primary site of language networks that process meaning. Later we shall see [...] that conscious words [...] gain the upper hand over much larger networks [...]" (Dehaene, 2014).

Figure 9.2. Figure of 8 and the attempt of R.F.S. to copy it.
Source: Schubert *et al.* (2020). Reproduced with kind permission of PNAS.

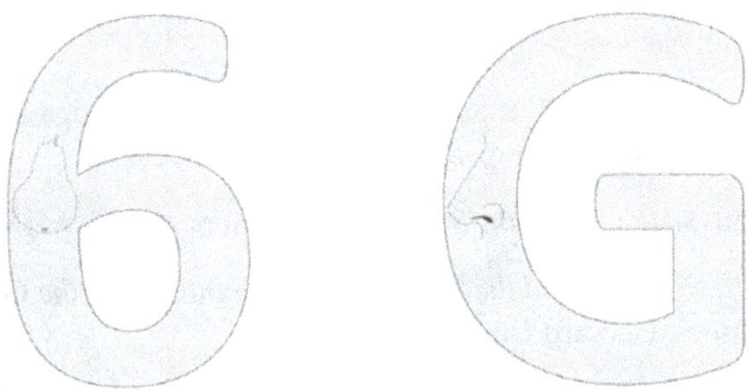

Figure 9.3. Figures of 6 and G.
Source: Schubert *et al.* (2020). Reproduced with kind permission of PNAS.

to copy the letter on the left of Figure 9.2, the patient began by drawing the black lines shown on the right of Figure 9.2, and then added an orange background. In addition, the ability of R.F.S. to consciously perceive stimuli placed in close proximity to the digits 2–9 is drastically impaired. For example, R.F.S. was able to identify the caricature in the letter G to the right of Figure 9.3 but did not consciously perceive the pear embedded in the digit 6 on the left of Figure 9.3. Despite the lack of awareness, imaging studies employing the imaging technique of EEG showed brain activation similar to that occurring in normal subjects during visual awareness.

Finally, it should be noted that consciousness is also affected by the process of *attention*. This implies that conscious attention must also be preceded by an unconscious phase. Indeed, in a relevant experiment, a subliminal stimulus was flashed in the corner of one of the two eyes. Following the identification of this corner as the spot of unconscious attention, it was then established that stimuli presented in this corner were processed faster and more attentively.

References

Baillargeon, R. & DeVos, J. 1991. Object permanence in young infants: Further evidence. *Child Development* 62(6), 1227–1246.

Bekinschtein, T., Manes, F., Villarreal, M., Owen, A., & Della Maggiore, V. 2011. Functional imaging reveals movement preparatory activity in the vegetative state. *Frontiers in Human Neuroscience* 5, 5.

Dehaene, S. 2014. *Consciousness and the Brain: Deciphering How the Brain Codes Our Thoughts.* Viking.

Dehaene, S. 2020. *How We Learn.* Random House.

Dehaene, S., Izard, V., Pica, P., & Spelke, E. 2006. Core knowledge of geometry in an Amazonian indigene group. *Science* 311(5759), 381–384.

Eagleman, D. 2015. *The Brain, the Story of You.* Canongate.

Everett, C. 2019. *Numbers and the Making of Us: Counting and the Course of Human Cultures.* Harvard University Press.

Kandel, E. R. 2006. *In Search of Memory: The Emergence of the New Science of Mind.* W. W. Norton & Company.

Koechlin, E., Dehaene, S., & Mehler, J. 1997. Numerical transformations in five-month-old human infants. *Mathematical Cognition* 3(2), 89–104.

Nehamas, A. 1994. *Plato's Meno in Focus.* Routledge.

Pascalis, O., Haan, M. D., & Nelson, C. A. 2002. Is face processing species-specific during the first year of life? *Science* 296(5571), 1321–1323.

Pica, P., Lemer, C., Izard, V., & Dehaene, S. 2004. Exact and approximate arithmetic in an Amazonian indigene group. *Science* 306(5695), 499–503.

Ridley, M. 2003. *Nature via Nurture.* Harper Collins.

Schubert, T., Rothlein, D., Brothers, T., Coderre, E., Ledoux, K., Gordon, B., & McCloskey, M. 2020. Lack of awareness despite complex visual processing: Evidence from event-related potentials in a case of selective metamorphopsia. *Proceedings of the National Academy of Sciences* 117(27), 16055–16064.

Schultz, W. 1998. Predictive reward signal of dopamine neurons. *Journal of Neurophysiology* 80(1), 1–27.

Seung, S. 2012. *Connectome.* Houghton Mifflin Harcourt.

van Gaal, S., Naccache, L., Meuwese, J., Van Loon, A., Leighton, A., Cohen, L., & Dehaene, S. 2014. Can the meaning of multiple words be integrated unconsciously? *Philosophical Transactions of the Royal Society B: Biological Sciences* 369(1641), 20130212.

Zaghloul, K., Blanco, J., Weidemann, C., McGill, K., Jaggi, J., Baltuch, G., & Kahana, M. 2009. Human substantia nigra neurons encode unexpected financial rewards. *Science* 323(5920), 1496–1499.

10 MEMORY AND LEARNING

New knowledge is acquired via learning. Memory is crucial not only for retaining knowledge over time but also for learning itself and for many other mental functions. The experiments with *aplysia* clearly demonstrate that synapses are malleable, a property called *synaptic plasticity*. These experiments also illustrate the occurrence of *intra-neural changes*, such as the activation of the gene CREB. The latter changes are referred to as *intrinsic plasticity*. These findings are consistent with the following neurophysiological postulate of the influential Canadian psychologist Donald Hebb made in the late 1940s:

"When an axon of a cell A is near enough to excite a cell B and repeatedly or persistently takes part in firing it, some growth process or metabolic change takes place in one or both cells, such that A's efficiency, as one of the cells firing B, is increased" (Hebb, 1949: p. 62).

This quote contains two concepts: synaptic plasticity and "some growth process or metabolic change" in the neuron, which is captured by the modern term of intrinsic plasticity.

Regarding the importance of continuity for the development of Hebb's ideas, it is noted that the Harvard psychologist and behaviourist Karl Lashley criticized Ivan Pavlov's approach to the physiology of learning (Pavlov, 1927, 1928a, 1928b). In particular Lashley claimed that Pavlov ignored the important concepts of attention, perception, thoughts, and mind (Lashley, 1932). Motivated by this criticism, Hebb, who worked with Lashley in the Yerkes Laboratories, supplemented the analysis of Pavlov with a comprehensive theory incorporating the concepts suggested by Lashley.

Hebb postulated three processes underlying the neuronal basis of learning and memory (Hebb, 1949): *synaptic changes*, the formation of a *cell assembly*, and the formation of a *sequence of assemblies* (this terminology was actually introduced by Valentino Braitenberg (Braitenberg, 1978) who was mentioned in Chapter 4). *A cell assembly is a set of neurons and a set of connections among these neurons, which allow them to act as a unit*; a stimulus activating one of the connections of this group, induces an activation of all other connections. *A sequence of assemblies is a collection of cell assemblies connected over time.* According to Hebb, the cell assemblies constitute the bridge from "the individual nerve cell to a psychological phenomenon".

For the formation of Hebb's sequences of assemblies, Braitenberg placed great emphasis on the *global activation or inhibition of pyramidal cells*, which takes place in many regions of the brain. Other authors emphasized temporal aspects of these assemblies, and specifically the *synchronization* of their activation, and called them *synfire-chains* (Abeles, 1982, 1991). Incidentally, Hebb's theory of activity-dependent synaptic plasticity, of the resulting formation of cell assemblies, and of the dynamic behaviour of a sequence of such assemblies, has led to the emergence of various types of useful mathematical models.

Using the imaging technique of *voltage-sensitive dye imaging*, introduced by Amiram Grinvald in the late 1990s, it is possible to study neuronal circuits on a scale of several millimetres. In such circuits, after a brief electrical stimulation, there is activation, which peaks at about 8 milliseconds. The neuroscientist, Baroness Susan Greenfield, claims that the activation of such circuits, which lasts for about 20 milliseconds and hence it is of the order of *meso-scale* with respect to the timescales involved in neuronal activation, is crucial for consciousness. In particular, she notes that anaesthesia abolishes consciousness by prolonging the duration and decreasing the size of such neural assemblies.[1]

---

[1]Baroness Greenfield claims that the main form of communication of the neurons forming such assembles is *not* via *synapses* or *volume transmission* achieved via bioactive molecules. Instead, according to the Baroness, the neurons of an assemble communicate via the

The experiments with *aplysia* indicate that *learning and memory are part of a continuum: learning is the process of acquiring new information or skills, as well as modifying existing ones. The duration and strength of the resulting memory depends on the frequency and intensity of learning.*

What is the mechanism for establishing continuity between learning and memory? The answer is synaptic plasticity. Indeed, the less frequent or intense the learning bouts, the weaker and more transient the synaptic changes induced by this learning. Such changes are called *short-term potentiation*. On the other hand, more frequent or intense learning bouts induce persistent synaptic modifications, a process known as *long-term potentiation*. Short-term potentiation is associated with *short-term memory*, typically lasting one to two hours, whereas long-term potentiation yields *long-term memory*, which can last for days, weeks, years, or even a lifetime. Incidentally, long-term potentiation is the result of high-frequency stimuli, whereas low-frequency stimulation produces synaptic weakening, often referred to as *long-term depression.*

The notion of continuity suggests that short-term and long-term memories must be intricately linked. This is indeed the case. For example, recent studies show that the neurocognitive procedures supporting short-term memories also enhance long-term memory stabilization (Ranganath *et al.*, 2014). In general, there does not exist a single cognitive mechanism, or a single brain region, supporting specifically short-term and long-term memories.

As noted in Chapter 4, Braitenberg provided support to Hebb's theory by showing that the anatomy and connectivity of the brain are consistent with Hebb's postulates. Also, he noted that the excitatory connections in the cortex, which are crucial for synaptic modification, greatly outnumber inhibitory ones. Interestingly, Braitenberg made indirect use of the notion of local versus global processes, which is one of the key notions employed in this volume. In this regard, he claimed that *local wiring is determined randomly* or more precisely individually, whereas *global wiring is inborn.*

---

so-called *gap-junctions*, which form a direct contact between adjacent neurons (Greenfield, 2016).

This claim is consistent with the summary of the nature of innate and acquired knowledge presented in the previous chapter.

SYNAPTIC PLASTICITY AND BEYOND

What are the mechanisms responsible for synaptic plasticity, namely for the changes in connectivity between neurons as a result of experience? An important such mechanism is *alterations in neurotransmitter release* (Ho *et al.*, 2011).[2] For example, a family of phosphoproteins, called *synapsins*, facilitate the *mobility of neurotransmitter-carrying vesicles*. Another important mechanism is *alterations in trans-synaptic signalling*. For example, a network of cell-adhesion molecules operating in the synaptic cleft is important for the *remodelling of the synapse*. Interestingly, *postsynaptic changes* can also affect synaptic plasticity.[3]

As discussed in the case of *aplysia*, *gene expression within neurons* is of fundamental importance for synaptic plasticity. In particular, several genes, including CREB and MEF_2, are vital for the *permanent changes associated with synaptic plasticity*.

Synaptic potentiation involves specialized receptors, including two types of receptors, called NMDA and AMPA. The regulation of these receptors, which are activated by the excitatory neurotransmitter glutamate, is astonishingly complicated. In particular, neurons employ a variety of proteins, including the so-called *transmembrane AMPA-receptor regulatory* proteins and the *cornichons*. These specialized proteins affect the trafficking and gating activity of AMPA.

In addition, neurons utilize the process of *endocytosis*, which is a very effective mechanism for degrading receptors. For example, this process is implemented regarding the KAR receptors, which are the third basic type of glutamate receptor, in addition to NMDA and AMPA. A subunit of KAR receptors binds to a small protein, called SUMO, which causes

[2]Different types of synapses employ distinct plasticity mechanisms. For example, the hippocampus of rodents consists of three sequential synaptic pathways, involving prefrontal, mossy fibre, and Schaffer collateral pathways.

[3]Long-term potentiation in mossy fibre synapses occurs primarily through presynaptic changes, whereas in Schaffer collateral synapses the main mechanisms are postsynaptic.

its rapid internalization into neurons through endocytosis, and then its destruction. This leads to a decrease in KAR-dependent synaptic activity (Martin *et al.*, 2007).

It was mentioned in Chapter 2 that the flow of calcium plays an important role in the release of neurotransmitters. Interestingly, this mechanism is also used for the regulation of the NMDA receptors. Indeed, during the induction of long-term potentiation, NMDA receptors mediate a rapid influx of calcium ions into spines, the tiny protrusions on dendrites. This initiates a cascade of events, finally leading to the insertion of additional AMPA receptors into spines, as well as to spine growth.[4]

Until recently, it was assumed that structural and functional plasticity is limited to the stimulated synapses. However, it is now understood that, remarkably, long-term potentiation at one synapse alters the properties of *its immediate neighbours*. Indeed, the threshold of induction of plasticity in the neighbouring synapses is dramatically reduced. This effect is seen in spines located within approximately 10 micrometres of the stimulated spine and lasts for about 5 minutes (Harvey & Svoboda, 2007).[5]

Synaptic changes and the subsequent intracellular cascades induced by the resulting alterations, give rise to further synaptic modifications, such as gene activation, *de novo* protein synthesis, and synaptic growth. In particular, strong stimulation of synapses leads to the synthesis of a particular atypical protein kinase, namely to the specific enzyme called PKMZ. This protein is known as the *memory molecule*, because it plays an important role in maintaining appropriate molecular and biochemical

---

[4]The process of recycling material from dendrites into spines, which is needed both for the construction of the AMPA receptors and for spine growth, is quite complicated: it starts with the influx of calcium ion signals, and involves *myosin V* proteins, which are actin-based motor proteins. These particular proteins specialize on organelle movements. Actually, they hijack endosomes from microtubules and move them along actin within spines (Wang *et al.*, 2008).

[5]It is worth noting that the occurrence of potentiation typically requires *simultaneous pre- and post-synaptic activity*. However, there exist some forms of potentiation that *start in the post-synaptic part*, including the following: the potentiation that occurs due to the presence of calcium ionic currents only at the post-synaptic neuron (Kato *et al.*, 2009), as well the potentiation due to the generation in the post-synaptic neuron of retrograde signals, such as nitric oxide, which travel to the presynaptic compartment where they elicit specific modifications (Padamsey & Emptage, 2013).

alternations at the synapse. In particular, it allows the memory trace to persist despite continued protein turnover (Langille & Brown, 2018).[6] Incidentally, the remodelling alterations, which include the growth of pre-existing and the formation of new dendritic spines, are correlated in magnitude, with the degree of training (Xu *et al.*, 2009; Yang *et al.*, 2009).

It should be noted that, although changes occurring at the synaptic connections are considered the main mechanism of neuronal plasticity, *experience* has additional important effects on learning. In particular, there occur alterations in the *intrinsic excitability* of a neuron, namely changes in the neuron's ability to generate an action potential. Indeed, it turns out that the action potential is initiated at the *axon initial segment*. This part of the axon is greatly enriched with ion-channels. Remarkably, the location of the site of this segment, as well as its length, *change* with experience. The closer this segment is to the cell body and the longer its length, the more excitable the neuron, i.e., the less stimulation is needed for its firing. For example, when researchers elevated the extracellular potassium levels of hippocampal neurons, mimicking the effect of increased neuronal activity, the neurons responded by shifting their axon's initial segment away from their cell body leading to a decrease in their excitability (Grubb & Burrone, 2010). On the other hand, following hearing loss, the total sodium current of neurons in *nucleus magnocellularis* increases. This is the result of the expansion of the axon initial segment of these neurons (Kuba *et al.*, 2010).

New mechanisms affecting the communication of neurons and hence their plasticity continue to be elucidated. In particular, it has emerged, that the prevailing model claiming that dendrites are passive structures simply receiving signals generated by action potentials, is too simplistic. Actually, dendrites themselves *can* release a variety of neurotransmitters. Such dendrites have been identified in many regions of the brain, including the cortex, thalamus, hypothalamus, and the hippocampus. They have also been found in the *substantial nigra*, where they release dopamine and in

[6]The relevant changes taking place at the synapses, as well as in the nucleus and the soma of neurons, are very complicated, involving many molecular and sub-molecular components, including: the activation of second messenger systems, nuclear binding proteins and protein kinases, as well as the transcription of DNA and translation of RNA.

the *raphe nucleus*, where they release serotonin (Margrie & Urban, 2007). When dendrites release a transmitter, the site of release is often close to the site where the dendrite forms a synapse with the presynaptic axon. This suggests that the released neurotransmitters may provide a local mechanism for controlling excitability. Incidentally, there are particular neurons which contain only dendrites and not axons. Such remarkable cells include the so-called *amacrine* cells, which are interneurons found in the retina.[7]

The glia cells provide an additional important mechanism affecting neuronal communication. These cells, whose role until recently was underestimated, can also excrete a variety of neurotransmitters.

It turns out that novel stimuli result in the release of *neuro modulatory transmitters*, including dopamine, serotonin, and acetylcholine that contribute to synaptic modification. For example, specific experiments have shown that during aversive learning, dopamine facilitates long-term potentiation (Bissière *et al.*, 2003; Marowsky *et al.*, 2005). In addition to these neurotransmitters, a variety of hormones, such as corticosterone and androgens, can affect synaptic plasticity[8] (Berger *et al.*, 2009). All these molecules may be released following various external and internal stimuli, including emotions, motivations, thoughts, and memories.

Hebb correctly argued that, since a memory is not localized in a synapse but is widely distributed, it becomes independent of the specific sensory modality that created it (Hebb, 1949: p. 129). For example, a memory formed by visual stimuli may be activated by auditory stimuli. Also, experiments have shown that, although the synaptic efficiency of a specific set of synapses is altered during a specific form of learning (Butler *et al.*, 2015, 2018), subsequent destruction of the modified synapses rarely eliminates the associated learning (Trettenbrein, 2016).

Depending on the physiological situation, learning may take place even after a single exposure to a novel environment. For example, such an

[7]Cajal gave them this name to emphasize that they lack long projecting processes. In Greek, *a* means without, *macros* means long, and *inos* means fibre.
[8]They include a variety of hormones such as oxytocin and ACTH, and even cytokines such as the tumour necrotic factor alpha.

exposure can lead to the activation of the specific gene *Arc* in neurons found in the hippocampus (Miyashita *et al.*, 2009).

Kandel's experiments mentioned in the earlier chapter have associated specific biochemical changes of neurons in *aplysia* with the *duration* of the induced learning and memory. These studies have been extended in humans. In particular, it has been shown that intracellular signalling cascades, protein synthesis, and morphological changes in those neurons that are affected by a specific training, are *slow* to occur. Thus, memory is initially encoded in the specific *electrical reverberations* of the cell assembly. This electrical activity induces *early* long-term synaptic potentiation that causes a series of temporary biochemical alterations, which are *independent* of protein-synthesis. These changes can maintain information until the process of protein-synthesis does take place, which depends on *late* long-term potentiation. It is through protein synthesis that memory is stabilized. In other words,

> specific electrical activity allows the brain to employ newly acquired information for immediate computational usage, bridging in this way the temporal gap between the acquisition of this information and its later stabilization achieved via de novo protein synthesis.

Short-term memory results from temporary modification of pre-existing proteins that are sustained for as long as the activity of certain enzymes, called *kinases*, dominates over the activity of certain other enzymes, called *phosphatases*. As soon as this dominance shifts, the biochemical substrate potentiating short-term memory is rapidly removed and this memory is lost (Langille & Brown, 2018). This process is of crucial importance in working memory.

The highly complex processes described above give rise to long-term changes in synaptic efficiency. Therefore, as a result of the associated synaptic modifications, subsequent activity in the neural network affected by these changes is substantially altered. This has the transformative consequence that when a stimulus similar to the one responsible for the above synaptic modifications appears, even in a fragmented or somewhat distorted form, then the entire cell assembly will be activated. In summary, a specific memory can be identified with a particular neural pattern of activity within a specific neural network. This pattern is the result of an intricate series of molecular, biochemical, and cellular modifications that have occurred within this network as a result of learning. Overall, the locus of a particular memory is a specific neural network that has been made more efficient because of learning (Costa *et al.*, 2017; Choi *et al.*, 2018).

As a result of the dynamic modifications of the brain, the re-activation of the specific network gives rise to the feeling of *recalling* a specific experience. Thus,

> access to a past event involves the reappearance of a similar neural pattern of activity that was present during the original experience. In other words, the brain achieves remembering by returning to a prior state that allows it to re-experience the past (Davachi & Preston, 2014).

MEDICAL IMPLICATIONS

Taking into consideration the enormous complexity of the structures involved in the synaptic transmission, it is not surprising that a variety of disorders can arise at various levels in the transmission chain. They include *channelopathies*, which are disorders of the ionic channels. Some of the synaptic transmission disorders are discussed hereunder.

Alzheimer's disease causes synaptic deterioration in the hippocampus. This includes reduction of dendritic spine density and loss of AMPAR receptors (Rodrigues *et al.*, 2016). As a result of these changes, patients suffering from Alzheimer's disease exhibit impaired long-term potentiation and serious defects in memory consolidation (Figueiredo *et al.*, 2013; Bilousova *et al.*, 2016; Yang *et al.*, 2017). This perhaps explains the so-called *anterograde amnesia* of Alzheimer's disease patients, namely their difficulty in forming new, lasting memories.

As an illustration of the complexity of synaptic channels and the potential for malfunction, it is noted that mammals have nine types of sodium-channels, denoted by $NA_V1.1$ to $NA_V1.9$. Specific ion-channel mutations can enhance (gain-of-function) or attenuate (loss-of-function) channel activity. For example, a particular gain-of-function mutation of $NA_V1.7$

increases the excitability of pain-signalling to dorsal root ganglion, causing *erythromelalgia*, namely the condition where exposure to mild warmth, causes severe burning pain and redness of the skin. A similar condition is seen in the *paroxysmal extreme pain disorder*, where patients experience episodes of pain in the lower body, eyes, and jaw. An example of an "opposite disorder", namely, one caused by of loss-of-function mutation, is the *associated insensitivity to pain*, where patients do not experience pain even after dental extractions, bone fractures, and burns (Waxman, 2011). Interestingly, the same mutation causes *anosmia*, namely, inability to sense odours (Weiss *et al.*, 2011).

It is worth noting that several different clinical syndromes occur as a result of the disturbance of the balance between excitatory and inhibitory neurotransmission. For example, the destruction, via specific antibodies, of the enzyme *glutamic acid decarboxylase*, which is responsible for the synthesis of GABA, or the destruction of glycine receptors, causes the so-called *stiff-person syndrome*. In December 2022, it was announced that Celine Dion was suffering from this syndrome. The decrease of the inhibition to neuronal excitation yields uncontrolled neuronal hyperexcitability resembling tetanus or strychnine poisoning. This is accompanied by severe muscle rigidity, painful muscle contractions, difficulty walking, seizures, or catatonia.

Similarly, in Alzheimer's disease and in several other circumstances, self-antibodies acting against NMDAR, AMPAR, GABA, or glycine receptors, *inactivate* GABAergic neurons. This process, which was elucidated only within the last 10–20 years, causes an increase in the extracellular glutamate resulting in a dis-inhibition of excitatory pathway. In turn, this causes psychosis, rigidity, dystonia, catatonia, and memory impairment.

Imbalance between excitatory and inhibitory neurotransmission can also be caused by a variety of self-antibodies against synaptic antigens. They include antibodies against a particular type of GABA receptors, the GABAB receptors, which are especially prevalent in the hippocampus.[9] These autoimmune conditions affect the proper function of inhibitory pathways,

[9]Antibodies have also been identified against NMDAR and AMPAR receptors.

and thus they cause neuronal hyperexcitability. Patients suffering from these disorders exhibit abnormal behaviour, agitation, excessive movements, muscular rigidity, paranoia, psychosis, hallucinations, intractable seizures, and impaired consciousness. The combination of these symptoms is a characteristic of the so-called *autoimmune limbic encephalitis*. Fortunately, this is a reversible condition, since the culprit antibodies can be suppressed with appropriate immunotherapies.

The recent remarkable progress in immunology, which will be reviewed in detail in the second volume, has led to the identification of several other classes of antibodies that cause neurological disorders. For example, recently, specific antibodies have been discovered which act against potassium channels. Similarly, antibodies against calcium channels can cause a myasthenic-type syndrome. In addition, a number of mutations in chloride or sodium channels cause myotonia, periodic paralyses, myoclonic epilepsy, or intractable pains. The aforementioned conditions are quite rare and fortunately most of them are treatable.

THE ROLE OF CONSCIOUSNESS IN LEARNING

As noted earlier, the hidden nature of unconscious processes is often underestimated. Interestingly, there exist a few investigators who apparently err in the opposite direction. They claim that, since the essence of mental functions takes place at a level below that of consciousness, being conscious does not offer any advantage. However, the fact that any experience is preceded by an unconscious process, does *not* imply that awareness is not useful. In my opinion, the important question is not whether consciousness is indeed useful, but the identification of its precise evolutionary advantage. As discussed earlier, *consciousness is of vital importance for the construction of a variety of metarepresentations*.

The emergence of metarepresentations has had a transformative impact on our survival. Related to this fact is another crucial contribution of consciousness: *its role in advanced forms of learning*. For example, it is inconceivable that an individual can learn algebra, which is a specific metarepresentation, let alone more advanced branches of mathematics, without awareness.

The Munduruku experiments discussed in the previous chapter suggest that humans have innate ability for exact arithmetic for numbers up to 3. It is interesting that the numbers 3 and 4 have appeared in a plethora of experiments concerned with the maximum number of percepts that can be handled consciously. In one such experiment, subjects were shown 10 crosses on a screen, and, initially, a subset of these crosses flashed. Then, the crosses began moving and the subjects were asked to follow only the subset that flashed. After the crosses became stationary, the subject had to identify those that were initially flashing. If only 3 crosses flashed, the subjects found the task of following these 3 crosses, relatively easy; for 4 crosses they were less accurate, and for 5 crosses they were unable to perform this task (Yantis, 1992). Interestingly, similar results are obtained for monkeys.

For exact arithmetic with numbers larger than 3, humans require the introduction of a specific word, or at least of an abstract symbol for each number, as well as the employment of specific rules. In other words, *humans must employ the power of metarepresentations.* Similarly, it appears that for performing a task involving more than 3 or 4 percepts, necessitates the use by humans of a specific algorithm. This suggests that,

conscious learning is intimately related to the ability of the brain to construct specific metarepresentations, such as words and numbers. Furthermore, conscious learning may require the use of algorithms, namely the use of appropriate rules specifying how specific metarepresentations are related.

For example, learning algebra requires the introduction of specific symbols and the use of specific rules for manipulating these symbols.

This discussion suggests that,

any form of learning that involves the systematic manipulation of metarepresentations, necessarily requires consciousness.

The relationship between metarepresentations and learning is implicitly noted in the introduction of Stanislas Dehaene's important book, *How We Learn.* Indeed, after presenting the accomplishments of a fully blind and paralysed child, as well as of a painter without a right hemisphere, Dehaene states:

"All these examples illustrate the extraordinary resilience of the human brain: even major trauma, [...] cannot extinguish the spark of learning. Language, reading, mathematics, artistic creations: all these unique talents of the human species, which no other primate possesses, can resist massive injuries [...]" (Dehaene, 2020: pp. xv–xvi).

Dehaene also notes that "Learning is the triumph of our species" (Dehaene, 2020: p. xx). Taking into consideration that advanced forms of learning would *not* be possible without the metarepresentations of "language, writing, mathematics and artistic creations", mentioned by Dehaene, it is perhaps fair to state that,

the singular characteristic of humans is not the ability to learn, but their capacity to construct metarepresentations.

Actually, starting with metarepresentations and employing the key concepts elaborated in this volume, including the concepts of associations, local versus global processes, abstraction, generalization, unification, and continuity, learning becomes *inevitable*. The relationship between learning, metarepresentations, and the above-mentioned vital concepts is briefly illustrated below with the aid of several examples.

The existence of concrete metarepresentations, such as language and mathematics, allows humans to learn from each other. This exceedingly important process has been formalized with the development of *pedagogy* and the invention of *specific educational institutions*. Several scholars have emphasized that two of the fundamental pillars of pedagogy are *active engagement* and *error feedback*. Active engagement facilitates the genesis of a plethora of *associations*, which are crucial for the elucidation of many aspects of the topic under consideration. Error feedback is necessary to counteract the innate tendency of consciousness to create a *complete picture of reality from insufficient knowledge and data*. This often gives rise to erroneous conclusions, which must be corrected via suitable feedback.

The importance of the notion of local versus global processes in visual perception was demonstrated in Chapter 4: a percept is *decomposed* to its constituents, which are reconstructed via the solution of *local* inverse problems, before *global synthesis* yields the perception of the given percept. This notion is also crucial for processes involving a variety of metarepresentations. In learning, this fundamental notion takes the form of *hierarchical decomposition*. For example, in language, a sentence is decomposed into words and sounds.

Regarding the importance of the notions of abstraction, generalization, and unification, it is shown in Chapter 18 that mathematics provides an example *par excellence* of the notions of abstraction and generalization. For physics, which will be discussed in the second volume, a similar role is played by the notion of unification.

The concept of continuity, together with the remarkable capacity of the brain for *plasticity*, imply that the specific brain domains associated with various *learned metarepresentations*, correspond to *enlargements of the domains associated with the innate versions of these metarepresentations.*[10] For example, the innate capacity for approximate arithmetic is associated with the parietal and prefrontal cortex. Imaging studies demonstrate that the same areas are also involved with advanced mathematical ruminations, such as those associated with complicated exact calculations or abstract mathematical reasoning. Interestingly, to enhance their mathematical capacity, professional mathematicians recruit additional adjacent brain areas. Namely, the lateral occipital regions of *both* hemispheres. These areas are part of the domains usually dedicated to face recognition, and hence as a result of this recruitment for mathematics, professional mathematicians have a *diminished capacity for remembering faces* (Amalric & Dehaene, 2016). On a personal note, despite the fact that my memory is in general very good, I have had a noticeable difficulty remembering faces. This often leads to the embarrassing situations of not being able to recognize individuals who greet me in a very friendly manner. I was always puzzled by this defect until I found out about the occurrence in mathematicians of the reduction of the brain areas dedicated to face recognition.

Another example of *adjacent recruitment* is provided by reading: this important linguistic function involves recognizing particular symbols, and

[10]Dehaene refers to this fact as "recycling the brain". The word recycling is used to emphasize that during learning "pre-existing neural circuits are reoriented in a new direction" (Dehaene, 2020, p. 122).

therefore is associated with a particular area of the visual cortex. This area, which finally becomes dedicated to word recognition, is directly connected with the language areas. In this way, words are recognized and at the same time give rise to sound and meaning. Again, as the capacity for reading improves, the brain recruits the adjacent areas in the *left* hemisphere, usually associated with face recognition. However, in this case, the brain *compensates* by enlarging the area in the *right* hemisphere dedicated to face recognition. Thus, in contrast to mathematics, literacy does *not* diminish the ability to recognize faces (Pegado *et al.*, 2014).

Incidentally, using imaging techniques, the recruitment of parts of the brain dedicated to face recognition can already be seen in children of ages 6–8, which is typically when learning of reading occurs. Interestingly, imaging studies showed that the enlargement of the reading visual area at the expense of the face recognition area did *not* occur in an adult learning to read, since by then the brain has reduced capacity for plasticity (Braga *et al.*, 2017).

The hugely important property of plasticity of the human brain is further supported by imaging studies demonstrating that the visual cortex of individuals who are born blind, can be recruited for mathematics and language. In this way, blind individuals can devote a larger part of their brain to enhance their mathematical and linguistic capabilities (Dehaene, 2020: p. 129). This is consistent with the fact that through the ages there have been several outstanding blind mathematicians. They include Nicholas Saunderson (1682–1738) who became the Lucasian Professor of Mathematics at the University of Cambridge (a prestigious Chair held by Newton and recently by my colleague Stephen Hawking); he became blind the first year of his life as the result of smallpox. Leonhard Euler (1707–1783), one of the greatest mathematicians of all time, became blind in the last 17 years of his life but continued to produce results of remarkable quality. The famous Soviet topologist Lev Pontryagin (1908–1988), became blind at the age of 14 due to an accident. The distinguished French geometers, Bernard Morin (1931–2018) who lost his sight at the age of six due to glaucoma, and Emmanuel Giroux, born in 1961, who lost his sight at the age of 11.

References

Abeles, M. 1982. *Local Cortical Circuits: An Electrophysiological Study.* Springer.

Abeles, M. 1991. *Corticonics: Neural Circuits of the Cerebral Cortex.* Cambridge University Press.

Amalric, M. & Dehaene, S. 2016. Origins of the brain networks for advanced mathematics in expert mathematicians. *Proceedings of the National Academy of Sciences* 113(18), 4909–4917.

Berger, S. L., Kouzarides, T., Shiekhattar, R., & Shilatifard, A. 2009. An operational definition of epigenetics. *Genes & Development* 23, 781–783.

Bilousova, T., Miller, C. A., Poon, W. W., Vinters, H. V., Corrada, M., Kawas, C., Hayden, E., Teplow, D., Glabe, C., Albay, R., Cole, G., Teng, E., & Gylys, K. H. 2016. Synaptic amyloid-b oligomers precede p-Tau and differentiate high pathology control cases. The *Americal Journal of Pathology* 186, 185–198.

Bissière, S., Humeau, Y., & Lüthi, A. 2003. Dopamine gates LTP induction in lateral amygdala by suppressing feedforward inhibition. *Nature Neuroscience* 6, 587–592.

Braga, L., Amemiya, E., Tauil, A., Suguieda, D., Lacerda, C., Klein, E., Dehaene-Lambertz, G., & Dehaene, S. 2017. Tracking adult literacy acquisition with functional MRI: A single-case study. *Mind, Brain, and Education* 11(3), 121–132.

Braitenberg, V. 1978. Cell assemblies in the cerebral cortex. In: Heim, R. & Palm, G. (eds.) *Proceedings Symposium on Theoretical Approaches to Complex Systems 1977. Lecture* Notes *in Biomathematics 21*. Springer, pp. 171–188.

Butler, C. W., Wilson, Y. M., Gunnersen, J. M., & Murphy, M. 2015. Tracking the fear memory engram: Discrete populations of neurons within amygdala, hypothalamus and lateral septum are specifically activated by auditory fear conditioning. *Learning and Memory* 22, 370–384.

Butler, C. W., Wilson, Y. M., Oyrer, J., Karle, T. J., Petrou, S., Gunnersen, J. M., Murphy, M., & Reid, C. A. 2018. Neurons specifically activated by fear learning in lateral amygdala display increased synaptic strength. *eNeuro* 5.

Choi, J.-H., Sim, S. E., Kim, J.-I., Choi, D., Oh, J., Ye, S., Lee, J., Kim, T., Ko, H.-G., Lim, C.-S., & Kaang, B.-K. 2018. Interregional synaptic maps among engram cells underlie memory formation. *Science* 360, 430–435.

Costa, R. P., Mizusaki, B. E., Sjostrom, P. J., & van Rossum, M. C. 2017. Functional consequences of pre- and postsynaptic expression of synaptic plasticity. *Philosophical Transactions of the Royal Society B* 372, 20160153.

Davachi, L. & Preston, A. 2014. The medial temporal lobe and memory. In: Gazzaniga, M. S. & Mangun, G. R. (eds.) *The Cognitive Neurosciences.* MIT Press, 5th ed., pp. 539–546.

Dehaene, S. 2020. *How We Learn.* Random House.

Figueiredo, C. P., Clarke, J. R., Ledo, J. H., Ribeiro, F. C., Costa, C. V., Melo, H. M., Axa P. Mota-Sales, A. P., Saraiva, L. M., Klein, W. L., Sebollela, A., De Felice, F. G., & Ferreira, S. T. 2013. Memantine rescues transient cognitive impairment caused by high-molecular-weight ab oligomers but not the persistent impairment induced by low-molecular-weight oligomers. *Journal of Neuroscience* 33, 9626–9634.

Greenfield, S. 2016. *A Day in the Life of the Brain.* Allen Lane.

Grubb, M. S. & Burrone, J. 2010. Activity-dependent relocation of the axon initial segment fine-tunes neuronal excitability. *Nature* 465(7301), 1070–1074.

Harvey, C. & Svoboda, K. 2007. Locally dynamic synaptic learning rules in pyramidal neuron dendrites. *Nature* 450(7173), 1195–1200.

Hebb, D. O. 1949. *The Organisation of Behaviour.* John Wiley & Sons.

Ho, V., Lee, J., & Martin, K. 2011. The cell biology of synaptic plasticity. *Science* 334(6056), 623–628.

Kato, H. K., Watabe, A. M., & Manabe, T. 2009. Non-Hebbian synaptic plasticity induced by repetitive postsynaptic action potentials. *Journal of Neuroscience* 29(36), 11153–11160.

Kuba, H., Oichi, Y., & Ohmori, H. 2010. Presynaptic activity regulates Na+ channel distribution at the axon initial segment. *Nature* 465(7301), 1075–1078.

Langille, J. J. & Brown, R. E. 2018. The synaptic theory of memory: A historical survey and reconciliation of recent opposition. *Frontiers of Systems Neuroscience* 12, 52.

Lashley, K. S. 1932. Studies of cerebral function in learning VIII. A reanalysis of data on mass action in the visual cortex. *Journal of Comparative Neurology,* 54, 74–84.

Margrie, T. & Urban, N. 2007. Dendrites as transmitters. In: Stuart, G., Spruston, N., & Hausser, M. (eds.) *Dendrites.* Oxford Academic. p. 401.

Marowsky, A., Yanagawa, Y., Obata, K., & Vogt, K. E. 2005. A specialized subclass of interneurons mediates dopaminergic facilitation of amygdala function. *Neuron* 48, 1025–1037.

Martin, S., Nishimune, A., Mellor, J., & Henley, J. 2007. SUMOylation regulates kainate-receptor-mediated synaptic transmission. *Nature* 447(7142), 321–325.

Miyashita, T., Kubik, S., Haghighi, N., Steward, O., & Guzowski, J. F. 2009. Rapid activation of plasticity-associated gene transcription in hippocampal neurons provides a mechanism for encoding of one-trial experience. *Journal of Neuroscience* 29, 898–906.

Padamsey, Z. & Emptage, N. 2013. Two sides to long-term potentiation: A view towards reconciliation. *Philosophical Transactions of the Royal Society* 369, 20130154.

Pavlov, I. P. 1927. *Conditioned Reflexes: An Investigation of the Physiological Activity of the Cerebral Cortex.* Oxford University Press.

Pavlov, I. P. 1928a. *Lectures on Conditioned Reflexes: Twenty-Five Years of Objective Study of the Higher Nervous Activity (Behavior) of Animals,* Vol. 1. International Publishers.

Pavlov, I. P. 1928b. Croonian lecture: Certain problems in the physiology of the cerebral hemispheres. *Philosophical Transactions of the Royal Society* 103, 97–110.

Pegado, F., Comerlato, E., Ventura, F., Jobert, A., Nakamura, K., Buiatti, M., Ventura, P., Dehaene-Lambertz, G., Kolinsky, R., Morais, J., Braga, L. Cohen, L., & Dehaene, S. 2014. Timing the impact of literacy on visual processing. *Proceedings of the National Academy of Sciences* 111(49), E5233–E5242.

Ranganath, C., Hasselmo, M. E., & Stern, C. E. 2014. Short-term memory: Neural mechanisms, brain systems, and cognitive processes. In: Gazzaniga, M.S. & Mangun, G.R. (eds.) *The Cognitive Neurosciences.* MIT Press, 5th ed., pp. 527–538.

Rodrigues, E. M., Scudder, S. L., Goo, M. S., & Patrick, G. N. 2016. Ab-induced synaptic alterations require the E3 ubiquitin ligase Nedd4–1. *Journal of Neuroscience* 36, 1590–1595.

Trettenbrein, P. C. 2016. The demise of the synapse as the locus of memory: A looming paradigm shift? *Frontiers of Systems Neuroscience* 10(88), 1–7.

Wang, Z., Edwards, J., Riley, N., Provance Jr, D., Karcher, R., Li, X., Davison, I., Ikebe, M., Mercer, J., Kauer, J., & Ehlers, M. 2008. Myosin Vb mobilizes recycling endosomes and AMPA receptors for postsynaptic plasticity. *Cell* 135(3), 535–548.

Waxman, S. 2011. Channelopathies have many faces. *Nature* 472(7342), 173–174.

Weiss, J., Pyrski, M., Jacobi, E., Bufe, B., Willnecker, V., Schick, B., Zizzari, P., Gossage, S., Greer, C., Leinders-Zufall, T., Woods, G., Wood, J., Zufall, F., & Woods, C. 2011. Loss-of-function mutations in sodium channel Na v 1.7 cause anosmia. *Nature* 472(7342), 186–190.

Xu, T., Yu, X., Perlik, A. J., Tobin, W. F., Zweig, J. A., Tennant, K., Jones, T., & Zuo, Y. 2009. Rapid formation and selective stabilization of synapses for enduring memories. *Nature* 462, 915–919.

Yang, G., Pan, F., & Gan, W.-B. 2009. Stably maintained dendritic spines are associated with lifelong memories. *Nature* 462, 920–924.

Yang, T., Li, S., Xu, H., Walsh, D. M., & Selkoe, D. J. 2017. Large soluble oligomers of amyloid ß-protein from Alzheimer brain are far less neuroactive than the smaller oligomers to which they dissociate. *Journal of Neuroscience* 37, 152–163.

Yantis, S. 1992. Multielement visual tracking: Attention and perceptual organization. *Cognitive Psychology* 24(3), 295–340.

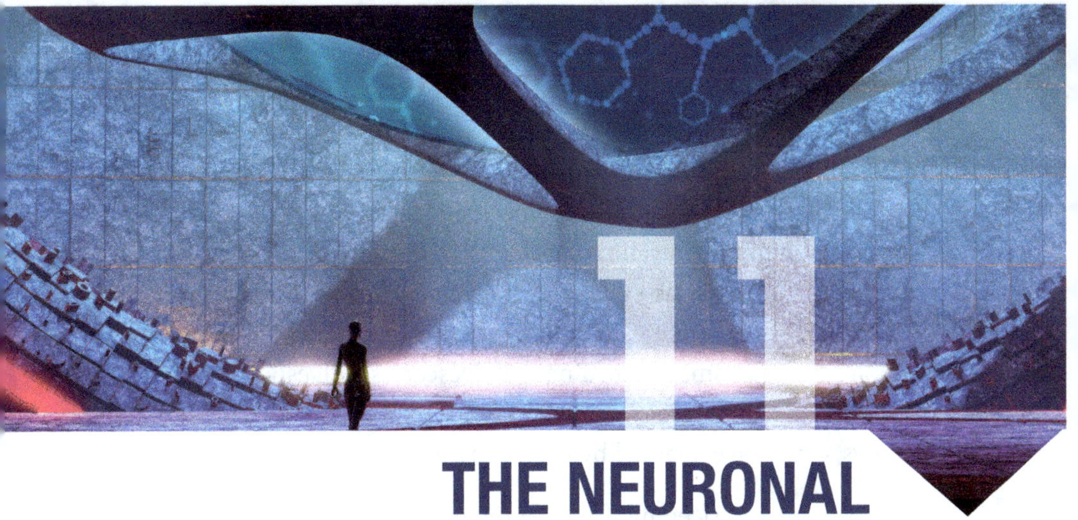

THE NEURONAL SUBSTRATES OF UNCONSCIOUS PROCESSES AND CONSCIOUS EXPERIENCES

The three great rationalist philosophers of the 17th century are René Descartes (1596–1650), Baruch Spinoza (1632–1677), and Gottfried Leibniz (1646–1716). Descartes confessed that, in his youth he "was in love with the poets"; from Ovid's *Fasti*, Cicero's orations, and Seneca's tragedies, to Virgil and Horace. His scientific and philosophical theories were written in Latin. Descartes' *Discourse on the Method* (Descartes, 1634) is an exception, written in French, because it was directed at the layman. His theory of *dualism* claims that the *mind* is made of a *non-material* substance that does not obey the laws of physics. This theory is regarded as the continuation of the theory of souls of Plato, summarized in Chapter 9, and of the theological doctrine of Thomas Aquinas written in *Summa Theologica* (Aquinas 1265–1274).[1] However, it should be emphasized that many aspects of the Cartesian dualism are based on *materialism*, and specifically on the doctrine of *reduction*.

---

[1]A distinction between body and soul also appeared in Homer, but this distinction is of a different nature: Homer employs the word *body* (*soma*), to refer to what is left in the battlefield. Also, *soul* (*psyche*) refers to what finally leaves from the lips of the dying warrior.

Indeed, he was the first to postulate a self-sufficient nervous structure capable of performing not only mechanical but also intellectual tasks. Consistent with the prevailing mechanistic attitude of his times, Descartes envisioned the nervous system as a *special machine.* Still, his approach was the first-ever attempt to employ theoretical modelling to a complex biological organism. According to Descartes, digestion, respiration, cardiovascular physiology, and sensations, as well as sleep, memory, and feelings, could be explained by his mechanical model:

"[...] waking and sleeping; [...] the retention or imprint of these ideas in the memory; the internal movements of the appetites and the passions; [...]. These functions follow in this machine simply from the disposition of its organs, as naturally as the movement of a clock or other automaton follow from the disposition of its counterweights and wheels" (*Description of the Human Body, Passions of the Soul, and L'hommer*).

Despite the impressive capabilities of his system, Descartes could not find a mechanistic solution to two problems: first, "using words or other signs by composing them, as we do, to declare our thoughts to others", and second, "How to generate an infinite variety of thoughts". Indeed, he claimed it was impossible,

"that there should exist in any machine a diversity of organs sufficient to enable to act in all the occurrence of life, in the way in which our reason enables us to act".

Thus, he felt, had no choice but to introduce the immaterial soul! Incidentally, Descartes understood that *reflexes* could be the subject of scientific investigation: in his book, *De Homine* (Descartes, 1662), he included a diagram of a reflex movement, depicting the withdrawal of a foot from fire. According to him, this involves the peripheral and central nervous system but not the pineal body, which he thought was the seat of consciousness. Although Descartes was convinced that the reflex-circuits could be studied scientifically, surprisingly, he thought that this was not possible for the associated pain.

Descartes knew early in his life that his progressive ideas were likely to create tension with the Catholic Church and monarchy in his native France.

Incidentally, following the propaganda of exponents of the French Revolution, the repressive impact of the church and the monarchy has been exaggerated. A well-known example is the falsehood claim that Queen Maria Antoinette, responding to people demanding bread, proclaimed "let them eat cake". Similarly, it is claimed that a year after Spinoza's birth, Galileo was placed by the Roman Inquisition under house arrest, as a result of his position that the Earth moves around the Sun, which was in contradiction to the Christian doctrine that the Earth is the centre of the universe. However, actually Galileo was a friend of Pope Urban VIII, who although supported financially Galileo's research, was reluctant to accept his conclusions, which were erroneously based on the analysis of circular orbits and hence gave less accurate predictions than the older calculations of the Ptolemian era (accurate predictions were obtained later via the use of Kepler's elliptical orbits). Independently of such exaggerations, it is a fact that Descartes withheld publication of his *Treatise of Man* anticipating attacks from the church. He moved to Amsterdam in 1630.

Amsterdam was the birthplace of Spinoza, who, incidentally, was taught algebra by Descartes. In his fascinating book *Looking for Spinoza* (Damasio, 2003), Antonio Damasio presents convincing evidence that Spinoza was a *proto-psychologist*. Indeed, Spinoza made several striking contributions, anticipating modern developments in psychology and neuroscience.

Foremost among them is his clearly formulated notion of *conatus, namely the relentless endeavour of organisms for survival*. For example, in his famous book, *The Ethics* (Spinoza 1661–1675), Part III, he writes that "Each thing, as far as it can by its own power strives to preserve its being". Furthermore, Spinoza traced fundamental aspects of both *ethical behaviour and human emotions to the endeavour for survival*. For example, in Part IV of his *Ethics* he writes that,

"the very foundation of virtue is the *conatus* to preserve the individual self, and happiness consists in the human capacity to preserve itself".

It should be noted that Spinoza, contradicting Descartes, did *not* believe that there is a dichotomy between body and mind. Also, remarkably, he intuited that perceptions have their origin in the *modifications of the body caused by external stimuli*. For example, in Proposition 23 of *The Ethics* it is stated that, "The Mind does not have the capacity to perceive [...] except in so far as it perceives the idea of the modifications of the body". This statement is remarkably modern: for example, as stated in Chapter 4, in visual perception the photons emitted from an object *modify* the sensory visual receptors, and this initiates a process that finally gives rise to the mental image of this object. Actually, the essence of perception is fully captured by this pioneering statement of Spinoza; this becomes clearer if the "idea of" is deleted. Similarly, the following statement in Proposition 26 is remarkably accurate: "The human Mind does not perceive any external body as actually existing, except through the ideas of the modification of its own body"; again, this becomes clearer by deleting the "idea of".

Another far-reaching achievement of Spinoza is his anticipation of the benefits of cognitive psychology, and in particular his emphasis on the vital importance of the *intellectual effort to fight a negative emotion with a stronger positive thought*: in Proposition 7, Part IV of *The Ethics* it is stated that,

"An affect cannot be restrained or neutralized except by a contrary affect that is stronger than the affect to be restrained".

Apparently, Spinoza understood the importance of unconscious processes. Indeed, taking into consideration that unconscious processes often express themselves via the notion of *intuition,* Spinoza's pioneering position that *intuition is the most advanced form of knowledge* can be considered as an endorsement of the primacy of the unconscious. Also, perhaps, he appreciated the interaction between conscious and unconscious processes. This is suggested by his claim that intuition can occur only after the accumulation of scientific knowledge, followed by reflection and subsequent deep analysis aided by reason.

Spinoza rejected the salvation proposed by his religious community. Instead, in the *Tractatus* (Spinoza 1677) and in *The Ethics* (Spinoza 1661–1675), he proposed a secular salvation (*salus*). This requires *a virtuous life mindful of God's nature that also abides with the laws of the society*. For Spinoza, God is an uncaused and eternal entity with infinite attributes. God is manifested in nature and most importantly in living creatures: *Deus sive Natura* (God or Nature). Spinoza interpreted the Bible as a source of valuable knowledge of proper human conduct and civil organization (Damasio, 2003: p. 274). A higher form of salvation and happiness can be achieved only via deeper intuitive understanding. A crucial role in the individual's salvation is played by the proper organization of society: salvation and happiness can take place only in a democratic and fair society that allows its citizens to live free from fear.

Spinoza was the epitome of a modest and stoic intellectual, seeking pleasure only via the pursuit of knowledge. He refused the attempts of his friend Simon de Vries to make him his heir, and only accepted the modest annuity of 500 florins.[2] As a result of his intellectual integrity, Spinoza, unfortunately, became an exile in his own community. This is ironic since this community itself consisted of exiles; he was born a *Marrano* Sephardic Jew. In 1492, Sephardic Jews fled from Spain to Portugal where later, some of them, became *Marrano*, namely, they behaved outwardly as Christians, but secretly observed Jewish rituals. Spinoza was ex-communicated from his synagogue in 1656 and consequently changed his name from Baruch to Benedictus.

Spinoza' positions were clearly ahead of its time. His *Ethics*, which was published a year after his death, was attacked by Jews, the Vatican, and the Calvinists. Moreover, it was banned by several European countries including Holland. Spinoza is now considered the main precursor of the movement of Enlightenment. But several participants of this movement avoided publicly crediting Spinoza's positions. Actually, although they were clearly influenced by Spinoza, publicly, they either ignored or made fun of him.[3] It is only after the Enlightenment that the far-reaching contributions of Spinoza began to be exalted. For example, the English poets Samuel Taylor Coleridge, William Wordsworth, Percy Shelley, and George Eliot, became his champions. Georg Hegel wrote, "To be a philosopher you must first be a Spinozist". Goethe, who was introduced to Spinoza via the writings of Gotthold Lessing, wrote that "[…] the man who was destined

[2]Spinoza promoted an ascetic model of living. Personally, I agree with Antonio Damasio's assertion in his book *Looking for Spinoza* that, true happiness can be achieved via "love and friendship".

[3]Voltaire writes in one of his poems: "[…] a little Jew, with a long nose and pale complexion, […]. A subtle but hollow spirit, […]. Hidden under the mantle of Descartes, his mentor […]".

to affect so deeply my entire mode of thinking, was Spinoza". Hermann von Helmholtz strongly supported the 1876 petition for erecting the statue of Spinoza that now sits in the Hague.

Leibniz, who had visited Spinoza in his house in Amsterdam, made the strongest claims among these three intellectual giants for the importance of unconscious processes: foreshadowing the existence of subliminal stimuli, Leibniz stated that there exist *petite perceptions which we perceive, but of which we do not become aware.* Taking into consideration that the concept of continuity is central in mathematics, it is not surprising that the great mathematician elevated continuity to a universal principle: *Natura non facit saltus* (Nature does not make jumps). Related to this principle, is Leibniz's theory of *pre-established harmony*, namely, his dogma that God has created the most harmonious world possible.

Employing the continuity principle, Leibniz concluded that there must exist intermediate steps between what we now call conscious and unconscious processes. Moreover, Leibniz intuited that the occurrence of perceptions of which we are unaware implies that there must exist part of the mind of which we are oblivious. The great philosopher influenced directly the anthropologist and physician Ernst Platner, who is credited with introducing the term unconscious, as well as the physician and physiologist Wilhelm Wundt, who is considered the founder of experimental psychology. Wundt was the first scholar to call himself a psychologist.

Contemplations of the German Romantics on the unconscious are summarized in Eduard von Hartmann's classic compendium, *Philosophy of the Unconscious* (von Hartmann, 1869). Perhaps the deepest understanding of the importance of the unconscious was reached by Hermann von Helmholtz (1821–1894), who is the author of the three-volume monumental work on the cognitive nature of perception, *Handbook of Physiological Optics* (von Hartmann, 1869). Helmholtz reached the conclusion that the brain attempts to reconstruct a *meaningful world picture by using information not only from sensations but also from prior knowledge, emotions, and unconscious inferences.* This truly modern understanding was the result of his extensive and meticulous experimental works. Indeed, in order to scrutinize how visual and auditory perceptions are achieved, he was the first scholar to combine theory and experiments. In other words, regarding perception, Helmholtz was the first scientist to elucidate the transition from the objective space (reality) to the subjective space (experience).

CONFABULATION

By definition, we become aware of only our conscious experiences. This implies that we are biased in favour of both the importance and truthfulness of conscious, as opposed to unconscious, processes. However, it turns out that unconscious processes represent objective reality more accurately than consciousness:

during the transition from an unconscious process to a conscious experience, there is always loss of information.

The study of *split-brain* patients, as well as experiments involving the phenomenon of *binocular rivalry* provide clear illustrations of this fact. Split-brain patients are discussed below, whereas binocular rivalry is discussed in the next section.

Consciousness strives to present a coherent picture of every situation. If there is an inadequate amount of information, consciousness does not hesitate to employ *confabulation*, namely, to *invent* a plausible scenario in order to try to fill gaps of its understanding of the given situation. Confabulation was illustrated in the 1970s, via famous experiments performed at Caltech in split-brain patients. Such patients had undergone a neurosurgical procedure, called *callosotomy*, which separates the two cerebral hemispheres, with the aim to control intractable epilepsy. The patients that underwent this surgery, not only began to live a normal life but also provided neuroscientists with a unique opportunity to study aspects of brain function.

For example, if such a subject is presented with an image to their left visual field, this image can be seen only by the subject's right hemisphere. For right-handed people the Broca's area, which is responsible for speech production, is located in the left hemisphere. Hence, such a split-brain patient cannot name this image. However, they can point to a card depicting this object using their left hand, which is controlled by the right hemisphere.

Incidentally, Descartes had wondered whether the existence of two hemispheres would affect the *unitary nature* of experience in normal subjects. For this reason, he postulated that the seat of the soul was the pineal gland, erroneously thinking that this structure was unified, namely, it did not have a left and a right half.

The callosotomy is an operation that divides the corpus callosum, the thick neural bundle consisting of about 200 million axons of pyramidal cells, connecting the left and right hemispheres. This surgical procedure was pioneered at Caltech in the 1960s, by the biologist Roger Sperry (Nobel Prize in Medicine, 1981) together with the neurosurgeon Joseph Bogen.

The most dramatic illustration of confabulation in split-brain patients was achieved in the late 1970s by Joseph LeDoux and by Sperry's PhD student, Michael Gazzaniga. In one such experiment, the right hemisphere was shown a scene with snow, whereas the left hemisphere was shown nothing. When the patient was asked to pick a relevant card from an array of cards, there were no surprises: the left hand, who knew about the snow, picked a card depicting a shovel; the right hand, which did not have knowledge of the snow, picked a random card. In the next experiment, the right hemisphere was again shown a scene with snow, whilst the left hemisphere was shown a chicken claw. When asked to pick an appropriate card, the left hand, which had knowledge of the snow, picked a shovel; the right hand, which had knowledge of the chicken claw, picked a chicken (see Figure 11.1).

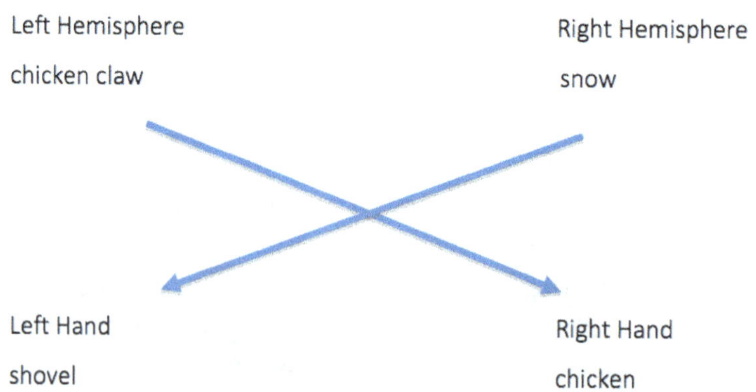

Figure 11.1. Illustration of an experiment in a split-brain subject.

When the patient was asked why his left hand had picked the shovel, his left hemisphere, with which the brain speaks and which had *no knowledge of the snow, but had seen the chicken claw*, responded: "you need the shovel to clean out the chicken shed". It is not surprising that the brain attempts to draw conclusions based on the data available to it. In that sense, since the brain knew of the chicken claw, it is not surprising that the brain attempts to provide a possible explanation for picking a shovel. However, remarkably, it does so with *the full conviction of certainty*. Indeed, Gazzaniga and LeDoux wrote that "the left did not offer its suggestion in a guessing vein but rather as a statement of fact".

As discussed in Chapter 5, emotions are more primitive than visual awareness. For this reason, emotions can cross the two hemispheres, not only via the corpus callosum but also via an evolutionarily older pathway in the lower temporal lobe. The existence of this additional pathway provides an explanation of the following study. When a split-brain patient was shown a picture of Hitler to her left visual field, her right brain became emotionally aware of this picture and was agitated. The right hemisphere is the dominant hemisphere regarding emotions and using the above older pathway, the right hemisphere imposed discomfort to the left hemisphere. When the patient was asked why she was upset, her linguistically dominant left hemisphere, which was *feeling upset without knowing why*, confabulated: "I was thinking about the time that someone made me angry".

NEURONAL CORRELATES OF CONSCIOUS AND UNCONSCIOUS PROCESSES

Are there specific cells or particular domains in the brain responsible for conscious experiences, as opposed to unconscious processes? Identifying such neurons or domains is referred to as finding the *neuronal correlates* or the *neuronal substrates*, of these processes. Remarkable experiments, using single neuron recordings, have made it possible to identify such domains.

Binocular rivalry

In split-brain patients there are no connections between the two hemispheres. Thus, consciousness does not have to face the potentially confusing problem of how to deal with two different images appearing simultaneously in its two different visual fields. But what happens if two different images are presented in a normal subject? This situation, which is referred to as binocular rivalry, was extensively investigated by Sir Charles Wheatstone in the early 1830s. Sir Charles was an ingenious inventor who constructed several new musical instruments. Also, he was a pioneer scientist who made important contributions to acoustics, optics, and electricity, in particular regarding the development of telegraphy. His extensive research in the area of binocular rivalry led him to discover the so-called *stereoscope.* This device is able to combine the views of the left and right eyes, and to create a single three-dimensional view. Using this device, Sir Charles observed the following. When the two eyes viewed two different alphabetic letters, there was a breakdown of the normal *binocular cooperation*, which occurs when the two eyes view the same letter. The viewer will see either one or the other. Wheatstone wrote:

"If *a* and *b* are each presented at the same time to a different eye, the common border will remain constant, while the letter within it will change alternately from that which would be perceived by the right eye alone, to that which would be perceived by the left eye alone. At the moment of change the letter which has just been seen breaks into fragments, while fragments of the letter which is about to appear mingle with them and are immediately after replaced by the entire letter. It does not appear to be in the power of the will to determine the appearance of either of the letters, but the duration of the appearance seems to depend on causes that are under our control: thus if the two pictures are equally illuminated, the alternations appear in general of equal duration; but if one picture be more illuminated than the other, that which is less so, will be perceived during a shorter time" (Wheatstone, 1838: p. 386).

From Wheatstone's studies the following characteristics of binocular rivalry emerged: First, consciousness completely suppresses one of the two stimuli. Second, this suppression changes throughout the given time period. Third, a spatial "fragmentation of the image" occurs during periods of transition. Fourth, the physical characteristics of the stimuli determine which of these stimuli is the dominant one. For example, a brighter stimulus dominates the one which has lower luminance.

Wheatstone's studies captured the attention of several distinguished scientists, including Hermann von Helmholtz, William James, and Sir Charles Sherrington. Hundreds of papers were published on this topic and several interesting results were obtained. In particular, it was shown that the successive periods of visibility and invisibility are indeed random (Levelt, 1966). These periods are referred to, respectively, as the dominance and suppression phases. The average dominance and suppression periods vary across subjects and also depend on the stimulus type. If the individual phase-durations are normalized, namely if they are divided by their mean, their distribution can be approximated with the use of a certain mathematical function, the *gamma function*. Interestingly, the relevant parameters appearing in these mathematical expressions show considerable inter-subject similarity for both humans and monkeys (Leopold & Logothetis, 1996; Sheinberg & Logothetis, 1997). Incidentally, an *increase* in the stimulus strength in one eye increases the predominance of this stimulus via the *decrease* of the mean dominance in the contralateral eye. Interestingly, the mean dominance of the eye receiving the stronger stimulus is essentially unaffected (Levelt, 1966).

Recent studies have shown that the autistic brain is slower in switching from one image to the other. This is apparently due to a defect in the action of the inhibitory neurotransmitter GABA. Using data from a single EEG electrode placed above the visual cortex obtained during binocular rivalry, researchers were able to predict with 87% accuracy whether or not an individual had autism (Spiegel *et al.*, 2019).

A deeper understanding of binocular rivalry necessitates the investigation of the underlying neural mechanism. This was achieved in the seminal work of Nikos Logothetis and his collaborators. Actually, the importance of this work goes much further than binocular rivalry.

It provides the first investigation at the neural level of the transition from an unconscious to a conscious process. Furthermore, it provides a clear determination of the neuronal substrates of seen and unseen visual images.

In these pioneering experiments, performed in 1996, David Leopold and Logothetis, first, trained monkeys to use a lever to report if they were seeing a house or a human face. After showing a house and a face in the two different visual fields of a monkey, these investigators established that, as expected, the monkey saw random alterations of the two different images, just as humans do. By employing single neuron measurements, Leopold and Logothetis established that in the primary visual cortex, which is the getaway of visual information in the cortex, neurons *encoded both images equally well.* Furthermore, the "left"-neurons fire when an image is presented in the left-visual field, and the "right"-neurons fire when an image is presented in the right-visual field. Also, this firing does *not* change when the monkey reports that its experience had switched from one object to the other. As visual processing progresses to more advanced levels, including the area V_4 and the inferior temporal lobe, the firing of more and more neurons begins to agree with the animal's report. Namely, *certain cells fire strongly when the animal reports seeing the face, and much less or not at all when the house is seen.* Actually, the firing of these neurons could be used by Leopold and Logothetis to *predict that the monkey will report seeing the face.* Similarly, with the neurons associated with the house.

As was noted in Chapter 4, different aspects of a given percept give rise to different unconscious structures. The integration of these representations yields a conscious experience. By appealing to the principle of continuity, it is natural to expect that,

the construction of a mental image is preceded momentarily by the unconscious integration of all relevant unconscious structures. Assuming that such an integrated set of neural circuits indeed exists, it will be referred to as a mental pre-image. I expect that mental pre-images can be identified experimentally.

It should be emphasized that when I refer to the unconscious structure associated with given mental images, I actually refer to the *set* of unconscious structures corresponding to the constituents of the given mental image. In any case, I hope that, by now, the reader has appreciated the neuronal mechanisms corresponding to both unconscious processes and conscious experiences, and this is far more important than the specific terminology used.

Here is a personal story, which I expect will resonate with many readers. This story supports the continuity of unconscious and conscious processes and hence the existence of a pre-image. Before leaving the house for my office, I usually take my notes, keys, and lunch. However, the morning just before Cambridge University's Christmas break of 2018, I needed to also take a present for my assistant. Because my brain is used to worrying about only three items, I was about to leave the house forgetting my lunch. However, I had *a strange feeling*, generated by my unconscious. Indeed, I had made the transient conscious decision to take three items, but I had also made the unconscious decision to take four, and the mental pre-image of this decision was revealing itself through the occurrence of an unspecified feeling. This pre-image had not become yet a mental image. This *did* happen when, finally, I realized I was forgetting my lunch. I believe that my strange feeling suggests the existence of the pre-image.

My personal experience described above, which is characterized by "vagueness", is consistent with the position advocated in this book that the transition from the unconscious to awareness is a *continuous* process. This is contrary to models proposed by several well-known neuroscientists, which are based on the assumption that information reaches awareness in a discrete manner associated with all-or-nothing changes in neural activity. I was very pleased to read a recent rigorous study at MIT, involving 112 individuals (Michael *et al.*, 2023), where it was shown that the set of data analyzed in this study (as well as the analysis of the investigators of earlier studies performed by other groups) is consistent with the continuous hypothesis.

Another important by-product of the binocular rivalry study is that it provided the first identification of neuronal correlates of both an unconscious process and a conscious experience. The activation of specific neurons associated with a house and specific neurons associated with a face, suggests that the brain constructs two separate mental pre-images for the house and the face before one of them generates a mental image.

The phenomenon of binocular rivalry has been studied in humans using the medical imaging technique of magnetoencephalography (MEG). Subjects were presented in one visual field with a blue horizontal grating and in the other a red vertical grading. In order to be able to follow the brain's responses to these two images, the stimuli were caused to flicker at two different specific frequencies. It turns out that the associated MEG data exhibited a sharp differentiation in these two frequencies. In this way, it was possible to follow the parts of the brain activated by each rival stimulus. As expected, each stimulus caused excitation in many parts of the brain even when the stimulus was not perceived consciously. However, the conscious perception of one of the two stimuli caused extensive activation (associated with the stimulus' specific frequency) in many areas of the brain, including the frontal, parietal, and occipital regions (Edelman & Tononi, 2001).

The fact that, at any given time, only one of the two rival images gives rise to a conscious experience, is consistent with the *unitary nature of consciousness*, namely to "awareness of one thing to the complete exclusion of everything else" (Edelman, 2006: p. 13). Also, both the experiments on split-brain patients and on binocular rivalry emphasize the predisposition of consciousness to present a

complete and coherent picture of a given situation, even if this is achieved at the expense of presenting an accurate picture of reality.

Concept neurons

As was briefly mentioned in Chapter 4 and will be discussed extensively in the next chapter, neurons in the primary visual cortex respond only to specific features of a percept: some neurons respond only to orientation, others only to colour, and others only to movement. In other words, these neurons concentrate on *local features* of a given situation. However, neurons in higher cortical areas, including the medial temporal lobe, encode *global features* of a percept. Such neurons have been recently termed by the leading neuroscientist Christof Koch, *concept neurons.*

As discussed in Chapter 10, the repeated exposure to the brain of a particular person, or object, or situation, gives rise to a *stable set* of neuronal circuits, which are activated by this specific percept. Taking into consideration that,

consciousness is the result of associations, it follows that this set will be activated by any characterization of this percept.

For example, for a subject who is familiar with the actress Jennifer Aniston, such a set of neuronal circuits will be activated any time the subject looks at a picture of Aniston, or hears her name, or even imagines her. Neurons belonging to such neural circuits were identified recently by the neurosurgeon and neuroscientist, Itzhak Fried, of the University of California, Los Angeles, along with Christof Koch of Caltech (now at the Allen Institute for Brain Sciences). Their discovery was facilitated by a standard neurosurgical procedure used for the treatment of intractable epilepsy. Before such a surgery, different techniques are used aiming to localize the place in the brain where the seizure originates, so that the extent of neurosurgery is minimized. For this purpose, Fried places in the brain around a dozen electrodes. These electrodes, which are inserted using small holes drilled through the skull, stay in the brain for about a week. This provides an opportunity for investigators to perform single neuron measurements in the brain tissue reached by the electrodes.

Most epilepsies originate in the medial temporal lobe, which includes the hippocampus. This is the area *par excellence* responsible for stabilizing a percept into long-term memories. In order to investigate how these memories are stored, after the placement of the electrodes, patients were shown pictures of familiar people, landmark buildings, animals, and objects. Simultaneously, neurons reached via these electrodes were monitored, with the hope of detecting a burst of action potentials. Most of the time there was no response, but occasionally some neurons responded to certain *categories*, such as the category of animals, or the category of faces.[4] Finally, after many trials, a breakthrough took place: neurons were discovered which were highly specialized. For example, a particular neuron fired only when the patient was looking at photos of Bill Clinton, who at that time was the President of the United States; this neuron did not respond to other famous people. Similarly, one hippocampal neuron responded to pictures of Jennifer Aniston but not to pictures of other famous actors or actresses. Importantly, the neuron responding to Aniston, was activated by any percept characterizing the concept Aniston. This is consistent with the statement made earlier regarding the vital role of associations for the generation of consciousness. Interestingly, a neuron of a mathematically oriented engineer responded to the famous Pythagorean equation $a^2 + b^2 = c^2$.

Neural circuits are enormously *redundant*. In particular, percepts belonging to the same category share a very large number of neurons with other members of this category. This implies that, if the particular neuron whose activation is monitored via a specific single neuron recording belongs to the very large number of neurons shared by a given category, then this neuron will be excited by any percept in this specific category. For example, if a neuron belongs to the category of *face*, it will be activated by the image of Aniston as well as by the image of any other face. On the other hand, the *unitary* nature of a particular conscious experience implies that,

within each category, there exists a particular neural network consisting of a unique subset of neurons that characterize only a specific percept or concept.

Thus, if the particular neuron monitored by one of the above electrodes belongs to the subset characterizing the concept *Aniston*, then this neuron will be activated only by any percept associated with this unique concept.

Regarding the total number of neurons corresponding to a particular concept, Koch writes (Koch, 2012) that "any one conscious percept is caused by a coalition of neurons numbering perhaps in the hundreds

[4]The ability of the brain for abstraction, namely, the ability to extract common, basic features among different percepts, leads to the formation of categories. For example, although different faces have different particular features, the brain extracts common characteristics and forms the category of face.

or thousands rather in the millions". This estimate seems reasonable. However, it is worth remembering that, in these fascinating experiments, the search for such neurons was localized only in a small part of the brain, whereas neural circuits containing these concept neurons could be widely distributed in an extended domain.

References

Aquinas, T. c.1265–1274. *Summa Theologica.*

Damasio, A. 2003. *Looking for Spinoza: Joy, Sorrow, and the Feeling Brain.* Harvest.

Descartes, R. 1634. *Discourse on the Method.*

Descartes, R. 1662. *De Homine.*

Edelman, G. 2006. *Second Nature.* Yale University Press.

Edelman, G. & Tononi, G. 2001. *A Universe of Consciousness: How Matter Becomes Imagination.* Basic Books.

Koch, C. 2012. *Consciousness: Confessions of a Romantic Reductionist.* MIT Press.

Leopold, D. A. & Logothetis, N. K. 1996. Activity changes in early visual cortex reflect monkeys' percepts during binocular rivalry. *Nature* 379(6565), 549–553.

Levelt, W. J. M. 1966. The alternation process in binocular rivalry. *British Journal of Psychology* 57, 225–238.

Michael A. C., Jonathan K., & Timothy F. B. 2023. Perceptual Awareness Occurs Along a Graded Continuum: No Evidence of All-or-None Failures in Continuous Reproduction Tasks, *Psychological Science*, 34, 1033–1047.

Sheinberg, D. L. & Logothetis, N. K. 1997. The role of temporal cortical areas in perceptual organization. *Proceedings of the National Academy of Sciences* 94(7), 3408–3413.

Spiegel, A., Mentch, J., Haskins, A. J., & Robertson, C. E. 2019. Slower binocular rivalry in the autistic brain. *Current Biology* 29(17), 2948–2953.

Spinoza, B. 1661–1675. *Ethics, Demonstrated in Geometrical Order.*

Spinoza, B. 1677. *Tractatus Theologico-Politicus.*

von Hartmann, E. 1869. *Philosophy and the Unconscious.* Duncker.

von Helmholtz, H. 1867. *Handbook of Physiological Optics (Originally published in German in three separate volumes between 1856 and 1866).*

Wheatstone, C. 1838. Contributions to the physiology of vision. Part the first. On some remarkable, and hitherto unobserved, phenomena of binocular vision. *Philosophical Transactions of the Royal Society of London* 128, 371–394.

12 DETAILS OF VISUAL PERCEPTION

After deriving his famous mathematical equations of electromagnetism in 1865, James Clerk Maxwell concluded that light is a particular type of *electromagnetic radiation*. This radiation travels in *wave form* with a constant speed, whose approximate value is 300 million metres per second. An important characteristic of any wave is its wavelength, defined as the distance between two consecutive crests. Cosmic rays, gamma rays, X-rays, ultraviolet light, visible light, and infrared light, are all different types of electromagnetic radiation, distinguished by their different wavelengths. The wavelength of visible light is in the range of 390–700 nanometres (a nanometre is one billionth of a metre).

Visible light of different wavelengths is perceived by humans in different colours. For example, light of a wavelength of 640 nanometres looks red, whereas light of 540 nanometres looks green. The ancient Greeks knew how to prepare all colours by using only four pigments, namely blue, green, yellow, and red. Leonardo da Vinci was also aware of this fact. Since green can be made by mixing blue and yellow, it follows that all colours can be obtained from blue, yellow, and red, which are known as the "primary colours". Many distinguished physicists, including Isaac Newton, attempted to explain the importance of primary colours by proposing theories in analogy with the theory of musical harmony. However, in 1802, Thomas Young proposed that there was nothing special about primary colours themselves, but rather that their apparent importance was the result of humans possessing three classes of photoreceptors. Maxwell was impressed:

"So far as I know, Thomas Young was the first who, starting from the well-known fact that there are three primary colours, sought for the explanation of this fact, not in the nature of light, but in the constitution of man".

ANATOMY AND FUNCTIONAL CHARACTERISTICS OF VISUAL PROCESSING

How does our eye process light? Light passes through the pupil and then the lens focuses the image on the retina, which is a sheet of layers of neural tissue lining the back of the eyeball.[1] For distant objects the lens generates a sharp image when it is in its flattest position, whereas for close objects the lens needs to be curved. These lens modifications are achieved by a circular muscle surrounding the lens, called the *ciliary* muscle. With age, the lens loses its plasticity, and this is the reason that the eye needs reading glasses in order to bring close objects into focus.

The outermost layer of the retina houses photoreceptors which contain light-absorbing pigments. As soon as these pigments absorb photons, their three-dimensional configuration changes. This *conformation transformation* initiates a chemical reaction that results in the opening of specific ionic channels, allowing the flow of ionic currents. The summation of these tiny currents causes the release of neurotransmitters that activate cells in the next retinal layer. In this way, *the distribution of the photons impinging on the retina gives rise to electrical signals.* These signals pass through three different types of cells, called *bipolar*, *horizontal*, and *ganglion* cells. By the time they reach the ganglion cells, these signals are sufficiently strong to generate action potentials. Finally, this neuronal information leaves the eye via the *optic nerve* and travels to the brain. The light that is not absorbed by the photoreceptors is absorbed by a layer of special cells located at the back of the eye, called *retinal pigment epithelium* cells.

---

[1]The importance of the retina is consistent with its high complexity; it contains approximately 60 different neuronal cell types.

Colour, luminance, acuity

The human eye contains two types of photoreceptors, called *rods* and *cones*. There exist about 120 million rods and only 6–7 million cones. There are three different kinds of cones, distinguished by their response to three different ranges of wavelengths: short (blue), middle (green/yellow), and long (red). Around 64% of the cones are *red*, 32% are *green*, and only 5% are *blue*. Cones have low sensitivity, which means that their function requires high levels of light. Rods are 500 times more sensitive than cones, and therefore are useful in dim-light conditions and especially in the night time; in bright light, rods saturate, and become inactive. In order to perceive colour, it is necessary to compare the information from at least two different types of photoreceptors. Since humans only have one type of rod, we cannot distinguish between colours in dim light.

Colour is perceived by comparing the activity occurring in the three different types of cones. Another crucial property of vision is *luminance*, which is perceived as *lightness*. This is the result of the summation of the activity occurring in different cones. It should be noted that the perceived luminance depends on the colour associated with it. Indeed, the response of the cones is not only a function of the number of photons received, which is determined by the brightness of the light but it also depends on the wavelength of these photons. All photoreceptors are sensitive to the green/yellow range of the spectrum, see Figure 12.1. It turns out that more photons are needed in the red or blue parts of the spectrum in order to achieve an equivalent luminance.

Since it is possible to perceive light by comparing the activity in only two different types of cones, it is natural to ask why a third type of cone cell is needed. The red cone pigment evolved via a minor mutation of the green cone pigment. This shifted the absorption peak towards the red end of the spectrum, allowing better discrimination between yellow and red. Animals possessing three types of cones have an enhanced ability to identify ripe (red) fruit in a green background of foliage and to distinguish tender yellowish leaves from tougher greener leaves. A crucial role in the function of the photo receptors is played by a class of proteins, called *opsins*.

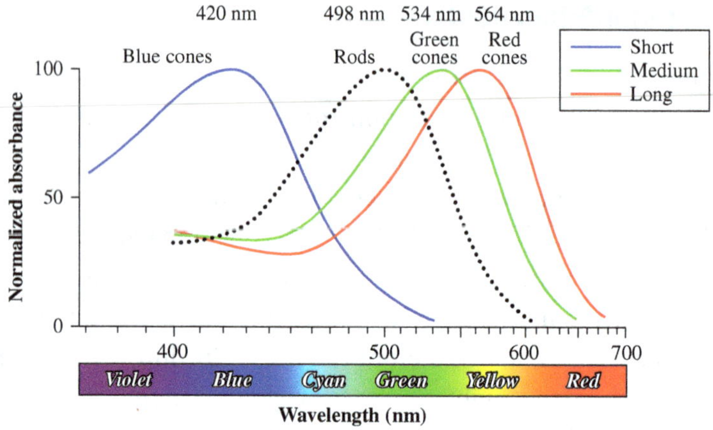

Figure 12.1. Photoreceptor response as a function of wavelength.
Source: Adapted from Anatomy & Physiology: Introduction, Rice University.

Green and red opsins are created via two similar genes that are located in the X chromosome.

A defect in one of these genes gives rise to the *two-colour vision (dichromacy)* passed from the mother to the son. About 1 in 12 men has this deficiency; in women, this is very rare. On the other hand, since women have two X chromosomes, a variation in one of the associated genes, can give rise to *four-colour vision (tetrachromacy)*. Such women can see a variety of shades and colours unseen by most people.[2]

It is interesting that many of our evolutionary predecessors have specific visual advantages in comparison to us. For example, cats and dogs, like most mammals, only have two types of cones (they cannot see red). However, they have more rods than us, and hence they have much better night vision. Bees and butterflies have four colour-receptor cones. They can see a much wider spectrum of colours than us, including ultraviolet colours, but we have a more detailed vision. Interestingly, certain nocturnal animals have colour night vision. Even more amazingly, the mantis shrimp

[2]The persistence in evolution of dichromacy suggests some evolutionary advantage. Indeed, studies focusing on the tribes of hunter-gatherers in the Amazon and South Africa show that the chief hunters have this "deficiency" because it allows them to better identify prey hidden in green foliage.

has 16 colour-receptive cones! They can see ultraviolet, infrared, and even polarized light (Levine, 2017).

High visual acuity occurs at the centre of the gaze, with much lower acuity at the periphery. This is due to the fact that the part of the retina at the very centre/back of the eye, called *fovea*, has special characteristics. First, the cones near the fovea are packed very tightly; actually, only two types of cones are found in this region. Second, the associated ganglion cells have a very small *receptive field*. This means that these ganglion cells receive information from an extremely small domain. As a matter of fact, there are as many foveal ganglion cells as corresponding photoreceptors. This means that, on average, the ratio between foveal ganglion cells and their photoreceptors is one–to-one. This is to be contrasted with peripheral ganglion cells that possess big, bushy dendritic arbours, and therefore receive input from a large number of photoreceptors. Since each foveal photoreceptor receives light from only a tiny part of the retina, it follows that foveal cells yield very high resolution. It turns out that the colour system operates at a low resolution, i.e., the cells which code colour has large receptive fields. This implies that our perception of colour is rather coarse.

Under an ophthalmoscope, the fovea appears as a tiny yellow spot. This is due to the fact that there are no blood vessels or other cell layers in front of the foveal photoreceptors, and light reaches these photoreceptors without being reflected or refracted. There is a yellow pigment layer that protects the foveal cones; this layer absorbs short wavelength light, causing the yellow appearance.

The external three-dimensional world projects a two-dimensional image onto the neurons of the two-dimensional retina of each eye. How does the brain reconstruct three-dimensionality using these two-dimensional images? The visual system employs various types of unconscious neural computations to estimate the distance and depth of objects in a given scene, and hence to determine the relevant three-dimensional spatial organization. For this purpose, it uses the following information: *Perspective*, namely, the changing of image size with distance; s*hading*, namely, the gradations with distance in the reflected light emitted from an object; o*cclusion*, namely, the fact that if an object occludes another object, then it must be in front of it. In addition, the visual system uses *relative* motion, also known as *parallax*, namely, the fact that the relative speed of an object with respect to the retina decreases as the distance to the object increases.[3] Also, it uses *stereopsis*, namely, the brain utilizes the slight differences in the images formed in the two eyes, which is the result of the slightly different viewing angles.

All these mechanisms except stereopsis are *monocular*. This means that the main clues used by the brain to estimate distance are provided via monocular considerations. It will be shown in the next chapter that artists, in order to depict three-dimensionality, also use techniques mimicking the approaches used by the brain.

Visual processing

After visual fibres leave the eye via the optic nerve, they travel to the *optic chiasm*, and from there to thalamus. Finally, from the thalamus, nerve fibres, known as *optic radiations*, travel either to the *primary visual cortex*, or to higher visual areas, see Figure 12.2.

By the late 1800s, neuroscientists knew that everything from the left visual field is transmitted to the right brain hemisphere, and from the right field to the left hemisphere.

More detailed information regarding the visual system was obtained by the Japanese ophthalmologist, Tatsuji Inouye, who examined several soldiers injured in the 1904 war, of the Tsarist Russia against Manchuria and Korea. The Russian soldiers used tiny bullets which had sufficient energy to penetrate the skull, but small enough to avoid shattering the brain. These bullets made small, clean wounds in the brain. If such wounds were at the back of the brain, namely, in the primary visual cortex, the victims developed blind spots in their vision. Inouye was able to establish a correlation between the position of a blind spot in the visual field and

[3]Regarding the effect of relative motion, an example is provided by seeing the moon, which to a walking observer appears immobile: an unconscious slight head movement causes the retina to move, creating a relative motion with respect to the moon.

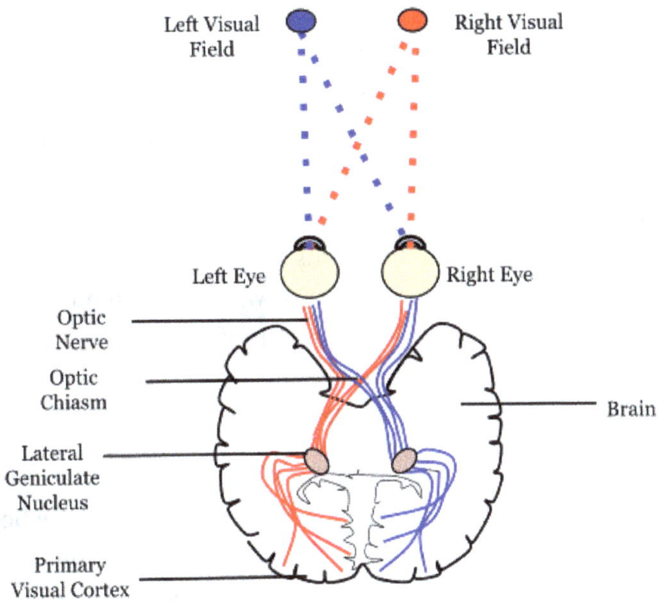

Figure 12.2. Left and right visual fields.
Source: Wikimedia Commons.

the position of the injury in the visual cortex. He concluded that *the more central the position of the blind spot, the larger the corresponding injury.* Indeed, it is now known that the visual processing of the central 1% of the visual field, requires about half of the 250 million neurons of the primary visual cortex. The results of Tatsuji are consistent with the findings of two English physicians, Gordon Holmes and William Lister, during the First World War, who were unaware of Inouye's publications.

The modern era of visual neurophysiology begins with the Cambridge physiologists Sir Edgar Adrian and Rachel Matthews, who successfully recorded electrical activity in the optic nerve of the eel. The next breakthrough was achieved by Haldan Keffer Hartline (Nobel Prize in Medicine, 1967). Hartline received his MD from Johns Hopkins University in 1927, but never went on to practise medicine.[4] It is interesting that, after being awarded an MD and after a brief exposure to experimental biology,

[4]Hartline used to joke that, "I was awarded an MD on the condition that I will never practice medicine". Adrian, who was a fellow of Trinity College at Cambridge, also had an MD degree.

Hartline decided to study mathematics and physics. He went to Germany to study under the great mathematical physicist Arnold Sommerfeld,[5] and also under one of the founders of quantum mechanics, Werner Heisenberg (Nobel Prize in Physics, 1932). However, it soon became clear that he lacked the background for such advanced studies, and after one year he returned to the United States to take his first appointment in biology. In spite of his disappointment with the outcome of this endeavour in Germany, Hartline's interest in mathematics and physics never waned throughout his career. In 1949, Hartline was appointed as the first professor of the newly formed Department of Biophysics at Johns Hopkins University. The first chairman of this department was Stephen Kuffler.

One of the most important contributions of Hartline was the successful recording of electrical activity from a single nerve fibre of the horseshoe crab *Limulus*. In these experiments he discovered the important phenomenon of *lateral inhibition*, also known as *surrounding suppression*: visual cells in the eye are excited by light impinging on their receptive-field *centres*, but they are inhibited, i.e., their activity is suppressed, by light on the immediate *surrounding region*. Kuffler extended Hartline's work by recording electrical activity from mammalian ganglion retina cells. David Hubel and Torsten Wiesel, who were mentioned earlier in this book, were trained by Kuffler and extended further their advisor's work, showing that visual neurons in the thalamus behave in a similar way.

These studies imply that visual neurons in the eye and the thalamus respond to abrupt, rather than gradual, changes in luminance. Hence, *the visual system is very efficient in encoding those parts of the image where there are discontinuities or abrupt changes.*

By 1958 scientists had built micro-electrodes sensitive enough to record the activity of single neurons. Employing these electrodes, Hubel and Wiesel studied visual neurons at the next level, namely, in the cortex. Their Nobel Prize calibre experiments in cats revealed that in the cortex there is higher specificity. Indeed, while retinal and thalamic cells are responsive to discontinuities in luminance regardless of the orientation of borders,

[5]Sommerfeld's first three PhD students, namely, Werner Heisenberg, Wolfgang Pauli, and Peter Debye, were separately awarded the Nobel Prize.

cortical cells are selectively responsive to *orientation*. There are *orientation cells*, which are excited only by a horizontal bar of light, and nearby cells which are excited only by a slanted or vertical bar of light. This means that cortical neurons are able to track *straight line-segments*.

In addition to discovering *line-detecting neurons*, Hubel and Wiesel also discovered *motion-tracking neurons*. They established that there are neurons that are excited by up-down motion, others by left-right motion, and still others by diagonal motion. Incidentally, taking into consideration that tracking movement is more difficult than mapping static images, and also that detecting movement can be lifesaving for animals threatened by attacking predators, it is not surprising that motion-sensitive neurons outnumber the orientation neurons.

Cajal's discovery that cortical neurons are organized in cellular columns was already noted in Chapter 2. The precise mechanism responsible for the genesis of the column-like structures of the human visual cortex remains unknown. The analogous physical process in the visual system of fruit flies was elucidated only recently. Three core neurons in the larval brain form concentric rings, held together by an adhesion protein called *N-cadherin*. As the fly matures, the differential expression of this protein between the three neuron types leads to different adhesive properties. These finally give rise to the columnar pattern (Trush *et al.*, 2019).

Hubel and Wiesel were the first to establish the *functional importance of the columnar organization*. They observed that cells of the same orientation are organized in a single column.[6] Indeed, they found a column of cells that are excited by horizontal lines, a column of cells that are excited by vertical lines, etc. The orientation of neighbouring cells in a column differs by only a few degrees, and the combination of all these cells forms a column that covers all possible orientations, i.e., 180 degrees. Furthermore, there is a column for the left eye and one for the right eye. Thus, by combining these two columns into a *hyper-column*, a one-millimetre strip of neuronal

[6]By employing two-photon imaging and large-scale electron microscopy, the neural anatomy and functions of specific stimulus-orientation neural networks in a mouse's primary visual cortex have been delineated. In particular, it has been established that inhibitory interneurons receive input from nearby excitatory neurons (Bock, 2011).

tissue in the brain area called V_1, it is possible to detect any line with any *orientation* at any position in the visual field. It was later discovered (Livingstone & Hubel, 1984) that hyper-columns have more structure. Namely, in the area called V_4, there are hyper-columns that contain *blobs*, which are cells sensitive to *colour*. Furthermore, in the area called V_5, there are hyper-columns containing cells that respond to *movement*. In this way, each hyper-column functions as an autonomous eye, able to detect orientation, colour, and movement, reminiscent of the compound eye of insects!

It has been established that at each successive stage in visual processing, neurons become selective for more precise features, such as corners, curvature, etc. Furthermore, more complicated neuronal processes give rise to *higher visual processing*. As noted in Chapter 4, all senses have *primary processing areas* in the cortex, where basic information encoded in a specific sense is extracted. In addition, each sense has one or more *association* areas, where further processing takes place. For vision, the relevant anatomical structures underlying the higher visual processing begin at the ganglion cells, see Figure 12.3, which come in two varieties: ganglion cells of high acuity, which are also able to detect colour, and ganglion cells of low acuity, that are colour blind. Both types of these cells project onto the visual cortex. However, only the low acuity type projects directly to the association areas.

The system formed by the projection of the high acuity-colour sensitive ganglion cells, known as the "what system", excels in image recognition, concentrating on shape and colour. The system formed by the projection

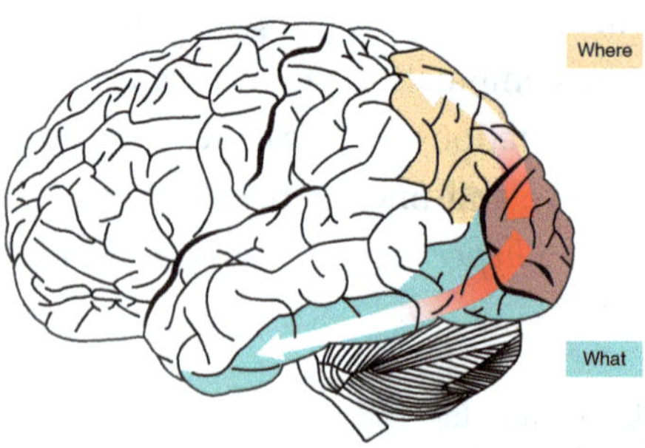

Figure 12.3. The what and where systems.
Source: Modified from Livingstone (2014, p. 123).

of the low acuity-colour blind ganglion cells, known as the “where system”, concentrates on the position and trajectory of the object. The where system implements a *gross classification* of complex visual scenes and objects, whereas the what system carries out a more detailed analysis. It should be noted that the where system is faster and more transient, i.e., its responses are of shorter duration, than the what system.

The pathways associated with these two fundamental systems have been further delineated: the fast pathway, also called the old pathway, goes through the part of the thalamus called the *superior colliculus*, and then arrives at the parietal lobe via the *pulvinar*, which is part of the visual cortex. The slower pathway, also called the new pathway, goes through the part of thalamus known as the *lateral geniculus*, and then, via the primary visual cortex, projects to the temporal lobe for further processing. From there, it sends sensory neurons to the prefrontal cortex. It has been verified that approximately 30 distinct areas specializing in visual processing are associated with this new pathway.

THE FACE AND EMOTIONS

Taking into consideration the importance of the face in social interactions, it is not surprising that a particular part of the brain, located in the temporal lobes, the *fusiform gyrus*, is crucial for the visual processing of the face. Winrich Freiwald and Doris Tsao have identified six interconnected brain regions of the fusiform gyrus in monkeys and humans, where neurons respond to faces much more strongly than to other objects; except to clocks, and to a few other objects that share common features with caricatures of a face (Freiwald & Tsao, 2010). These neuroscientists, using functional MRI, studied the activation of these regions caused by important facial features, such as eye size, inter-eye distance, mouth, nose, and hair. They concluded that the higher the deviation of such features from a *typical face*, the stronger the neuronal response. In other words, the visual system concentrates on the analysis of *how a particular face differs from an average face.*

A caricature emphasizes precisely these individual differences. In his book on caricatures (Grose, 1788), Francis Grose writes:

"a slight deviation [from typical forms and proportions] by the predominance of any feature, constitutes what is called Character, and serves to discriminate the owner thereof, and to fix the idea of identity. The deviation, or peculiarity, aggravated, forms a caricature".[7]

According to this description, a caricature provides an ideal illustration of the difference between a particular face and a typical face. This suggests that *a facial caricature should elicit a stronger neuronal response than a photograph of a face.* This is indeed the case.[8]

Humans have a universal ability to perceive emotions expressed by facial expressions. In this regard it is noted that in the late 1700s, the pioneer neurologist, Duchenne de Boulogne (1806–1875), whose name is associated with the dreaded neuromuscular disease, Duchenne muscular dystrophy, was able to elicit emotions by exciting individual facial muscles via the placement of electrodes in the face of volunteers. These expressions were systematically studied by Darwin, who after analyzing 11 pictures of such faces, concluded that the mechanisms of defence and protection are *universal.*

Actually, humans not only have the ability to perceive facial expressions but can also decipher a variety of subtle features, such as the emotional content of the gait of an individual. This is a consequence of the occurrence of direct neuronal connection between the areas of the brain that handle visual and emotional processes. There is a pathway from the fusiform gyrus to an area in the temporal lobe called *superior temporal sulcus*, and from there to the amygdala, which is the area *par excellence* for gauging the emotional significance of the context of an image. If an image is threatening, then the organism prepares for action. This is achieved with the help of the *hypothalamus*, which orchestrates the proper hormonal response and activates the autonomic nervous system. This causes the release of

[7]Francis Grose is known for his books on the antiquities of England, Wales, and Scotland.
[8]The importance of the face for social interactions suggests that evolution has assigned a particular module of the brain to the task of its visual processing. Although it is possible that such an area does not exist at birth but emerges as a result of repeated experiences during early development, the experiments of Tinbergen mentioned in Chapter 9 suggest that this is unlikely.

adrenaline, an increase in heart and respiratory rates, an increase in blood pressure, sweating, and secretion of cortisol, the key stress hormone. There also occurs a contraction of the skin blood vessels, of the musculature of the gut, and of the face muscles, causing the adoption of a characteristic "mask of fear". These dramatic physical changes *precede the experience of the feeling of fear*. The correct sequence of these events was first understood by William James and published in his classical paper "What is an Emotion", which appeared, in 1884, in the journal *Mind* (James, 1884). As noted in Chapter 5, the unconscious phase of the *feeling* of fear has been named by Antonio Damasio the *emotion* of fear. In contrast to typical unconscious states, which remain hidden, the unconscious emotion of fear reveals itself via the above dramatic bodily changes.

An emotional response can be measured objectively by using the *galvanic skin response* (GSR). This involves attaching two electrodes on the skin of a subject and connecting these passive electrodes to a device called *ohmmeter*. A strong emotional response causes the skin to sweat, and this lowers the skin's resistance to the flow of electricity measured by the ohmmeter. For example, if neutral pictures, such as a chair or a table, are shown to a subject, there is no GSR response. But, the picture of a familiar face elicits a strong GSR response.

MALFUNCTIONING OF VISUAL PROCESSING

The inability to see motion (akinetopsia) and the inability to see colour (achromatopsia)

As noted earlier, specific areas of the medial temporal lobes, called V_5, are activated by the visual perception of movement. Patients with a stroke in this area, as well as healthy volunteers whose medial temporal area is inactivated via transcranial magnetic stimulation (this technique is discussed in Chapter 16), behave in a characteristic way consistent with their inability to see motion. Such patient is described in Ramachandran (2011: p. 66):

"She could read newspapers and recognise objects and people. But she had great difficulty seeing movement. When she looked at a moving car, it appeared like a long succession of static snapshots. She could read the number plate and tell you the car's colour, but there was no impression of motion. She was terrified of crossing the streets because she didn't know how fast the cars were approaching. When she poured water into a glass, the stream of water looked like a static icicle [...]. Talking to people was like talking on the phone because could not see the lips moving".

Evidence for the crucial importance of this part of the brain for perceiving motion is further supported by studies that stimulate electrically this area of monkey's brains using microelectrodes. This prompts the monkeys to start moving their eyes as if they are tracking moving objects, which apparently appear in their visual fields as a result of the electrical stimulation. In addition, studies using functional MRI in volunteers looking at moving objects show a clear activation of this area.[9]

Regarding the visual processing of colour, it is noted that the complete loss of colour perception in individuals with otherwise normal vision is rare. Patients suffering from this condition, called *cerebral achromatopsia*, experience the world in black and white.[10] This condition should not be confused with *complete colour blindness*, where the genes for all three cone pigments are missing. People suffering from this condition are *achromats* and are highly sensitive to light because only their rods function; in daylight they need to wear dark red glasses or contact lenses.

Blindsight and Anton's syndrome

The remarkable phenomenon of *blindsight* was first reported during the Great War, by the physician George Riddoch. Riddoch, who is known for his pioneering studies on spinal cord injury, demonstrated that *patients could still perceive motion despite blindness caused by damage to their visual*

[9] As expected, when taking evolutionary considerations into account, animals are particularly sensitive to detecting movement, so that they can identify predators. For example, a snake's visual acuity is based primarily on movement; snakes have difficulty detecting still objects (Vyshedskiy, 2014: p. 70).

[10] A patient who had lost colour vision in the upper left quadrant of his visual field is described in Gallant *et al.* (2000). This defect followed a cerebral artery infarct that created a pea-sized lesion on the right side of his higher visual centres beyond the primary visual cortex.

cortex. This phenomenon was rediscovered independently by Lawrence Weiskrantz in the late 1970s. This Oxford physician had a patient who had suffered substantial damage to his left visual cortex, causing blindness to his right visual field. When Weiskrantz asked the patient to touch a tiny spot of light in his right visual field, the patient protested that this was pointless, since he could not see anything. However, to his own amazement, the patient was able to touch this spot. Repeated trials established that this was not an accident. Part of the old pathway was functional, and apparently this allowed the patient to have unconscious visual access.

Typically, patients suffering from blindsight are reluctant to accept that they do possess the ability for some visual processing. It is interesting that there are patients with the opposite behaviour. Namely, such patients deny their blindness despite objective evidence of visual loss. These patients confabulate, namely, they make up stories to support their position. In this rare condition, known as *Anton's syndrome*, in addition to bilateral injury in the occipital cortex, other cortical areas must also be malfunctioning (Maddula *et al.*, 2009). Remarkably, the Roman stoic philosopher, famous for his tragedies, Seneca (4 BC-AD 65), knew such a patient. In his *Letters to Lucilius* (*Liber V, Epistla IX*), he writes:

"This foolish woman has suddenly lost her sight. Incredibly as it may appear, what I am going to tell you is true. She does not know she is blind. Therefore, again and again she asks her guardian to take her elsewhere. She claims that my home is dark".

Anton's syndrome is a particular form of a condition known as *anosognosia*, which means that there is unawareness of a particular deficit.

Prosopagnosia and Capgras syndrome

Prosopagnosia, i.e., face-agnosia, refers to a difficulty in perceiving familiar faces. In its extreme form, a patient cannot recognize a face as a face. A patient suffering from total prosopagnosia perceives distinct features of the face, such as eyes and mouth, but cannot combine them to perceive the percept of a face. In its more common form, a patient can recognize a face as a face, but is not able to associate a specific face with a particular person.

This leads to social isolation, because people afflicted with prosopagnosia have difficulty recognizing and naming familiar persons, and thus they avoid social interaction.[11]

The neurologist Oliver Sacks, in his famous book, *The Man Who Mistook His Wife for a Hat* (Sacks, 1985), discusses a man who tried to shake hands with a grandfather clock, when he mistakenly identified the face of the clock as a human face. As stated earlier, face-cells are known to be activated by face-like objects, such as clocks.

It is interesting that people suffering from prosopagnosia have a normal galvanic skin response. In particular, they have an increased response when they see a familiar face. *This means that the unconscious processing of emotions evoked by the given face remains intact, despite the fact that the face is not consciously recognized.*

Precisely the opposite occurs in *Capgras syndrome*. This fascinating condition is named after a French psychiatrist who, in 1923, described a female patient who complained that imposters had taken the place of her husband and some of her friends. Until the 1980s, this syndrome was classified as a psychiatric disease, but is now viewed as a neurological disorder that can appear in schizophrenia, in some cases of epilepsy, and after brain injury. In his book *Phantoms in the Brain* (Ramachandran, 1998), V. S. Ramachandran describes a young man suffering from this syndrome. This patient was comatose for two weeks following a car accident, but then he made a remarkable recovery:

"He could think clearly, was alert and attentive, and could understand what was said to him. He could also speak, write, and read fluently, even though his speech was slightly slurred. He had no problem recognising objects and people. Yet he had a profound delusion. Whenever he saw his mother, he would say Doctor, this woman looks exactly like my mother but she isn't-she is an imposter pretending to be my mother. He had a similar delusion for his father but not for anyone else".

[11]These patients are often helped by concentrating on a characteristic feature, such a mole or a large nose, to help them identify a familiar face.

Apparently, in this syndrome, the connection between visual processing and emotional response malfunctions. Hence, although the patient could recognize the faces of his parents, these faces did not evoke the expected emotional warmth. Consciousness, which always attempts to reach a complete understanding of every situation, concludes that the people claiming to be his parents are imposters.

This explanation is further supported by Ramachandran's report that the photographs of the parents of the above patient did *not* evoke a galvanic skin response. Interestingly, the patient *could* recognize his parents via a telephone conversation. This apparent paradox can be perhaps explained as follows: the visual and auditory systems use different neuronal connections to the *limbic system*, namely to the part of the brain responsible for an emotional response. Apparently, in this patient only the visual connection malfunctions. If the patient sees and hears his parents, because vision overwhelms the other senses, the emotional response of the auditory system is vetoed by the lack of response of the visual system. Incidentally, there is a particular form of Capgras syndrome where a patient looking in the mirror thinks that there is a sinister stalker attempting to replace them.

Some investigators consider *Cotard's syndrome*, mentioned in Chapter 3, "as Capgras syndrome turned inwards". Perhaps, in these devastating cases, patients feel no emotional response about *themselves*, and this complete lack of feeling convince them that they are dead. For the occurrence of the Capgras as well as of the Cotard syndrome it is apparently necessary for the right, as opposed to the left, hemisphere to be damaged.

References

Bock, D. D. *et al.* 2011. Network anatomy and *in vivo* physiology of visual cortical neurons. *Nature* 471(7337), 177–182.

Freiwald, W. & Tsao, D. 2010. Functional compartmentalization and viewpoint generalization within the macaque face-processing system. *Science* 330, 845–851.

Gallant, J. L., Shoup, R. E., & Mazer, J. A. 2000. A human extrastriate area functionally homologous to macaque V4. *Neuron* 27(2), 227–235.

Grose, F. 1788. *Rules for Drawing Caricaturas.*

James, W. 1884. What is an emotion? *Mind.*

Levine, J. 2017. 5 things you didn't know about how animals see color. https://crosstalk.cell.com/blog/5-things-you-didnt-know-about-how-animals-see-color.

Livingstone, M. S. 2014. *Vision and Art: The Biology of Seeing.* Abrams.

Livingstone, M. S. & Hubel, D. H. 1984. Anatomy and physiology of a color system in the primate visual cortex. *Journal of Neuroscience* 4(1), 309–356.

Maddula, M., Lutton, S., & Keegan, B. 2009. Anton's syndrome due to cerebrovascular disease: A case report. *Journal of Medical Case Reports* 3, 9028.

Ramachandran, V. S. 1998. *Phantoms in the Brain: Probing the Mysteries of the Human Mind.* Harper Collins.

Ramachandran, V. S. 2011. *The Tell-Tale Brain: A Neuroscientist's Quest for What Makes Us Human.* W. W. Norton & Company.

Sacks, O. 1985. *The Man Who Mistook His Wife for a Hat.* Summit Books.

Trush, O., Liu, C., Han, X., Nakai, Y., Takayama, R., Murakawa, H., Carrillo, J. A., Takechi, H., Hakeda-Suzuki, S., Suzuki, T., & Sato, M., 2019. *N*-cadherin orchestrates self-organization of neurons within a columnar unit in the drosophila medulla. *Journal of Neuroscience* 39(30), 5861–5880.

Vyshedskiy, A. 2014. *On the Origin of the Human Mind.* Greatspace Independent Publications.

13 LIGHT, VISION, AND ART

What is light? This issue was hotly disputed in Isaac Newton's time, especially the particular question of whether light consists of particles or is formed by waves. Newton performed ingenious experiments supporting both hypotheses, but in his famous work *Optics* (1704) he presented arguments slightly in favour of the particle theory. Thomas Young revisited Newton's arguments in 1802, and although he expressed in his writing admiration for Newton, concluded that the great man was wrong to support the particle as opposed the wave hypothesis. As stated in the previous chapter, James Clerk Maxwell concluded in 1865 that light is a particular type of electromagnetic radiation. The theoretical results predicted by the mathematical analysis of Maxwell's elegant equations were confirmed in 1887 in a series of seminal experiments by Heinrich Hertz. Hence, everyone accepted that *light is waves.*

However, Hertz, during his experiments to confirm the wave nature of light, observed the so-called *photoelectric effect*. In this phenomenon, after light illuminates a metal, electrons may jump out of this metal.[1] Further experiments performed by others showed that the kinetic energy of the electrons released from the metal is *independent of the intensity of the striking light.* Moreover, the electron emission starts almost *simultaneously* with the illumination of the surface of the metal. These results could not be explained by employing classical physics. Indeed, according to the classical

[1]It is remarkable that the same experiment that confirmed one of the most important theories in Physics, simultaneously produced an effect that could be used to raise questions about the general validity of this theory!

understanding, a body absorbs or emits energy in a *continuous manner*. Hence, the maximum kinetic energy of the electrons should be proportional to the amount of energy they receive, and therefore of the intensity of the incident light. Furthermore, the metal should be illuminated for some time before an electron is released.

Albert Einstein understood that this paradox could be resolved by employing an earlier proposal by Max Planck (Nobel Prize in Physics, 1918), the leading physicist of the time. Planck had proposed that the energy of an electromagnetic wave is *not smoothly distributed throughout the wave front but is carried in discrete amounts as tiny packets*. Specifically, Plank proposed that the smallest amount of energy that could be released is given by the constant h, later named Plank's constant. In order to explain the photoelectric effect, Einstein postulated, in 1905, a hypothesis, which he himself considered a "revolutionary idea": light must consist of particles!

In fact, Einstein argued that each light particle gives its energy to one and only one electron. A light particle would cause the release of an electron from the metal, provided that this electron receives a sufficient amount of energy to overcome its attraction to the nucleus. Taking into consideration that photons of frequency f carry energy which equals hf, it follows that the maximum kinetic energy, KE_{max}, of an ejected electron, is given by $KE_{max} = hf - \phi$, where ϕ is the minimum energy required to remove an electron from the surface of the metal. Indeed, an *accurate* plot of the maximum kinetic energy for different light frequencies *does* give rise to a straight-line graph, as predicted by this equation (Figure 13.1).

Astonishingly, Einstein was aware that the experiments performed during that time did *not* produce a straight line, and hence did *not* verify his hypothesis. However, his phenomenal self-confidence allowed him to still propose this theory, speculating that more accurate experiments would produce a straight-line graph! Naturally, his paper was met with scepticism even by his strongest supporters. For example, Max Planck did not think that his own theory, as employed by Einstein, could be used to explain the photoelectric effect. However, a decade after Einstein's paper was published, the American physicist Robert Millikan confirmed, to a great degree of accuracy, Einstein's prediction. Einstein was awarded the Nobel

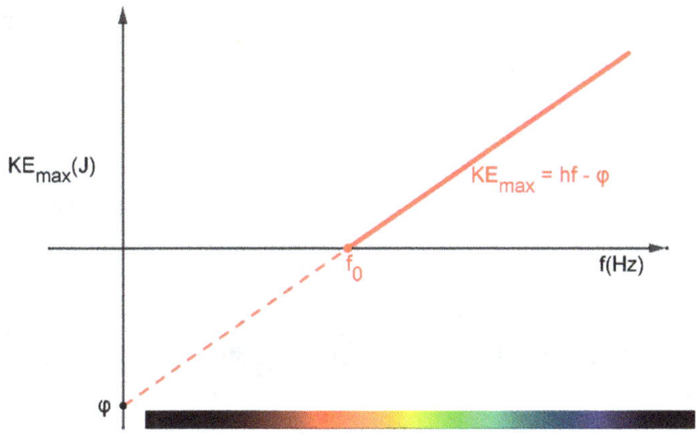

Figure 13.1. Diagram of the maximum kinetic energy (KE_{max}) as a function of the frequency of light (f).

Prize in Physics for the photoelectric effect in 1922, and Millikan the Nobel Prize in Physics in 1923 for his experimental contributions to the electron and to the "light-particles" (this is further discussed in the second volume). By then, these particles were named *photons*.

Interestingly, a few years later, the paths of Einstein and Millikan crossed again: Millikan had become the President of Caltech. Einstein visited Caltech for a year, staying in the Athenaeum, the guest house of Caltech, in what is now called "the Einstein Suite". Einstein wanted to stay at Caltech permanently, but Millikan insisted that no one at Caltech should have a higher salary than its Nobel Laureate President. As a result of Millikan's refusal to satisfy Einstein's salary demands, Einstein became the first appointee of the Institute of Advanced Studies at Princeton. In 2014, on the occasion of delivering at Caltech the Paco Lagerstrom Lecture, I had the immense honour of staying in the Einstein Suite and to enjoy the historical photos of the great man.

The final confirmation of Einstein's theory took place in 1923 when Arthur Compton (Nobel Prize in Physics, 1927) demonstrated that the behaviour of light when it scatters off electrons, cannot be explained via Maxwell's equation, but *can* be explained via the existence of photons.

In summary, Einstein was able to elucidate the deeper structure of light:

light consists of photons which travel in the form of electromagnetic waves.

The question "What is light?" is historically related to the question "What is sight?" According to the historian A. Mark Smith,

"For the vast majority of its history, the science of optics was aimed primarily at explaining not light and its physical manifestations, but sight in all its aspects from physical and physiological causes to perceptual and cognitive effects".

Smith claims that a paradigm shift took place in the 17th century when the analytical focus of optics shifted "from sight to light" (Smith, 2015). He argues that, while modern optics concentrates on the study of light and its properties, ancient optics was the study of sight.[2]

However, as noted by the priest and theologian, Isidoros Katsos, light was *the* key element in ancient theories of vision. Indeed, this is clearly demonstrated, by Plato: according to his dialogue *Timaeus*, sight is explained in terms of the interaction between the eye and physical light. Plato attempted to establish the fundamental importance of physical light by arguing that it is a *variant of fire*, which together with air, water, and earth, were the constituent elements of his physical world (Katsos, 2019). Incidentally, Katsos following earlier works and especially Runia (1986), argues that the physical and metaphysical positions presented in *Timaeus* greatly influenced the Jewish-Christian exegesis of the six-day narrative in *Genesis*. There again the role of light is central: "And God said, let there be light: and there was light".

It is important to note that the writings of Maxwell and Helmholtz on the nature of light had a significant impact on the Arts. In addition, the French chemist Michel Chevreul influenced painting, as a result of creating his "colour wheel". Remarkably, in contrast to Newton's classic prism experiments of 1704, where only 7 colours were identified, Chevreul's wheel consisted of 72 distinct colours: the 3 primary colours red, blue and yellow (green, which often replaces yellow as a primary colour, can be formed from mixing yellow with blue); three secondary mixtures of

---

[2]This point of view is consistent with the etymology of *optics*, which derives from the Greek word *όψις* meaning view, sight, vision.

orange, green, and violet; and 66 additional mixtures generated from the primary and secondary colours.

Chevreul's colour wheel, published in 1838, influenced many artists: from Seurat and Mondrian to several leading contemporary painters, including: Frank Stella, for example, *Harrach II* (1967); Ellsworth Kelly, for example, *Colours for a Large Wall* (1951); and Bridget Riley, for example, *Nataraja* (1993).[3]

The intuition of great artists allows them to use several of the visual mechanisms employed by the brain, independently of their knowledge regarding properties of light. As noted in the earlier chapter, the brain starting with only a two-dimensional retinal information is able to fully reconstruct a three-dimensional world. Artists aspiring to representational painting face an additional formidable challenge: they must construct a two-dimensional painting that appears three-dimensional to viewers looking at it. As noted in Chapter 4, the brain constructs a mental image by first solving many separate inverse problems and then integrating the associated reconstructions. If an artist could have access to these representations, it would be easier for the artist to depict them directly, avoiding altogether the problem of three-dimensionality. These remarks provide a clear illustration of the importance of the unconscious in arts.

The more access an artist could have to unconscious processes, the easier it would be for the artist to depict reality.

THE INTUITION OF GREAT ARTISTS REGARDING CARICATURES AND THE IMPORTANCE OF DIRECTING THE GAZE AT THE FACE

Remarkably, great artists make unconscious use of some of brain's visual mechanisms. For example, Livingstone has shown that the impressionists made extensive use of the interplay of luminance and colour. As discussed in the previous chapter, the global nature of a scene is determined by the

[3]Incidentally, the importance of *relationships* in art has been eloquently expressed by Riley: "The marks on the canvas are sole and essential agents in a series of relations which form the structure of the painting".

"where" system, and this relies on luminance (and not on colour). The less the contrast in luminance between different parts in an image, the less clear it is. This fact is perfectly illustrated in Claude Monet's painting, *Vetheuil dans le Brouillard* (1879). In this painting, the low luminance contrast stimulates the "where" system. Hence the beholder, although becomes aware of the holistic features of the scene, is unable to immediately identify the specific objects in the image.

The use of *synergy* of luminance and colour is beautifully illustrated in Margaret Livingstone's highly insightful book *Vision and Art: The Biology of Seeing* (Livingstone, 2014: pp. 120–121): Livingstone computes the luminance of Claude Monet's painting, *Impression: Sunrise* (1872). This painting has a uniform luminance, except for the darker boat. In particular, this computation shows that the luminance of the sun and its shadow is the *same* as the background's luminance. The glorious effect exerted by the sun is the result of the synergy of this uniform luminance and of the blurring of the different colours.[5]

In what follows, an effort will be made to show that artists have unconscious appreciation of a variety of the basic characteristics of visual perception. In particular,

all great portraitists have made use of the fact that the highest resolution is achieved at the centre of the gaze.

This is illustrated, for example, in one of Pablo Picasso's earliest portraits, *Gustave Coquiot* (1901). Coquiot was a prolific writer known for his titillating stories. This characteristic is clearly captured in this portrait. Especially, by the combination of Coquiot's lustful smile and the frieze behind his portrait, which depicts women in oriental costume performing an erotic dance. The face is painted in detail and is in focus, whereas everything in the periphery is rather blurred, especially the black suit and Coquiot's hand (Figure 13.2).

As explained in the previous chapter,

gazing at the periphery gives rise to an overall impression of the image that cannot be achieved by concentrating on the details of the image.

Figure 13.2. Pablo Picasso, *Gustave Coquiot* (1901).
Source: © Succession Picasso/DACS, London 2024.

This is demonstrated in a dramatic manner in Leonardo da Vinci's famous *Mona Lisa* (1503), exhibited at the Louvre Museum. Concentrating the gaze on Mona Lisa's mouth causes her enigmatic smile to disappear; gazing away from the mouth the smile reappears. Focusing the gaze on the mouth makes it impossible to perceive gestalt features, and perceiving a smile requires precise awareness of such features since it requires appreciating a variety of different elements.

As elucidated in the previous chapter,

a caricature of a person elicits a stronger visual response than the picture of a person.

This fact has been intuitively exploited by many artists. Perhaps no artist employed this fact more dramatically than the Viennese expressionist Egon Schiele, who used caricature not only for the face but also for the entire body. A typical example can be seen in *Self-Portrait with Raised Bare Shoulder* (1912) (Figure 13.3).

Figure 13.3. Egon Schiele, *Self-Portrait with Raised Bare Shoulder* (1912).

Figure 13.4. Picasso, *Portrait of Sebastià Junyer Vidal* (1903).
Source: © Succession Picasso/DACS, London 2024.

Picasso was both an admirer and user of caricatures. The latter is clearly illustrated in his painting, *Portrait of Sebastià Junyer Vidal* (1903). The face depicted in this portrait has caricaturist features, including Sebastia's bulbous forehead which is "framed" by the curly hair and a thick moustache (Figure 13.4).

Figure 13.5. Picasso, *Portrait of the Poet Sabartés* (1901).
Source: © Succession Picasso/DACS, London 2024.

Regarding the role of continuity, it is noted that Picasso's portrait of his friend, sculptor, and writer, Jaume Sabartés, *Portrait of the Poet Sabartés* (1901), reveals the influence on Picasso's early works of van Gogh, Gauguin, Degas, and Toulouse-Lautrec. By combining caricature and solemnity with different shades of blue (a defining characteristic feature of Picasso's *blue period*), this portrait provides an eloquent meditation on melancholy (Figure 13.5).

Caricature is also used by Picasso in his work *Casagemas in His Coffin* (1901). In death, a person loses their individuality, therefore Casagemas' exact physical resemblance is not important; this makes caricature an appropriate style. Picasso painted many caricatures, including those of his friends and writers, André Salmon and Guillaume Apollinaire, of the composers Igor Stravinsky (1917, 1920) and Francis Poulenc (1957), as well as of the multi-talented artist Jean Cocteau (1917).[4]

---

[4]In 1917, Cocteau introduced Picasso to the legendary Russian ballet impresario, Sergei Diaghilev. This led to the collaboration of Picasso and Diaghilev in the short *avant-garde* ballet *Parade*, with scenario by Cocteau, designs by Picasso, music by Erik Satie, and choreography by Léonide Massine. In the same year, Picasso travelled with Cocteau to Rome for the production of *Parade*, and there he befriended Stravinsky.

Figure 13.6. Picasso, *Gertrude Stein* (1905–1906).
Source: © Succession Picasso/DACS, London 2024.

Picasso's portrait of the American writer Gertrude Stein provides an early indication of Picasso's brilliance. Picasso was impressed by the strong-willed, erudite Stein. He was determined to convey profound truths about her inner character. After many sittings, Stein has claimed 80 or 90, and after long periods of introspection, Picasso created a *new type of portrait*: he retained the schematic style of a typical caricature, but the predominant feature became a mask-like, Iberian face, with an intense fixed stare[5] (the origin of the Iberian influence will be discussed in the third volume). This work justifies Picasso's maxim that "All good portraits are in some degree caricatures" (Penrose, 1958: p. 126) (Figure 13.6).

After the creation of his iconic painting *Les Demoiselles d'Avignon (The Young Ladies of Avignon)*, Picasso became preoccupied with depicting a head seen simultaneously from many different directions, a key feature of Cubism that will be analyzed in the third volume. This led to the creation of several portraits where caricature is combined with Cubism. An example

[5]This style was used by Picasso the following year in his masterpiece *Les Demoiselles d'Avignon* (1907).

Figure 13.7. Picasso, *Portrait of Ambroise Vollard* (1910).
Source: © Succession Picasso/DACS, London 2024.

is provided by the fabulous illustration of the sleep disorder *narcolepsy* in the *Portrait of Ambroise Vollard* (1910).[6] Vollard was notorious for dropping off to sleep in company, which suggests that he was suffering from narcolepsy. Picasso depicts perfectly this condition: Vollard's eyes are closed and his body assumes a sleeping posture. The portrait captures the sitter's main features, including his balding head, large nose, and trimmed beard (Figure 13.7).[7]

Another example is the portrait, *Jaume Sabartés with Ruff and Cap* (1939), depicting Sabartés, who, in 1935, became Piacasso's personal secretary (Figure 13.8).

[6]At the beginning of the 20th century, Vollard was one of the most important art dealers in Europe. He provided exposure and support to numerous artists, including Pierre-Auguste Renoir, André Derain, Paul Gauguin, Vincent van Gogh, and Picasso.
[7]Painting Stein was an onerous experience for Picasso. In addition to the psychological pressure born from his determination to create a novel type of portraiture, Stein's intensely intrusive character, apparently unsettled Picasso to the extent that he later avoided using live models. Vollard's portrait was made mostly via the use of photographs.

Figure 13.8. Picasso, *Jaume Sabartés with Ruff and Cap* (1939).
Source: © Succession Picasso/DACS, London 2024.

Regarding this portrait, Sabartés notes that "My portrait has truly all the characteristics of my physiognomy, and only the most essential ones",[8] and continues:

"A caricature is a kind of a 'minimum' portrait done with the avowed purpose of ridiculing a person, whereas a portrait is the 'maximum' expression of personality, the qualities of which the painter emphasises by means of lines, colours or both, as in this case, without however, overlooking certain features which might seem ridiculous to anyone else" (Sabartés, 1949: p. 27).

Regarding the almost grotesque appearance of this portrait, it is perhaps useful to note the following remark made by Picasso to Christian Zervos, who was the compiler of the complete catalogue of Picasso's works, consisting of 33 volumes:

"Academic training in beauty is a sham. Art is not the application of a canon of beauty, but what the instinct and the brain can conceive beyond any canon" (Zervos, 1935).

[8]Sabartés had severe myopia, curiously shaped lips, a pointed nose, a jutting chin and a bald, domed head.

The statement "what the instinct and the brain can conceive" is a reflection of the great artist's deep understanding that art is created by the interplay of unconscious (instinct) and conscious (mind) processes.

Incidentally, Christian Zervos, an art critic, connoisseur of modern art, and founding editor of the important art magazine *Cahiers d'art*, was born on the Greek island of Cephalonia. In Paris, he met his compatriot Gerasimos Sklavos, and was fascinated by the sculptures created by this remarkable innovator, who surprisingly remains largely unknown. In 1966, Zervos claimed that "After Giacometti's death, Sklavos is the greatest living sculptor". I am proud to share the same birthplace with Zervos and Sklavos.

It is worth emphasizing that Picasso was perfectly capable of conveying classical beauty. For example, the portrait of his wife, *Olga Picasso* (1923), exemplifies Picasso's ability to employ subtle combinations of colours to create a portrait characterized by beauty and gentility. The overall technique expresses the elegance and grace that were apparently entrenched within Olga's personality. In terms of beauty, this portrait is on the same level as works by great exponents of aesthetic perfection, such as Jean Auguste Dominique Ingres. Interestingly, Olga's portrait reveals her detachment from Picasso. The absence of any interaction with the observer betrays that the couple had grown apart. Indeed, a short time after this portrait was made, Picasso and Olga divorced (Figure 13.9).

Francisco de Goya, one of the greatest and most prolific portraitists of all time, painting around 160 portraits, was well aware of the importance of directing the gaze to the face. This can be seen, for example, in the portrait of a young aristocrat, *Don Valentín Bellvís de Moncada y Pizarro* (1795). The "unfinished" quality of the uniform prompts the viewer to focus on the more clearly defined face, where the eyes look directly at the viewer, enhancing the emphasis on the face. The sitter's white costume creates an antithesis with the red sash, the prominent black cuff displaying three silver strips, and an oak leaf motif in gold braid that is repeated on the sash (proclaiming the rank of Brigadier). The proto-impressionistic brushstrokes reveal Goya's ability to give form to the body below the sash (Figure 13.10).

Figure 13.9. Picasso, *Olga Picasso* (1923).
Source: © Succession Picasso/DACS, London 2024.

Figure 13.10. Goya, *Don Valentín Bellvís de Moncada y Pizarro* (around 1795).

Figure 13.11. Goya, *Time and the Old Women* (1812–1815).

Goya employed caricature to create some of the most disturbing images of conflict ever made, such as the etching *The Disasters of War* (1810). He also used caricature for allegorical purposes, as seen in *Time and the Old Women* (1812–1815). In this painting, two elderly women, whose dress and pose reveal an aristocratic pedigree, hold a plaque inscribed with *Que tal* (How are you). Behind them the dominant figure of "Father Time" completes the composition (Figure 13.11).

How do humans respond emotionally to art? This important question was extensively analyzed by Alois Riegl (1858–1905). This Viennese art historian was the first to focus on the *emotional involvement of the viewer*, claiming that the viewer interprets the artist's intentions in a unique and highly personal way. Following Riegl's position, the Viennese art critique and psychoanalyst, Ernst Kris (1900–1957), emphasized the *importance of ambiguity*, noting that every powerful work of art arises from conflicts in the artist's life, and therefore is ambiguous. The viewer adds to this ambiguity by interpreting the work in terms of their own experiences and conflicts. Sir Ernst Gombrich (1909–2001), who fled Austria in 1936 and spent the rest of his life in the UK, combined the above ideas with newer understanding of cognitive psychology and visual perception. He presented a more complete theory for the emotional involvement of the viewer, which he called *beholder's share.*

Riegl and Kris, as well as Gombrich, who is considered the leading art historian of his generation, are products of Vienna's golden age (this remarkable period will be discussed in detail in the third volume). Gombrich's father was a schoolmate of the writer Hugo von Hofmannsthal, and his mother was a distinguished pianist who was a student of Anton Bruckner. Gombrich studied at Vienna University, completing his PhD in 1933. During World War II, he worked in the BBC World Service, monitoring German radio broadcasts. When, in 1945, an upcoming announcement was prefaced by Bruckner's Seventh Symphony, Gombrich aware that this symphony was written for Richard Wagner's death, correctly speculated that Hitler was dead and immediately informed Churchill.

It has been repeatedly emphasized in this volume that consciousness strives to achieve a complete and precise understanding of reality. This implies that consciousness avoids ambiguity. On the other hand, ambiguity is a highly sought virtue of arts and letters. This antithesis is explained by the fact that, in my opinion,

> Arts and Letters are strongly influenced by emotions, which are unconscious processes and hence free from the constraints imposed by consciousness.

Ambiguity in art, regarding both the artist and the beholder, is clearly revealed in Édouard Manet's iconic painting *Le Déjeuner sur l'herbe* (1863). This painting is considered by several scholars, including Eric Kandel, as the first truly modern painting, both stylistically and thematically (Kandel, 2012). Stylistically because there is little depth or perspective, a feature also prominent in painting during the golden age of Vienna. Thematically because it expresses the interplay between fantasy and reality, as well as the ambivalent nature of sexual interactions. The nude woman exists only in the fantasies of the two men, who appear to speak with each other, completely indifferent to her presence. At the same time, the nude woman looks somewhat provocatively at the viewer (Figure 13.12).

Figure 13.12. Manet, *Le Déjeuner sur l'herbe* (1863).

Of course, sexuality is only one aspect of nudity, and certainly not the most important one. For example, nudity symbolizes the universal freedom of human beings, and also symbolizes equality. This is beautifully illustrated in El Greco's *The Opening of the Fifth Seal* (1608–1614), also known as *The Fifth Seal of the Apocalypse*, or *The Vision of Saint John*, which was painted in the last years of the Greek-born artist. In this highly influential work, nakedness is presented as an essential quality of both men and women, who joyfully remove their tunics to emphasize their equality before God.

Goya, a meticulous student of the masters that preceded him, had a clear understanding of the interplay between the nude and the clothed. This duality is evident in his paintings *The Naked Maja* (1795–1800) and *The Clothed Maja* (1800–1807).

True artistic ambiguity is based on subtlety, and in this sense perhaps the apogee of the nude-clothed duality is expressed in Goya's *The Marchioness of Santa Cruz* (1805), see Figure 13.13. Remarkably, this masterpiece was

Figure 13.13. Goya, *The Marchioness of Santa Cruz* (1805).

painted six decades before Manet's *Le Déjeuner sur l'herbe.*[9] As discussed in the outstanding book (Bray, 2015), the erotic gaze, the muslin dress worn without corsets or other undergarments that reveals the anatomical position of the navel, and her prominently displayed breasts to which attention is drawn by a seductive tendril of dark hair, emphasize the ability of the beholder to view this beautiful aristocrat unclothed. It is worth noting that, by the time that this work was made, Goya had replaced the quick flicks of the brush of his early years with thicker and heavier brush marks. Here, he uses thick white paint to create the form of the glowing white dress. He also employs reds, blacks, whites, and pinks to give the folds of drapery a sense of motion. This further emphasizes the immobility of the reclining Marchioness, adding to the seductive nature of the painting.

Interestingly, this painting also provides a striking example of Goya's response to the art of antiquity and to the Neo-classical style of his time. The Marchioness, who is crowned with a headdress of wine leaves and grapes, supports a lyre-guitar with her left hand. She is depicted as a Muse,

[9]The influence of Goya on Manet is also revealed by comparing Goya's *The Naked Maja* with Manet's famous paintings, *Olympia* (1865).

perhaps as Terpsichore, the muse of dancing and song, or Erato, the muse of lyric and love poetry.

ASSOCIATIONS IN THE ARTS AND THE INSIGHT OF GIACOMETTI

Suppose that the eyes view a face. Then, immediately begins the activation of the local neural circuits involved in the construction of the unconscious structures of different aspects of this face. However, as mentioned earlier, neuronal circuits are highly redundant. Namely, the same neural circuit participates in the formation of many other faces. Thus, concurrently with the initiation of the reconstruction of the perceived face, there begins the unconscious reconstruction of a plethora of other faces, among which, a couple may even become conscious. This is the reason that the viewing of a particular face often brings into a memory another one. Here is a personal example: upon meeting my wife, Regina, many people have remarked that she looks like the actress Penelope Cruz. Although neither Regina nor I can see such resemblance, it is interesting that no one ever mentioned any other person.

Associations are created not only in the process of perceiving a face but also in the process of thinking about one. This is beautifully illustrated by one of Picasso's late portraits, *Seated Old Man* (1970–1971) (Figure 13.14). This portrait is the result of the associations created in Picasso's brain, while he was contemplating the recreation of the *Self-Portrait* by Rembrandt (1658) (Figure 13.15). These associations were a consequence of Picasso's knowledge of van Gogh's *Self-Portrait* (1887).

The origin of these association can be easily explained: During the last part of his life, Picasso spoke often about van Gogh, describing him as "the one painter whose life was exemplary, up to and including his death" (Parmelin, 1969: pp. 37).

The overall pose of Picasso's creation, including the right hand resting on the arm of the chair, is close to Rembrandt's portrait. However, Rembrandt's large black velvet beret has been replaced by van Gogh's straw hat. Moreover, the gold of the robe has been transferred to the background. In addition,

Figure 13.14. Picasso, *Seated Old Man* (1970–1971).
Source: © Succession Picasso/DACS, London 2024.

Figure 13.15. Rembrandt, *Self-Portrait* (1658).

Picasso uses van Gogh's green instead of Rembrandt's brown. This portrait is a remarkable achievement, in particular taking into consideration that it was completed a month after the artist's 90th birthday. Picasso fought the process of ageing right to the very end of his life, using work as his main defence weapon. He never stopped painting, drawing, and making prints. Despite this defiance, this portrait reveals that Picasso was well aware of the inevitable effects of ageing. This explains the expression of despondence in the old man's face.

Consistent with the ancient Greek aphorism, "there is no good devoid of ill", together with the crucial role of associations in creativity, it should also be noted that,

the relentless bombardment of the brain by associations can hinder the artist's attempt to perceive reality.

No one expressed this potential difficulty more eloquently than Alberto Giacometti:

"You begin by seeing the person who is posing, but gradually, every possible sculpture interposes itself between the sitter and you [...]. You are no longer sure of its appearance, its size or anything at all".

Giacometti is considered one of the pre-eminent artists of the 20th century. His thin, distinctly elongated sculptures have acquired an iconic status. For example, in the *Woman of Venice VIII* (1954) (see Figure 13.16) the body is eroded by the surrounding void. Its inert material appears antithetical to human vitality. At the same time, despite the use of abstraction, the ability of this creation to both stand and walk appears intact.

Regarding the origin of these iconic sculptural creations, it is noted that Giacometti spent part of the Second World War exiled in Geneva. During that period, he made some tiny sculptures of the human figure, which were brought to Paris in mailboxes. These figures provided the basis for his tall, thin, signature sculptures. Discussing these tiny figures, Giacometti said: "Figures only seem slightly real if they are minute". This statement, as well as the statement, "I see things small", may reflect the fact that mental images occupy a small volume in the cognitive space of the brain.

Figure 13.16. Giacometti, *Woman of Venice VIII* (1954).
Source: © Succession Alberto Giacometti/DACS 2024.

Several critiques have interpreted these works as a sculptural reaction to the Second World War and its darkness. Thinness symbolizes physical and spiritual starvation, which reached its nadir in Auschwitz, the Warsaw Ghetto, and other abominable Nazi creations.

Artistic emphasis throughout the 20th century was on the engagement with *mental* and *psychological* aspects of the model. This general attitude of modernity is consistent with the phantasmagorical and perplexed works of Giacometti in his first one-man exhibition, in 1932, in Paris. At that time, he felt the need "to abandon the real". Picasso was among the first to visit this exhibition, and Jean-Paul Sartre wrote his appreciation for the originality of these works. During that period, Giacometti was considered a Surrealist.

Incidentally, Surrealists attempted to depict the mysterious world of dreams, of fantasies and of the unconscious, in general. André Breton, the leader of Surrealism, was a medical school dropout, who was strongly influenced by Pierre Janet, the heir of the distinguished psychologist and neurologist Jean-Martin Charcot. Well-known exponents of Surrealism were film-maker Luis Buñuel, and artists like Salvador Dalí, Max Ernst, and René Magritte who famously depicted *free associations*. Surrealism was influenced by mathematics, cosmology, and especially by Einstein's work. For example, in 1938 Breton and Christian Zervos published the article *Mathematics in Abstract Art*, in the art magazine *Cahiers d'art*. Also, Magritte's *Time Transfixed*, 1938, provides a metaphor for the relativity of time as understood by Einstein. A detailed analysis of this remarkable movement will be presented in the third volume.

Within two years after his first exhibition, Giacometti distanced himself from the French poet André Breton and his followers. He actually moved in the opposite direction. His new goal became "just to reproduce on canvas or clay what I see". In contrast to other great artists of the 20th century, he did not concentrate on the inner psychology of an individual, justifying this decision by the phrase, "I have enough trouble with the outside without bothering with the inside".

As mentioned in Chapter 4, neuroscience has established that,

> a very brief glance is subliminal, i.e., it does not lead to awareness. On the other hand, the longer the visual engagement, the richer the emerging associations, and therefore the greater the potential that the image will be distorted.

Giacometti's unquenched desire to see "that which at first glance is invisible", led him to repetitive scrutiny of the model. However, this gave rise to a plethora of different impressions. Therefore, the more he scrutinized the model, the fuzzier the sitter became: "The more I looked at the model, the more the screen between reality and myself thickened".

Figure 13.17. Giacometti, *The Artist's Father* (1927).
Source: © Succession Alberto Giacometti/DACS 2024.

Artists are aware that the *a priori knowledge of a subject affects their creation*. For example, Nicolas Poussin has stated that "I paint what I know, not what I see". However, Giacometti was adamant that *one must ignore this knowledge and reproduce purely what one sees*: "One must create a vision and not merely what one knows to exist". Giacometti's homage to his father, *The Artist's Father* (1927), expresses Giacometti's conviction that if one looks at the head frontally and one is able to consciously suppress the knowledge that the head is three-dimensional, then the head would appear two-dimensional (Figure 13.17).

In visual perception,

> prior knowledge plays a crucial role in achieving a fast reconstruction of the relevant object and thus minimizes the time available for making associations.

By refusing to use such prior knowledge of the subject's features, Giacometti was exposed to multiple impressions. Each time he looked at a model, a new and different visual experience was born, each giving rise to a new creative response. Depending on his own relative position, the size, shape, and appearance of the head of the sitter looked different. Each new visual impression obscured the earlier one, prompting a new attempt by Giacometti at recreating this new reality.

Unable to cope with this multitude of alternatives, Giacometti tended to destroy his work in progress, and to start again from scratch. Faced with the impossibility of completing anything, he abandoned the sittings and attempted to work from memory. However, this did not resolve his conundrum, because independently of whether he was working from observation or from memory the identity of the specific model remained elusive. He wrote:

"When I draw, or sculpt, or paint a head from memory, it always turns out to be more or less Diego's head, because Diego's head is the one I have done most often from life. And a woman's head tend to become Annette' head, for the same reason".

The description by Giacometti himself of the influence of his brother and wife on his creative process, suggests that the artists' main concern was not to represent the individual identity, but to decipher the mystery of "seeing". This goal becomes clear by his statement that,

"I do not work to create beautiful paintings or sculptures. Art is only a means of seeing. No matter what I look at, it all surprises me and eludes me, and I am not too sure what I see. It is too complex. So, we must try to copy, simply in order to begin to realise what we are seeing. It is as if reality continuously hides behind curtains, one tears a curtain […]. But there is always another […] always more".

Giacometti's true obsession was not the understanding of the visual experience in general, but

the engagement with perception in the presence of another human being.

This engagement, described by Sartre as "pure presence", makes this visual experience even more elusive: "The distance between myself and the model tends to increase continuously; the closer I get, the further away it moves".

Giacometti continued to struggle with his perception of the human presence, "I need to find out why I can't reproduce what I see", until 1946, when a revelation took place, recorded by him as "the day I begun to see". That special day, he wrote, "My view of the world was a photographic view, as I think everyone's is […], but actually, the world is dazzling and at the same time impossible to render". He understood that the reason that he had been tormented by his attempts to depict reality is that life unfolds continuously, relying on the interplay of conscious and unconscious processes. In this regard, it should be noted that,

the reflection of the reality on the brain, as well as the associations made by the brain, change endlessly. Hence, since a portrait attempts to record life, a portrait can never be fixed or definitive.

Giacometti, finally became aware that there is a huge difference between the representation produced by a camera, the so-called literal appearance, and the representation constructed by himself, based on his subjective apprehension of reality:

“Each time I look at a glass, it has an air of re-making itself, its reality becomes doubtful because its projection in my brain is doubtful, or partial. I see it as if it disappeared […] reappeared […] disappeared […] reappeared […]. In other words, it really always is in between being and not being. And this reality is what we want to copy […]”.

Perhaps for great artists the interplay between unconscious and conscious processes is much more dynamic!

After this crucial understanding, Giacometti renewed his engagement with sculpture and especially with portraiture. Actually, after 1950, portraiture became his main preoccupation. His new *modus operandi* became sculptures created from memory, and portraiture made from life. In the *Bust of Annette* (1954), he depicted his revealing understanding of *the continuously changing visual perception* via a flurry of lines and abbreviated nervous brush marks (Figure 13.18).

Another clear illustration of the importance of continuity is provided by the impact on Giacometti of the ancient Egyptian art, as well as of Tintoretto. In 1920, Giacometti accompanied his father to Venice, where he encountered, for the first time, the powerful works of Tintoretto. In addition, he saw ancient Egyptian art in the Museo Archeologico of Florence. Several researches have pointed out the direct influence of Egyptian art on Giacometti. For example, the *Head of Isabel* (1936), combines Egyptian royal elegance with divine serenity (Figure 13.19).

In this connection, it is noted that the individual faceted planes used by Giacometti in some of his works are considered elements of Cubism.

Figure 13.18. Giacometti, *Bust of Annette* (1954).
Source: © Succession Alberto Giacometti/DACS 2024.

Figure 13.19. Giacometti, *Head of Isabel* (1936).
Source: © Succession Alberto Giacometti/DACS 2024.

However, taking into consideration that the Egyptians were the precursors of aspects of Cubism, in the sense that they were the first to attempt to depict reality as seen simultaneously from different angles, there is, in my opinion, another possibility: Could it be that Giacometti's use of this technique was influenced at least as much from the Egyptians as from Picasso and Braque? Also, could it be that the remarkable elongated forms used by Giacometti in his iconic sculptures have their origin in Tintoretto, who also exerted a profound influence on El Greco?

Concluding this section, it is worth noting that Giacometti interacted with several important personalities of his time, including André Breton, Jean-Paul Sartre, Simone de Beauvoir, and Samuel Beckett. Therefore, it is not surprising that many of his ideas were affected by these intellectuals. For example, under the influence of Sartre's existentialism, summarized in the maxim, "A man chooses and makes himself". Beckett wrote: "Ever tried. Ever failed. No matter. Try again. Fail better". Echoing this phrase, in an interview in 1961, Giacometti stated:

"I tried to do the same thing that I found impossible to do thirty years ago. It seems as impossible to me now, as it was then, totally impossible, all it can lead to is failure",

but he also writes,

"If I am able to see better [...], even if the picture doesn't make sense, or it is ruined, in any event I have won. I have won a new sensation, I never had before".

References

Bray, X. 2015. *Goya: The Portraits of Goya*. National Gallery Company.

Kandel, E. 2012. *The Age of Insight: The Quest to Understand the Unconscious in Art, Mind, and Brain, from Vienna 1900 to the Present*. Random House.

Katsos, I. 2019. Chasing the light — What happened to the ancient theories?" *Isis* 110, 270–282.

Livingstone, M. 2014. *Vision and Art the Biology of Seeing*. Abrams.

Parmelin, H. 1969. *Picasso Says...* Translated by Christine Trollope. Allen and Unwin.

Penrose, Ronald. 1958. *Picasso: His Life and Work*. Victor Gollancz.

Runia, D. 1986 (repr. 2016). *Philo of Alexandria and the Timaeus of Plato*. Brill.

Sabartes, J. 1949. *Picasso: An Intimate Portrait*. Translated by Angel Flores. W. H. Allen.

Smith, M. 2015. *From Sight to Light: The Passage from Ancient to Modern Optics*. Chicago University Press.

Zervos, C. 1935. *Conversations Avec Picasso*.

IV

FUNCTIONAL IMAGING, BRAIN STIMULATION, AND MEDICINE

In 1971, the first scan obtained via computed tomography (CT), performed in Wimbledon, England, via a machine constructed by EMI, revealed a tumour in the frontal lobe of a 41-year-old woman. Eight years later, the Nobel Prize in Medicine was awarded to two pioneering researches who had been instrumental for developing this machine, Godfrey Hounsfield and Allan Cormack. Perhaps, the earliest and most important medical development that allowed physicians to localize lesions within the body was the invention, in 1816, of the stethoscope by René Laennec. The transformative invention of the X-ray machine allowed for the direct visualization of several tissues, specifically bone. However, the brain remained invisible until the revolutionary discovery of CT. Incidentally, the invention of the CT scanner marked an additional radical change in medical diagnosis: for the first time, digital computers were indispensable for acquiring and analyzing medical data.

The development of *functional* imaging techniques provided a new, higher level development in medical imaging. Namely, the passage from the imaging of anatomical structures to imaging functional properties of various tissues. Functional imaging has had a major impact on medicine. In particular, psychiatry has been one of the major beneficiaries of the clinical use of these techniques. Indeed, functional imaging had a significant effect in the emergence of the new understanding that psychiatric disorders including schizophrenia, major depression, and bipolar disorder, are diseases of the brain, as opposed to behavioural disorders, as erroneously postulated by Freud and others.

Chapter 14 introduces the important functional imaging techniques of positron emission tomography (PET) and single photon emission computed tomography (SPECT). Details are presented on the use of PET for the diagnosis of Alzheimer's disease and other dementias. Chapter 15 introduces the imaging technique of functional magnetic resonance (fMRI). In addition, the work of Adrian Owen is reviewed, where fMRI was miraculously used for communicating with patients who were thought to be in a vegetative state.

Chapter 16, first, summarizes the work of the surgeon Wilder Penfield, who used brain stimulation to locate the specific parts of the brain involved in the processing of the sensation of touch arising from different parts of the body. Then, transcranial electric and magnetic stimulations are reviewed, and the use of deep brain stimulation for the treatment of Parkinson's disease and major depression is discussed in detail.

Famous individuals who suffered from depression include Newton, Beethoven, Goethe, Schumann, Huygens, and van Gogh. The latter's struggle with this disorder is evident in his painting, *Portrait of Dr. Paul Gachet* (1890). It depicts the homeopathic physician and artist, Paul Gachet, who took care of van Gogh during the final months of his life, following van Gogh's brief spell in an asylum at Saint-Rémy-de-Provence (Figure PIV.1).

Incidentally, other well-known depictions of melancholy include Pablo Picasso's *Melancholy Woman* (1902), Edvard Munch's *Melancholy* (1894), and Artemisia Gentileschi's *Mary Magdalene as Melancholy* (1625–1626). A less widely known work is Paul Gauguin's *Melancholic* (1891). Incidentally, it appears that Gauguin's fabulous portraits have been underrated. Astonishingly, the first exhibition concentrating on his work as a portraitist took place only in 2019 (Homburg and Riopelle, 2019).

As discussed in Parts I–III, the function of the brain depends crucially on the dynamic interaction between *local* and *global* neural processes.

Figure PIV.1. Vincent van Gogh, *Portrait of Dr. Paul Gachet* (1890).

This interaction takes various forms. In particular, the employment of *local processes* drives the tendency of the brain to scrutinize many *different local (particular)* aspects of a subject. This necessitates the employment of a *variety of disciplines*, giving rise to *interdisciplinarity*. On the other hand, the use of *global mechanisms* motivates attempts to achieve a *global view*, which leads to *unification*. Chapter 17 shows that medicine provides an example *par excellence* of the process of interdisciplinarity (unification in medicine will be discussed in the second volume). The interdisciplinary nature of medicine is illustrated by discussing the impact on medicine of mathematics, computer science, physics, chemistry, bacteriology, pharmacology, and molecular biology. Additional important applications of physics in medicine and molecular biology, as well as important developments in epigenetics, immunology and the impact of molecular biology, immunology, and epigenetics on the new understanding and treatment of cancer will be discussed in the second and fourth volumes.

Reference

Homburg, C. & Riopelle, C. 2019. *Gauguin Portraits, National Gallery of Canada, Ottawa and the National Gallery, London*. Yale University Press.

THE FUNCTIONAL IMAGING TECHNIQUES OF PET AND SPECT AND THE DIAGNOSIS OF DEMENTIA

The functional imaging techniques of Positron Emission Tomography (PET) and Single Photon Emission Computed Tomography (SPECT) offer the unexpected opportunity to observe the activation of the brain as it occurs in real-time. The predecessor of PET and SPECT is Computed Tomography (CT), which is briefly reviewed in the following section.

COMPUTED TOMOGRAPHY

The advent of CT made possible the construction of direct images of brain tissue. Furthermore, the subsequent development of Magnetic Resonance Imaging (MRI) facilitated discrimination between grey and white matter. These achievements have had a tremendous impact on neurology and neurosurgery. Although the first applications of CT and MRI were in these two areas, later there were applications in almost all of medicine. Indeed, it is impossible to think of medicine today without CT and MRI.

New applications continue to emerge. For example, *coronary computed tomography angiography* (CCTA) is the imaging of the main arteries supplying blood to the heart, which are called coronary arteries. This new technique provides a non-invasive alternative to *coronary angiography*, which uses a catheter and a contrast dye to delineate the anatomy of the coronary arteries. Although this technique remains the standard way for determining if there is significant blockage in these arteries, it is invasive and is associated with a small risk of complications. It was shown in 2018 that, if CT angiography is added to standard medical therapy, then, after 5 years of follow-up there is a 41% reduction in death from coronary artery disease or non-fatal heart attack (compared to the standard medical therapy alone) (Hoffmann & Udelson, 2018). More precisely, there was a significant reduction of non-fatal *myocardial infarctions* which is the technical terminology for the death of part of the *myocardium*, which is the muscle forming the wall of the heart. Presumably, this significant benefit was the result of changing the medical management of these patients following the information obtained from CCTA.

It is important to note that approximately 60% of patients referred for angiography have no haemodynamically significant coronary artery stenoses. Furthermore, among patients with abnormal results in *functional* testing, such as changes in electrocardiography during exercise, SPECT studies, and the analysis of fractional flow reserve, only about half of the patients had evidence of stenosis on angiography. For these reasons, alternative approaches have been sought to identify obstructive lesions of *functional* importance. In the last 20 years it has been argued that, at least in patients with stable chest pain and an intermediate pre-test probability of obstruction, CCTA should be preferable to angiography. Indeed, in a recent double-blind study of such patients that were followed for 3.5 years and who had either angiography or CCTA, it was found that there were no differences in cardiovascular death or cardiac complications (Loscalzo, 2022).

Almost all medical imaging techniques are based on the solution of certain mathematical problems, known as *inverse problems*. In particular, the mathematical formulation of CT involves determining the *X-ray attenuation coefficient* of the tissue. This coefficient, which characterizes the amount of energy absorbed when X-rays transverse the tissue, reflects the density of the given tissue. The CT apparatus has the capacity to measure this absorbed energy. This provides the data needed for the solution of the associated inverse problem. The mathematical formalism needed for this solution was developed in 1917 by the Austrian mathematician, Johann Radon.

Both CT and MRI provide high spatial resolution information about the *anatomy* of the brain and of other tissues, but they cannot provide any information regarding *functional properties* of these tissues. Such information can be obtained via functional imaging techniques. These techniques have many clinical applications, and in addition have had a transformative impact on the understanding of mental functions. Indeed, the triptych of *neuroscience* (the science of the brain), *cognitive psychology* (the science of the mind), and *functional imaging techniques*, provide the basis for the empirical investigation of various cognitive processes. Functional imaging techniques play a crucial role in the diagnosis and management of a plethora of diseases in many areas of medicine, including oncology, cardiology, neurology, and psychiatry.

POSITRON EMISSION TOMOGRAPHY

Modern functional imaging techniques evolved from the seminal studies of the American neuroscientist, Seymour Kety, and his colleagues. In 1945, these researchers developed the first effective way of measuring blood flow in the living brain. In these studies, it was established that there is a difference between the blood flow (technically, of the haemodynamic response) in awake and asleep subjects. Subsequently, in 1955, it was demonstrated that it is possible to visualize local blood flow in 28 different regions of a cat's brain. In later studies it was shown that visual stimulation increases the blood flow only in the areas of the brain involved with vision (Landau, 1955). Moreover, in 1977, the American neuroscientist Louis Sokoloff pioneered a technique for measuring *local* metabolic activity in the brain (Sokoloff, 1977). The combination of these developments led to the emergence of PET and SPECT.

These powerful imaging techniques can be used to observe activation in live brains in real-time, by utilizing the fact that the brain uses only glucose, as opposed to proteins and fats (except in conditions of food deprivation). Since the more active parts of the brain consume more glucose, the measurement of glucose consumption provides an indirect measurement of brain activity. The spatial distribution of glucose consumption can be determined using the following procedure. An individual is injected intravenously with a dose of ^{18}F-2-deoxy-2-fluoro-D-glucose (FDG), which is a deoxy glucose analogue where the normal hydroxyl group in the glucose molecule is substituted with the radioactive isotope ^{18}F. This isotope undergoes radioactive decay, emitting the antimatter version of the electron, called positron (in contrast to the negatively charged electron, a positron has an equal positive charge). The positron collides with a nearby electron. As a result, they annihilate each other, emitting two gamma rays (photons) travelling in opposite directions. These photons are detected by PET cameras.[1] In this way, the measurement of the positron-emitting ^{18}F provides a quantitative measure of the glucose consumed in the tissue. The distribution of the radioisotope, which will be denoted by g, can be measured by utilizing the *gamma-ray attenuation coefficient*. This is defined in a similar manner to the X-ray attenuation coefficient, but with the term "X-rays" replaced with the term "gamma-rays" (actually, these two coefficients are numerically related). A mathematical analysis implies that, while CT computes a certain mathematical entity of the X-ray attenuation coefficient, called *Radon transform*, PET involves a more complicated mathematical entity of g, known as the *attenuated Radon transform*.[2] However, the fact that the two gamma rays travel in opposite directions gives rise to certain cancellations. In this way, the attenuated Radon transform is reduced to the usual Radon transform. This means that PET measures the Radon transform of the gamma-rays attenuated coefficient. Therefore, the inverse mathematical problem needed in PET is identical with the inverse mathematical problem in CT.

Medical applications

As noted in Chapter 2, neurotransmitters, which bind to specific neuro-receptors, play a crucial role in the function of the brain. There are specific

[1]FDG is metabolically trapped by the cells that use glucose via an enzyme called *hexokinase*.
[2]In the attenuated Radon transform, the function g is first multiplied by a certain exponential function involving the X-ray attenuation coefficient, and then it is integrated along the line connecting the tissue to the detector.

molecules, the *labelled ligands*, which bind to a variety of important receptors, including receptors for dopamine, serotonin, opiates, and benzodiazepine (these receptors are targeted by the medication called Valium). By using such specific ligands, it is possible to determine the density of the corresponding receptors. This procedure is particularly useful for drug development. Usually, the therapeutic dose of a drug blocks 50% of the relevant receptors; after a drug that blocks these receptors is injected, the density of the free receptors is determined again. In this way, it is possible to determine the dose needed to block 50% of the receptors, as well as to determine the duration of the blockade.

Parkinson's disease is characterized by dopamine's deficiency (details are discussed in Chapter 16). By employing the isotope ^{18}F Fluro dopa, it is possible to measure dopaminergic presynaptic receptors in the basal ganglia. Similarly, by using ^{11}C Flumazenil, it is possible to measure the density of GABA receptors in the basal ganglia (which, incidentally, are severely depleted in Huntington's disease). These types of neuropharmacological studies, as well as the use of small-animal PET, have transformed PET into an essential imaging modality for preclinical research, as well as for drug development and discovery.

In addition, a plethora of rigorous clinical studies have established the importance of PET in detecting, staging, and monitoring the progress of several neoplasms (cancers). Most tumours exhibit glucose hyper-metabolism, so they can be easily localized by PET. Oncologic imaging is needed in order to provide accurate pretreatment staging, monitor the response to therapy, and to provide surveillance after curative treatment. An example relevant to neurology is provided by high-grade cerebral neoplasms, such as gliomas. Moreover, post-radiation necrosis (death) can clearly be distinguished from recurrent gliomas, or other metastatic tumours.

It is well known that, in cancer, the involvement of lymph nodes is an adverse prognostic factor, namely it increases the probability for a negative outcome. For a long time, biopsy of the nodes was the only accepted method for assessing nodal involvement. But now, both PET and SPECT can rule out cancer in enlarged reactive lymph nodes. Furthermore, utilizing the fact that glucose metabolism in cancerous tissue is higher than in normal tissue, the newly developed apparatus CT-PET, which is a device combining CT and PET, is highly effective in the early detection of cancer.

For example, among 91 patients with non-small-cell lung cancer, 38 pathologically confirmed mediastinal lymph-node metastasis were missed by CT, whereas among 98 similar patients, only 21 metastases were missed by CT-PET (Fischer & Lassen, 2009). Since the occurrence of a metastasis is a contraindication for surgery, this means that the use of CT resulted in 38 futile thoracotomies, whereas using CT-PET there were only 21. In another application, PET targeting the protein *specific-prostrate membrane antigen* detected local or metastatic tumour growth in almost all patients who were treated with castration for prostatic cancer. In contrast, there was no detectable metastasis with conventional imaging.

Using more specific positron emitters it is possible to obtain highly specific information for a variety of disease processes. For example, using F-uro-17-estradiol, it is possible to access *in vivo* the density of estrogen receptors, as well as to monitor the response of the treatment for estrogen receptor positive breast cancer. Similarly, using ^{18}F-annexin V, it is possible to follow tumour cell apoptosis (death) *in vivo*, as well as to monitor treatment response of various types of cancer.

ALZHEIMER'S DISEASE AND FRONTOTEMPORAL DEMENTIA

The discussion in the previous section suggests that PET has the advantage, in comparison to fMRI, of targeting specific tissues by employing specific isotopes. This is exemplified in the disease named after the German psychiatrist, Alois Alzheimer, who was a colleague of the distinguished psychiatrist Emil Kraepelin. The most common form of the disease occurs late in life, and it presents with defects in recent memory and word-finding difficulties. There is also a rarer form with early onset, which presents with personality changes. In fact, the first patient studied by Alzheimer was a 51-year-old woman who became irrationally jealous of her husband. She was admitted to a psychiatric clinic, and following her death 5 years after the onset of symptoms, Alzheimer performed an autopsy where he observed the following three anatomical hallmarks of the disease: there was dense material *outside* her neurons that is now called *amyloid plaques*; there was accumulation of certain tangled protein fibres *inside* her neurons, now called *neurofibrillary tangles*; the brain was atrophied.

Amyloid plaques consist of a protein called *amyloid-β peptide*, which is part of a larger protein found in the cell membrane of dendrites. Two different enzymes cut this larger protein, releasing the *amyloid-β peptide*. In Alzheimer's patients, this peptide accumulates, either due to its accelerated production or its reduced clearance. The second hallmark of Alzheimer's disease, namely the formation of neurofibrillary tangles, is due to *misfolding of the protein* τ (tau), which is found inside neurons. As a result of an unknown defect, the process of forming the three-dimensional structure of these proteins, instead of yielding a normal protein, gives rise to a pathological "misfolded protein" (misfolding is further discussed in Chapter 19).

Using PET with a specific ligand, the *Pittsburgh compound B* (PiB), it is possible to *quantify* the concentration of β-amyloid (Aβ) plaques in the brain. Furthermore, several tracers have recently been developed for τ-protein PET imaging.[3] PET studies using these tracers have had a significant impact on delineating the relevant roles of β-amyloid and τ-proteins as precursors to this devastating disease. In particular, using β-amyloid PET imaging, it has been established that, for the late-onset Alzheimer's disease, the correlation between the occurrence of β-amyloid plaques and loss of cognition is rather *weak*. Moreover, although such plaques are correlated with brain atrophy, it is possible to have atrophy without these plaques. Consistent with this new understanding, until very recently, the use of monoclonal antibodies aiming at the *destruction* of β-amyloid did not benefit Alzheimer's patients. Similarly, the use of an oral medication, called verubecestat, that *inhibits* the production of β-amyloid, did not slow progression in patients with *mild-to-moderate* Alzheimer's disease, despite causing a reduction of β-amyloid in the cerebrospinal fluid and some regression of amyloid plaques in the brain (Egan *et al.*, 2018). Furthermore, in another disappointment, a similar failure was observed

[3]These include, the Aryquino-line derivatives THK5117 and THK5351, the Pyrido-indole derivative AV-1451, and the derivative PBB3 (this is obtained from the same tracer family as the Aβ ligand Pittsburgh Compound B).

in patients with *prodromal* Alzheimer's disease, namely, patients that had some early features but not the full-blown disease (Knopman, 2019).

In contrast to these early disappointments, a study published in May 2021 provided some encouragement. In this double-blind study, 257 participants with *early* Alzheimer's disease were randomly assigned to receive every 4 weeks for approximately 1.5 years, intravenously, either an anti-amyloid monoclonal antibody, called donanemab, or placebo. There was a very slight (barely statistically significant) improvement in the Integrated Alzheimer's Disease Rating Scale; a 3-point difference in a scale ranging from 0 to 144. Unfortunately, there was *no* clinical improvement in dementia severity, cognition, and functional abilities (Mintun *et al.*, 2021). In other words, there was no discernible clinical benefit.

A similar study, using another anti-amyloid monoclonal antibody, aducanumab, also did *not* show any clinical benefit. For this reason, FDA's controversial approval of aducanumab, in June 2021, was highly surprising. Actually, members of the evaluating committee used by FDA were critical of this decision, noting that the approval was *not* based on any clinical improvement but on the reduction of β-amyloid observable on PET (Alexander *et al.*, 2021). Since the cost of this treatment is $56,000 per year, fears were immediately expressed that this approval could cost the USA federal government many billions of dollars (Sachs & Bagley, 2021).

It is worth noting that this approval was made possible due to the existence of FDA's "accelerated approval program". Originally founded for HIV drugs, this programme has been used mainly for cancer medications. It is worth noting that in addition to the controversial approval of aducanumab, other medications have also been approved despite the opposition of the external advisory committee. This includes, eteplirsen, approved in 2016, for Duchenne's muscular dystrophy (the full results of a rigorous trial regarding the possible benefits of this medication are expected late in 2024 and of the aducanumab trial late in 2029).

The first monoclonal antibody to show a small but statistically significant clinical benefit for *early* Alzheimer's disease is lecanemab. In a recent study, 898 patients received this medication and 897 received placebo. After 18 months, all patients decline, but those receiving lecanemab had a slightly less decline in measures of cognition and function. However, almost 13% of the patients receiving lecanemab developed brain edema or effusion (van Dyck *et al.*, 2023).

The impact of donanemab was further analyzed in a study published in July 2023 (Sims *et al.*, 2023). In this rigorous study, 1,736 participants with mild cognitive impairment or mild dementia were followed for 19 months. Donanemab was very effective in eliminating cerebral amyloid. Remarkably, by the end of the treatment 80% of the amyloid plaques were eliminated. However, cognition and daily function continued to decline in all participants. The treatment group showed a *delayed progression* by 4 months. In addition to the disappointment of this rather weak clinical benefit, there were some serious side effects. In particular, several treated patients developed swelling or brain bleeding, and there were three deaths attributed to the treatment.

Regarding these recent developments I agree with the judgement expressed in the editorial accompanying the most recent donanamab study, which states that, "Similar to previous trials of lecanemab and aducanumab, this donanemab trial does not provide sufficient evidence of safety or efficacy [...]" (Manly & Deters, 2023). However, I am quite hopeful that this is the beginning of a long process that *will* lead to the discovery of a new treatment with strong clinical benefits.

In this connection, I should add that I am looking forward to pending trials that use medications capable of the destruction of τ-proteins. It has been speculated that τ-proteins play a more important role in the development of late-onset Alzheimer's disease than the amyloid plaques. Indeed, τ-amyloid PET imaging has established that the deposition of τ-protein begins at the entorhinal cortex. This part of the brain (as stated in Chapter 3) plays a crucial role in transforming short-term to long-term memories, and one of the first symptoms of late-onset Alzheimer's disease is a deficit precisely in this ability. For example, Alzheimer's patients may remember stories from their childhood, as the consolidation of these memories took place when their entorhinal cortices were intact, but they may not remember what they had for breakfast, since now these patients are not able to transform short-term memories to long-term ones.

It is possible that, for rare familial cases of Alzheimer's disease, the β-protein is more important. Indeed, particular genes have been identified causing Alzheimer's disease affecting many members of related families. One of these genes encodes a protein called *presenilin*, which helps the elimination of amyloid-β peptides found in the extra-neuronal space (Sherrington *et al.*, 1996). Amyloid plaques initially appear in the prefrontal cortex, and this, perhaps, explains the initial occurrence of personality changes observed in early-onset Alzheimer's disease.

Some studies suggest that Alzheimer's disease may be primarily a disorder of synaptic failure (Selkoe, 2002). For example, in Alzheimer's patients, glutamate receptors are chronically activated, and this causes neuronal toxicity and hence ensuing neuronal damage. Also, there is a very strong correlation of synaptic loss with cognitive decline.

Several pathological processes occur in the neurons of Alzheimer's patients, which include reduced production of *neurotrophies*, which are molecules that promote the proliferation, differentiation, and survival of neurons and glia. In addition, it turns out that β-amyloid plaques are poisonous to *mitochondria*, the "energy engine" of the cell. In turn, this increases the production of free radicals, which are highly reactive atoms or molecules due to the fact that they contain *unpaired* electrons.

Incidentally, free radicals are natural by-products of various biochemical processes, and can cause damage to proteins, DNA, RNA, and cell membranes, by "stealing" their electrons. This process is called *oxidation* and causes "oxidative stress". The body produces its own antioxidants, which are "free-radical scavengers". Namely, compounds that reduce the formation of free radicals or react with them causing their neutralization. The inability of the body to counteract the damaging effects of the oxidation of free radicals damages the brain. It is worth noting that fried foods, tobacco smoke, alcohol, as well as air pollutants and pesticides, are substances that generate free radicals. On the other hand, fruits, vegetables, and other plant-based, whole foods, are sources of antioxidants. Vitamins E, C, and several other vitamins are also effective antioxidants.

In Alzheimer's disease, there is also a progressive loss of calcium regulation, which in turn affects the function of neurons. In addition, microglia, which, as stated earlier in this book, is a particular type of glia cells, release a number of substances, including *interleukins* and *tumour necrosis factor*, which cause inflammation (Querfurth & LaFeria, 2010).

The most significant risk factor identified so far for late-onset Alzheimer's disease is the presence of the gene APOE4, a variant of the gene APOE. About half of the patients with late-onset Alzheimer's disease carry this gene. APOE codes for a protein that combines with fats to form the molecule apolipoprotein E, that is used to transport cholesterol and other fats through the bloodstream. This suggests that Alzheimer's disease may be related to a defect in cholesterol metabolism. This hypothesis is consistent with the finding of observational studies that the use of statins is associated with reduced risk for Alzheimer's disease. Another important risk factor for Alzheimer's disease is type 2 diabetes.

It is important to mention that in the last five years there has been major progress in the development of a blood test for the diagnosis of Alzheimer's disease in the very early stages. In particular, in a very recent publication (Shea *et al.*, 2022) it is shown that this can be achieved by measuring a specific brain-protein, the "toxin Aβ oligomers". Researchers at the University of Washington, after developing a particular assay for measuring this protein (known by the acronym SOBA), used it in the blood plasma of 310 participants. Some of these participants showed mild cognitive impairment or were suffering from Alzheimer's disease, whereas others had no cognitive defect. By measuring this protein, they were able to identify all 53 participants with Alzheimer's, who were later confirmed to have the disease post-mortem. In addition, this protein was found in 11 participants with normal cognitive function. Remarkably, 10 of them were later diagnosed with Alzheimer's disease. Assuming that these results are confirmed by additional studies, this is a major development, since any future breakthrough in the treatment of this devastating disease would be most effective if it attacks the disease as early as possible.

Frontotemporal dementia

Alzheimer's disease should be differentiated from *frontotemporal dementia*, earlier called Pick's disease, in honour of Arnold Pick, the psychiatrist who first identified this disease in 1892 (a decade before Alzheimer's pioneering studies). As suggested by its name, this dementia results from damage to the frontal and temporal lobes. It usually occurs in people under the age of 65. It is usually straightforward to differentiate it from Alzheimer's disease because patients present with profoundly affected moral reasoning and overall strange social behaviour, which is *not* the usual presenting symptom of patients with Alzheimer's disease.

If the defect starts on the left side of the brain (for right-handed people), patients present with progressive *aphasia*, i.e., impairment in language. If it starts on the right side, patients present with psychiatric symptoms, among which are the following: disinhibition, apathy, loss of sympathy and empathy, repetitive behaviour, loss of executive functions, and hyper-orality, namely, excessive chewing, sucking, lip smacking, or food craving. Initially, it is easy to mistake such patients' illness for a psychiatric disorder. Frontotemporal dementia is also caused from misfolded proteins.[4]

It is worth noting that some patients with dementia exhibit dementia-related psychosis. The treatment of such patients is quite challenging. It appears that the medication pimavanserin may be beneficial (Tariot *et al.*, 2021). This is a serotonin-receptor *modulator*, whose action is different than the conventional antipsychotic medications (which bind mostly to the D^2 dopamine receptors).

Traumatic encephalopathy

PET can also be used to monitor *chronic traumatic encephalopathy*. This is a neurodegenerative disease associated with exposure to repetitive head impacts, such as those occurring in contact sports including American football and soccer. At early stages, this disease is characterized

[4]In frontotemporal dementia, in addition to the neurofibrillary tangles created by misfolded τ-peptides, there also occur other aggregates, called TDP-43. They are caused by mutations in the gene C90RF72 and in the gene coding for the *progranulin* protein. There is hope that drugs which elevate the concentration of normal progranulin will be useful for treating this disease (Miller, 2013).

by τ-aggregates observed in the frontal, temporal, and parietal cortices. In a recent study in a group of former National Football League players with cognitive, mood related, and behavioural symptoms, PET identified elevated τ-aggregates in the bilateral superior frontal, medial temporal, and left parietal cortices. There was no significant elevation in β-amyloid deposition (Stern *et al.*, 2019). In another study, the records of 7,676 former elite soccer players and matched controls were examined, regarding the cause of death and the use of medications for dementia. Mortality from common non-neurological diseases was lower among former soccer players than among the controls. On the other hand, mortality from neurodegenerative diseases was higher, and prescriptions of dementia-related medications were more common among the former soccer players (Stern, 2019).

SINGLE PHOTON EMISSION COMPUTED TOMOGRAPHY

SPECT is the most widely available diagnostic imaging technique that uses short-lived radiographic isotopes. The isotopes used in SPECT have a longer half-life than those used in PET. For example, the half-life of FDG, the main isotope used in PET is 110 minutes, whereas that of Technetium, which is extensively used in SPECT, is 6 hours. Thus, in contrast to PET, it is not necessary to have an elaborate infrastructure, called *cyclotron*, near the hospital to produce the relevant radioisotopes. The isotopes used in SPECT emit a single photon, and hence the cancellation occurring in the inverse problem in PET does not take place. Therefore, the basic mathematical problem for SPECT is the construction of the inverse attenuated Radon transform. This mathematical problem is rather complicated, and was solved analytically only in 2002 (Novikov, 2002), by using a new method for inverting integrals developed jointly by Israel Gelfand and the author (Fokas & Gelfand, 1994). My group in the Centre of Mathematics at the Academy of Athens has implemented this formula numerically using the so-called attenuated Spline Reconstruction Technique (Protonotarios *et al.*, 2018). This is a variant of a numerical technique developed earlier by our group for PET (Kastis *et al.*, 2014, 2015).

It was shown in our PET publications that this technique has certain advantages with respect to the standard commercial technique of filter back projection.

In neurology, SPECT can be used to obtain images of the distribution of the blood flow in the normal brain. Furthermore, by employing specific radioisotopes, it is possible to obtain useful information about several diseases. For example, the most effective early treatment for Parkinson's disease is the medication levodopa, which is a pro-drug, namely it is a precursor of dopamine which is converted to dopamine in the brain via the enzyme DOPA decarboxylase. Levodopa, in contrast to dopamine, can cross the blood-brain barrier and hence can reach the brain. Using SPECT, it has been possible to establish that, although levodopa is indeed highly effective as dopamine-replacement in Parkinson's disease, this treatment decreases the density of brain's dopamine receptors. This effect is highlighted in the movie *Awakening*.

SPECT has many additional clinical applications. In particular, it is used extensively in cardiology, oncology, endocrinology, orthopedics, and pediatrics. An update of clinical applications of SPECT-CT, which is an apparatus combining CT with SPECT, is presented in Israel *et al.* (2019), where the results of over 400 papers published in this area are reviewed. An example of the use of SPECT in oncology is the study reported in Gershenwald & Ross (2011). In this study, SPECT-CT was used for *lymphoscintigraphy*, namely for establishing a preoperative road map of lymph nodes that are at risk for metastatic melanoma. In particular, it is used for the intraoperative identification of the so-called *sentinel nodes*. The rationale for sentinel-node biopsy relies on the fact that different regions of the skin have specific patterns of lymphatic drainage to the regional nodes. Moreover, for a given region of the skin there exists a specific node, the sentinel-node, in which lymphatic vessels drain first. This node is the most likely first site of metastasis.

In cardiology, SPECT is part of the common stress-testing procedures for the evaluation of chest pain (Abrams, 2005). In the USA, more than 12 million myocardial perfusion scintigraphy imaging studies are performed annually. Myocardial perfusion studies employ either thallium or technetium as radiotracers. Two sets of images are obtained, one at peak stress and one at rest. In a normally perfused myocardium, radiotracers are distributed uniformly. Areas which are supplied by a coronary artery with a functionally significant stenosis, exhibit a stress defect, which improves upon rest. If this defect does not improve, it represents a necrotic area.

References

Abrams, J. 2005. Chronic Stable Angina. *New England Journal of Medicine* 352, 2524–2533.

Alexander, G., Knopman, D., Emerson, S., Ovbiagele, B., Kryscio, R., Perlmutter, J., & Kesselheim, A. 2021. Revisiting FDA approval of aducanumab. *New England Journal of Medicine* 385(9), 769–771.

Egan, M., Kost, J., Tariot, P., Aisen, P., Cummings, J., Vellas, B., Voss, T., Mukai, Y., Aisen, P., Cummings, L., Tariot, P., Vellas, B., van Dyck, C., Boada, M., Zhang, Y., Li, W., Furtek, C., Mahoney, E., Mozley, L., Mo, Y., Sur, C., & Michelson, D. 2018. Randomized trial of verubecestat for mild-to-moderate Alzheimer's disease. *New England Journal of Medicine* 378(18), 1691–1703.

Fischer, B. & Lassen, U. 2009. Preoperative staging of lung cancer with combined PET–CT. *New England Journal of Medicine* 361, 32–39.

Fokas, A. S. & Gelfand, I. M. 1994. Integrability of linear and nonlinear evolution equations and the associated nonlinear Fourier transforms. *Letters in Mathematical Physics* 32(3), 189–210.

Gershenwald, J. & Ross, M. 2011. Sentinel-lymph-node biopsy for cutaneous melanoma. *New England Journal of Medicine* 364, 1738–1745 .

Hoffmann, U. & Udelson, J. 2018. Imaging coronary anatomy and reducing myocardial infarction. *New England Journal of Medicine* 379, 977–978.

Israel, O., Pellet, O., Biassoni, L., De Palma, D., Estrada-Lobato, E., Gnanasegaran, G., Kuwert, T., la Fougère, C., Mariani, G., Massalha, S., Giammarile, F., & Paez, D. 2019. Two decades of SPECT/CT–the coming of age of a technology: An updated review of literature evidence. *European Journal of Nuclear Medicine and Molecular Imaging* 1–23.

Kastis, G., Gaitanis, A., Samartzis, A., & Fokas, A. 2015. The SRT reconstruction algorithm for semiquantification in PET imaging. *Medical Physics* 42(10), 5970–5982.

Kastis, G. A., Kyriakopoulou, D., Gaitanis, A., Fernández, Y., Hutton, B. F., & Fokas, A. S. 2014. Evaluation of the spline reconstruction technique. *Medical Physics* 41(4).

Knopman, D. 2019. Lowering of amyloid-beta by β-secretase inhibitors — Some informative failures. *New England Journal of Medicine* 380, 1476–1478.

Landau, W. 1955. The local circulation of the living brain; values in the unanesthetized and anesthetized cat. *Transactions of the American Neurological Association* 80, 125–129.

Loscalzo, J. 2022. Evaluating stable chest pain — An evolving approach. *New England Journal of Medicine* 386(17), 1659–1660.

Manly, J. & Deters, K. 2023. Donanemab for Alzheimer disease — Who benefits and who is harmed? (Editorial). *JAMA*.

Miller, B. 2013. *Frontotemporal Dementia.* Oxford University Press.

Mintun, M., Lo, A., Duggan Evans, C., Wessels, A., Ardayfio, P., Andersen, S., Andersen, S., Shcherbinin, S., Sparks, J., Sims, J., Brys, M., Apostolova, L., Salloway, S., & Skovronsky, D. 2021. Donanemab in early Alzheimer's disease. *New England Journal of Medicine* 384(18), 1691–1704.

Novikov, R. G. 2002. An inversion formula for the attenuated X-ray transformation. *Arkiv för Matematik* 40(1), 145–167.

Protonotarios, N. E., Fokas, A. S., Kostarelos, K., & Kastis, G. A. 2018. The attenuated spline reconstruction technique for single photon emission computed tomography. *Journal of the Royal Society Interface* 15(148), 20180509.

Querfurth, H. & LaFeria, F. 2010. Alzheimer's disease. *New England Journal of Medicine* 362, 329–344.

Sachs, R. & Bagley, N. 2021. Medicare coverage of Aducanumab — Implications for state budgets. *New England Journal of Medicine* 385(22), 2019–2021.

Selkoe, D. 2002. Alzheimer's disease is a synaptic failure. *Science* 298(5594), 789–791.

Shea, D., Colasurdo, E., Smith, A., Paschall, C., Jayadev, S., Keene, D., Galasko, D., Ko, A., Li., G., Peskind, E., & Daggett, V. 2022. SOBA: Development and testing of a soluble oligomer binding assay for detection of amyloidogenic toxic oligomers. *PNAS* 119(50), e2213157119.

Sherrington, R., Froelich, S., Sorbi, S., Campion, D., Chi, H., Rogaeva, E., Levesque, G., Rogaev, E., Lin, C., Liang, Y., Mar, l., Brice, A., Agid, Y., Percy, M., Clerget-Darpoux, F., Piacentini, S., Marcon, G., Nacmias, B.,

Amaducci, L., Frebourg, T., Lannfelt, L., Rommens, J., & St George-Hyslop, P. 1996. Alzheimer's disease associated with mutations in presenilin 2 is rare and variably penetrant. *Human Molecular Genetics* 5(7), 985–988.

Sims, J. R., Zimmer, J. A., Evans, C. D., Lu, M., Ardayfio, P., Sparks, J., & Kaul, S. 2023. Donanemab in early symptomatic Alzheimer disease: The TRAILBLAZER-ALZ 2 randomized clinical trial. *JAMA* 330(6), 512–527.

Sokoloff, L. 1977. Relation between physiological function and energy metabolism in the central nervous system. *Journal of Neurochemistry* 29(1), 13–26.

Stern, R. A. 2019. Soccer and mortality — Good news and bad news. *New England Journal of Medicine* 381(19), 1862–1863.

Stern, R., Adler, C., Chen, K., Navitsky, M., Luo, J., Dodick, D., Alosco, M., Tripodis, Y., Goradia, D., Martin, B., Mastroeni, D., Fritts, N., Jarnagin, J., Devous Sr, M., Mintun, M., Pontecorvo, M., Shenton, M., & Reiman, E. 2019. Tau positron-emission tomography in former national football League players. *New England Journal of Medicine* 380(18), 1716–1725.

Tariot, P. N., Cummings, J. L., Soto-Martin, M. E., Erten-Lyons, D., Sultzer, D. L., & Foff, E. P. 2021. Trial of pimavanserin in dementia-related psychosis. *New England Journal of Medicine* 385(4), 309–319.

van Dyck, C. H., Swanson, C. J., Aisen, P., Bateman, R. J., Chen, C., Gee, M., & Iwatsubo, T. 2023. Lecanemab in early Alzheimer's disease. *New England Journal of Medicine* 388(1), 9–21.

THE FUNCTIONAL IMAGING TECHNIQUE OF FMRI AND AN ASTONISHING MEDICAL APPLICATION

FUNCTIONAL MAGNETIC RESONANCE IMAGING

The Italian physiologist, Angelo Mosso, was apparently the first scientist to appreciate that there exists a relationship between emotional and cognitive activity on the one hand, and blood flow in the brain on the other. While Mosso was examining the pulsations on the cortex of a peasant who had a permanent pulsating soft spot in his skull as a result of an injury, the church bells rang noon. Mosso noted that this caused an increase in the rate of pulsations. A similar increase was observed when the peasant said his mid-day prayers. Then, in what can be considered as the first-ever experiment correlating cognition and brain flow, Mosso asked the peasant to multiply 8 by 12. This also caused an increase in the pulsations, leading Mosso to conclude that there is a strong correlation between cognition and changes in brain blood flow (Raichle & Shepherd, 2014). Mosso proposed a concrete method known as the "human circulation balance" to quantify this relationship. Although this technique was mentioned in William James' book published in 1890, the work of Mosso remained largely unknown until the recent discovery of the original instruments used by Mosso and related reports (Catani & Sandrone, 2015). It is not clear whether Mosso was able to measure changes in cerebral flow during cognition using his instruments. However, using an apparatus of the type described by Mosso, along with modern imaging techniques, it is indeed possible to detect changes in cerebral blood flow caused by cognition (Field & Inman, 2014).

Experiments performed in 1890 at Cambridge University by Sir Charles Sherrington and Charles Roy provided direct evidence for a link between brain function and cerebral blood flow. In 1936, Linus Pauling (Nobel Prize in Chemistry, 1954 and Nobel Peace Prize, 1962) along with Charles Coryell, discovered that the magnetic properties of *haemoglobin* depend on its oxygenation. Haemoglobin is the crucially important blood protein that carries oxygen to the various tissues, as well as carbon dioxide from the tissues to the lungs. The next decisive step was taken in 1990 by Seiji Ogawa, who recognized that the dependence of the magnetic properties of haemoglobin on its oxygen content, resulted in measurable changes in the Magnetic Resonance Imaging (MRI) signal.

Magnetic resonance imaging

The predecessor of functional MRI (fMRI) was the MRI. This type of imaging is based on the fact that, if a hydrogen atom is placed in a strong magnetic field, its nucleus can both absorb and emit radio waves. Since hydrogen atoms are abundant in water and fat, by using this property, an MRI can map the location of water and fat in the body. More precisely, a patient is positioned within an MRI scanner that creates a magnetic field of 1.5 tesla, which is more than 15,000 times stronger than that of a microwave (commercial scanners can create fields of strength ranging from 0.2 to 7 tesla). Then, this equilibrium field is perturbed, which causes the excited hydrogen atoms to emit a radio frequency signal, measured by a receiving coil. By varying the applied field using appropriate gradient coils, the emitted radio signal can provide information about the location of the emitting cells.[1] This gives rise to the following *inverse problem*:

[1]The acquired data are displayed in the spatial-frequency domain on a rectangular grid with principal axes k_x and k_y, which denote the frequencies in the x and y directions. These data encode frequency and phase information, which are directly affected by the applied magnetic field.

reconstruct the intensity of the emitted signal from the knowledge of the defining features of the applied magnetic field. It turns out that the associated basic mathematical problem is the determination of a function from the knowledge of a certain mathematical entity, called *Fourier transform*. This mathematical problem was first solved in 1807 by the French mathematician Joseph Fourier.[2]

The remarkable fact that Radon and Fourier's work in pure mathematics has had such important applications reminds us of the statement of the physicist and Nobel Laureate, Eugene Wigner: "The miracle of the appropriateness of the language of mathematics in physical sciences [...] is a wonderful gift which we never understood or deserved". This "miracle" is discussed in Chapter 18.

Following a magnetic excitation, each tissue returns to its state of equilibrium via two independent processes, called T_1 and T_2. T_1-weighted images are used for assessing the cerebral cortex and for obtaining focal morphological information, whereas T_2-weighted images are useful for revealing white matter lesions and for detecting edema and inflammation.

There exists a useful variant of MRI called *Diffusion Tensor Imaging*. This imaging technique measures the diffusion of water through fibrous tissue. So, it provides an ideal approach for assessing the *connectivity* between different areas of the brain.

Functional magnetic resonance imaging

MRI, like CT, cannot image functional properties of a given tissue. This *can* be achieved by fMRI. As noted earlier, the more active areas of the brain consume more glucose. Since the brain does not store glucose, higher blood flow is required to transport more glucose. The higher blood flow also carries more oxygen, and this provides the physiological basis of fMRI. The oxygen is carried in the form of oxygenated haemoglobin molecules in red blood cells. This form of haemoglobin does not have magnetic properties

[2]Fourier succeeded in constructing an analytical formula for the inverse Fourier transform via a limiting process of the *Fourier series*. The Fourier transform and the Fourier series are extremely useful mathematical tools employed in a variety of different areas.

(diamagnetic), as opposed to deoxygenated haemoglobin that is magnetic (paramagnetic). By utilizing this difference in the magnetic properties of haemoglobin, fMRI can provide images of those areas of the brain which are more active. For the identification of such areas, fMRI provides a better alternative to PET, because it does not expose the individual to radiation. On the other hand, it takes much longer to produce an fMRI image than producing a PET image, and also the process is very noisy.

The precise relationship between fMRI images and the underlying neural activity was investigated in Logothetis *et al.* (2001). In this landmark study, fMRI measurements were performed simultaneously with microelectrode recordings in anaesthetized monkeys. It was shown that, in typical situations, fMRI recordings directly reflect increased neural activity, which starts 2–3 seconds after the stimulus onset. Most of the haemodynamic signal changes reflect the energetically expensive activity taking place at the synapses.[3]

It is worth noting that there occur circumstances where the haemodynamic response is *not* directly correlated with neural activity. Indeed, in preparation for expected tasks, there may be additional blood flow in specific parts of the brain *without* any neural activity. For example, in a relevant experiment, monkeys were required to fixate their gaze on a tiny spot on a computer monitor. As expected, there was the usual correlation between blood flow in the primary visual cortex and neural activity. However, when this task was carried out in complete darkness, there was *no* neural activity but the haemodynamic signal was as strong as during the earlier experiments, apparently in *anticipation* of the visual task (Sirotin & Das, 2009).

The imaging technique of fMRI has had a major impact on medicine, neuroscience, and psychiatry. Since fMRI studies can assess the brain's activity without the use of exogenous radioactive tracers, they can be used

[3]More precisely, the haemodynamic response is related with a specific neural activity giving rise to *local voltage potential* signals. These signals, which reflect the neuronal input and intra-cortical processing in a given area, vary very slowly with time. Microelectrode recordings reveal that, superimposed on these signals, there also occur high-frequency oscillations caused by action potentials. The latter spiking output is *not* reflected in the fMRI recordings.

to delineate differences between children and adults regarding the location, magnitude, and volume of activation following a particular mental task (Casey, 2002).

It should be emphasized that functional imaging techniques in general and fMRI in particular are the main reasons that psychiatry is now considered part of neurology. Indeed, pathologists cannot identify any specific anatomical defects associated with schizophrenia, major depression, bipolar disease, anxiety states, and addictive disorders. This fact, together with the unfortunate impact of the Freudian assertion that mental illnesses are due to traumatic sexual experiences that supposedly often occurred in early childhood, led many psychiatrists to erroneously consider psychiatric disorders as behavioural disorders, and not disorders of the brain. The modern point of view, put forward by Philippe Pinel and Emil Kraepelin, was firmly accepted only after it was established, via functional imaging techniques, that the above disorders involve subtle, but significant, functional changes in specific neuronal circuits. Additional support for the neurological basis of mental diseases was provided by the development of animal models, and the elucidation of the genetic origin of some of these disorders. The investigation of the genetic origin of psychiatric disorders was pioneered in the early 1930s by the German-born American psychiatrist Franz Kallmann (1897–1965).

COMMUNICATING WITH PATIENTS THOUGHT TO BE IN A VEGETATIVE STATE VIA FUNCTIONAL MAGNETIC RESONANCE IMAGING

The most dramatic illustration of the importance of fMRI for studying the functional properties of the brain is provided by the remarkable work of Adrian Owen at Cambridge, concerning patients who appeared to be in a vegetative state. Typically, patients in this state breathe on their own, have sleep–wake cycles, and all the functions that depend on the autonomic system remain intact, including regulation of pulse, body temperature, and vascular tone. Their eyes are open, occasionally they move parts of their body, and at times appear to look fleetingly around the room. However, they do not respond to prompting by family members or physicians.

This state should be differentiated from the minimally conscious state, where patients respond to simple commands, such as to move a finger, and they appear to move in and out of awareness. The vegetative state should also be contrasted with coma (from the Greek word *κώμα*, meaning deep sleep), defined clinically as a prolonged loss of capacity to be roused; these patients are completely unresponsive and lie with eyes closed. Comatose patients retain some high-level reflexes. For example, such patients gag when their throat is stimulated, and also their pupils contract in response to bright light, indicating that some neural circuits located in the brain stem remain functional. In contrast to the case of coma, such brain stem reflexes are totally absent in the state of brain death, where the EEG is completely flat, and the patient is unable to initiate breathing.

Owen's seminal discovery is based on preliminary results obtained via PET. In the period of 1992–1995, Michael Petrides along with Owen used PET to investigate various mental functions at the Montreal Neurological Institute. This included elucidating processes in the frontal cortex where working memory is employed to make specific "memory decisions". For example, remembering in which parking space the car is parked on a given day, avoiding confusing this space with the parking slot used in earlier days. Upon his return, in 1996, to the UK, Owen decided to use his expertise in PET imaging to study vegetative patients at the Wolfson Brain Imaging Centre at Cambridge Addenbrooke's Hospital.

The first patient in this series of studies had extensive damage to her white matter, i.e., damage to mainly axonal connections. This was the result of acute disseminated encephalomyelitis (widespread inflammation of the brain and the spinal cord) caused by a viral infection. This patient was shown 10 photos of familiar people on a computer screen, each presented for 10 seconds. As a "control" the patient was shown 10 images constructed from blurring these photos, so they contained no discernible faces. PET images showed that both sets of images activated the visual cortex, implying that visual information was reaching the patient's gateway to vision. However, only the first set led to activation of the fusiform gyrus, which is the part of the brain responsible for recognizing faces. Furthermore, the relevant activity pattern was similar to that observed in normal people.

Remarkably, a few months later the patient recovered. Her recollections are shocking:

"they said I could not feel pain; they were so wrong", "they thought I wasn't me; they thought I was just a body. It was horrendous. I still had feelings. I was still a person! I was incredibly angry inside. The main thing is I had no idea where I was or why I was there. I thought I'd forgotten how to walk", "No one told me where I was; I couldn't hear anyway. I could only hear noise. No words".

The patient tried to take her own life by holding her breath, but "I could not stop my nose from breathing. My body did not seem to want to die".

Taking into consideration that the patient could not comprehend words, it was not clear if she was able to "comprehend" faces, namely, if she was *aware* of the perceived faces. However, the PET images clearly indicated that the 10 photos were perceived by the patient, at least unconsciously.

The second patient studied by Owen had damage to her third cranial nerve and the upper part of the brain stem. She appeared to be in an even worse state than the first patient. She had dilated, unreactive pupils, and she was completely unresponsive to any questions. She flinched when exposed to painful stimuli, but such responses are considered reflexive. Imaging studies were performed in 2000 at the Applied Psychology Unit of the Medical Research Council (MRC). Since the pupils of the patient were unresponsive, it was decided to perform studies involving sound. The investigators chose simple, sufficiently familiar words so that these words could immediately evoke memories related to their content. They avoided abstract words, such as the word "uncertainty", which are difficult to visualize. Every two seconds the words "sofa…candle…table…lemon" were heard, delivered in a steady voice, followed by bursts of noise. Remarkably, the activation of the patient's brain was similar to that of normal people: the part within the auditory cortex that is specifically responsible for processing speech sounds, the *planum temporale*, was activated by the sounds of the words but not the noise. These studies supported the conclusion that the patient was perceiving, at least unconsciously, the words she was hearing. However, as with the first patient, the question of awareness remained open.

This patient also began recovering after a few months and became more and more aware of her environment.

The last patient studied using PET was someone who had a massive stroke in his thalamus and in part of the brain stem. This study, performed in 2003, was based on earlier studies carried out at Wolfson. These studies established that the activation of Wernicke's area, observed via PET, is correlated with the level of "intelligibility" of the spoken sentence under consideration. In a particular study, sentences were acoustically delivered to a group of healthy volunteers, and the amount of static noise was adjusted so that some of the sentences were easily understood, some could be deciphered with some effort, and some were very difficult to understand. The more difficult the sentences were to understand, the higher the amount of activation, consistent with the fact that the brain had to work harder. Remarkably, a similar pattern of activation was observed in the Wernicke's area of the stroke patient, suggesting that he could process linguistic meaning. But was the patient aware of this meaning?

The patient in this study was also the first patient to be studied using fMRI. Taking into consideration that the frontal lobes are important for resolving ambiguities, it is not surprising that following the acoustic presentation of sentences containing ambiguous words, the fMRI images of healthy volunteers, in addition to activation in Wernicke's area, also show activation in the lower part of the frontal lobes. The same pattern of activity was observed in the third patient, suggesting that by integrating the context of ambiguous words, the patient *could understand* the meaning of complex sentences.

The frontal lobes are also important for making decisions. Indeed, in an fMRI study performed by Owen's group in 2004, it was established that, when healthy volunteers were shown images of paintings, there was activation in the temporal lobes but not in the frontal lobes. However, when the volunteers were instructed to remember the next image, there was activation in the frontal lobes, with no additional increase in the activation of the temporal lobes. Owen realized that the *decision* to "remember" a painting as opposed to simply "look at it", was evidence of consciousness. This provided the motivation for a study described below, which established beyond any doubt that certain patients thought to be in a vegetative state, were actually conscious!

It was noted in Chapter 9 that John O'Keefe was awarded the Nobel Prize in Medicine for establishing that the place cells of the hippocampus of a rat's brain construct a unconscious structure of the spatial environment. Similarly, when humans look at a familiar spatial environment, a specific area of the brain near the hippocampus, the para-hippocampal gyrus, gets activated. Furthermore, this part of the brain is also activated when humans simply *imagine* that are moving in a familiar environment. By using fMRI imaging on several healthy volunteers, Owen's group established that this activation is reliable. Similarly, they established that when volunteers were *imagining* playing tennis, there was reliable activation in their premotor cortex.

The identification of these two areas of brain activation, corresponding to two different imaginary processes, provided the basis for the following seminal imaging study: a 23-year-old female patient with bilateral frontal damage after a traumatic injury, was placed in the scanner and was asked to imagine playing tennis. Then, 30 seconds later, she was asked to relax and "empty her mind". This sequence was repeated several times. Remarkably, every time the patient was asked to imagine playing tennis, her premotor cortex was activated in a way similar to the activation observed in normal volunteers. When she was asked to relax, the activity in the premotor cortex disappeared. Moreover, when the patient was asked to imagine navigating in her house, her para-hippocampal gyrus was activated (just like in the healthy volunteers), and this activation disappeared when she was asked to relax. These studies, which were published in *Science* in 2006 (Owen *et al.*, 2006), paved the way for *communicating directly* with such patients. For example, in 2010, they instructed a patient to imagine playing tennis if his mother was alive and to imagine navigating his house if his mother was dead. Observing the clear activation of his para-hippocampal gyrus, they concluded that his mother was dead, which was indeed the case. Similarly, they could ask patients many "yes" or "no" questions, such as whether they had a sister, a brother, if they were in pain, or if they would like to watch a particular programme on television.

Owen's group introduced an additional approach to their search for establishing awareness in patients appearing to be in a vegetative state. They performed fMRI studies while patients were watching Alfred Hitchcock's movie, *Bang! You're Dead.* This approach is based on the observation that when healthy volunteers watch a movie, their brains "synchronize". For example, in response to loud sounds, the auditory cortex of *all* viewers is activated, whereas when the camera concentrates on a face, their fusiform area gets activated. More importantly, healthy volunteers can *experience* the movie. This is facilitated by utilizing a significant feature of the brain usually referred by using the unfortunate name "theory of mind". This refers to the ability to understand how others think and feel.[4] In the movie *Bang! You're Dead*, there is a boy who plays with a gun thinking it is simply a toy. The viewer *knows* that the boy thinks this is not dangerous, thus the viewer appreciates the associated danger, and this creates suspense. Of course, this presupposes that the viewer is aware of the nature of guns. Therefore, consciousness is a prerequisite for the effect of the theory of mind to be materialized. Remarkably, there were patients considered to be in the vegetative state, who, while watching this movie, had a similar pattern of activation to healthy volunteers. Furthermore, surprisingly, there were cases when the earlier test of "imagining playing tennis" or "imagining navigating in your house" did not show reliable activation, whereas the "movie test" did establish awareness.

Using these imaging studies, it has been shown that approximately one in five patients thought to be in a vegetative state has awareness. Unexpectedly, almost no patient classified as minimally conscious exhibited reliable brain activation with any of the above tests. Currently, there are ongoing efforts to find even more reliable imaging techniques to establish awareness, as well as to use EEG, which is more widely available and can be used at the bedside.

References

Casey, B. 2002. Windows into the human brain. *Science* 296(5572), 1408–1409.

Catani, M. & Sandrone, S. 2015. *Brain Renaissance. From Vesalius to Modern Neuroscience.* Oxford University Press.

[4]Many autistic people lack theory of mind.

Field, D. T. & Inman, L. A. 2014. Weighing brain activity with the balance: A contemporary replication of Angelo Mosso's historical experiment. *Brain: A Journal of Neurology* 137(2), 634–639.

Logothetis, N., Pauls, J., Augath, M., Trinath, T., & Oeltermann, A. 2001. Neurophysiological investigation of the basis of the fMRI signal. *Nature* 412(6843), 150–157.

Owen, A., Coleman, M., Boly, M., Davis, M., Laureys, S., & Pickard, J. 2006. Detecting awareness in the vegetative state. *Science* 313(5792), 1402.

Raichle, M. E. & Shepherd, G. M. 2014. *Angelo Mosso's Circulation of Blood in the Human Brain.* (Angelo, Ed.). Oxford University Press.

Sirotin, Y. & Das, A. 2009. Anticipatory haemodynamic signals in sensory cortex not predicted by local neuronal activity. *Nature* 457(7228), 475–479.

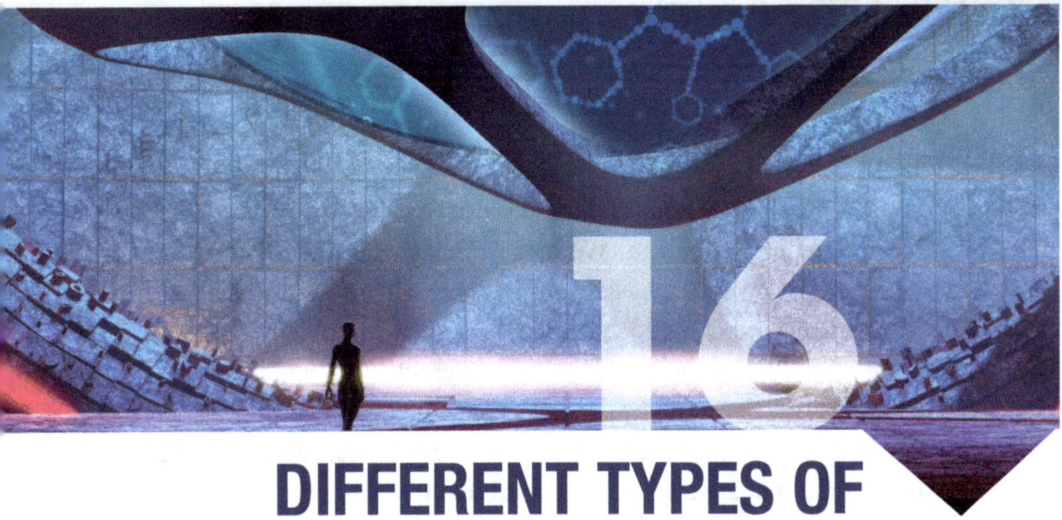

16 DIFFERENT TYPES OF BRAIN STIMULATIONS, PARKINSON'S DISEASE, MAJOR DEPRESSION, AND BIPOLAR DISORDER

A *functional modulus* is a particular brain area involved with a specific mental task. For many years, delineating such modules in the human brain was based on following patients with a specific neurological dysfunction and then localizing the corresponding anatomical defect via brain autopsy. The most dramatic discovery of this type is due to Pierre Paul Broca, who proclaimed in 1864: "we speak with the left hemisphere". Indeed, he was able to establish that damage in an area of the left posterior frontal lobe, which was later called Broca's area, yields the so-called *Broca's aphasia*. Patients with this defect lose the ability to speak fluently but retain the ability to comprehend. Broca had followed a patient who was suffering from syphilis; at a late stage, this venereal disease can infect the brain, giving rise to neurosyphilis. Broca's patient could understand and follow instructions, but when he attempted to speak, he could only produce unintelligible mumbles. After the patient died, Broca identified, via autopsy, the area of his brain that was affected. He eventually studied eight additional patients with a similar language difficulty.

In 1875, Carl Wernicke observed a complementary language defect. He studied a patient whose words flowed freely, but who lacked the ability to comprehend language. Wernicke tracked the underlying anatomical defect to an area in the back of the left hemisphere, later called Wernicke's area. Furthermore, he established that language expression and language comprehension are processed in two different areas, which are connected by a pathway, the *arcuate fasciculus*. Generalizing these findings Wernicke realized that complex mental functions are not localized in a single region of the brain but are distributed in different interconnected brain regions.

Further progress in identifying functional modules was made by studying the effect of the activation of specific areas of the brain during neurosurgery. A decisive step in this direction was taken by Wilder Penfield. After studying neuropathology at Oxford University under Sir Charles Sherrington, Penfield studied medicine at Johns Hopkins University. Then, he returned to Oxford where he met the distinguished physician, William Osler. Afterwards, Penfield took a surgical apprenticeship in Boston with the famous surgeon Harvey Cushing, who was the first to show that electrical stimulation of the sensory cortex causes the patient to feel tingling sensations. In 1934, Penfield became the co-founder and first director of the Montreal Neurological Institute at McGill University. Under the influence of Cushing, and by extending the techniques of the neurosurgeon Otfrid Foerster, which he had learned while on sabbatical in Germany (Penfield, 1958), he introduced, in collaboration with Herbert Jasper, the so-called Montreal procedure. This was executed before performing a brain operation for intractable epilepsy and while the patient was conscious on the operating table. Penfield stimulated the brain with an electrical probe and observed the response. In this way, he could delineate the epileptogenic focus, which allowed him to reduce the brain tissue affected by the operation. This technique, in addition to minimizing side effects, allowed Penfield to construct functional maps of both the sensory and motor cortices.

Regarding the sensory cortex, it was mentioned earlier that cells in the visual cortex responsible for the early stages of visual perception are topographically organized. Interestingly, the neurons responsible for the sense of touch are also organized in a topographic manner. Nerve fibres

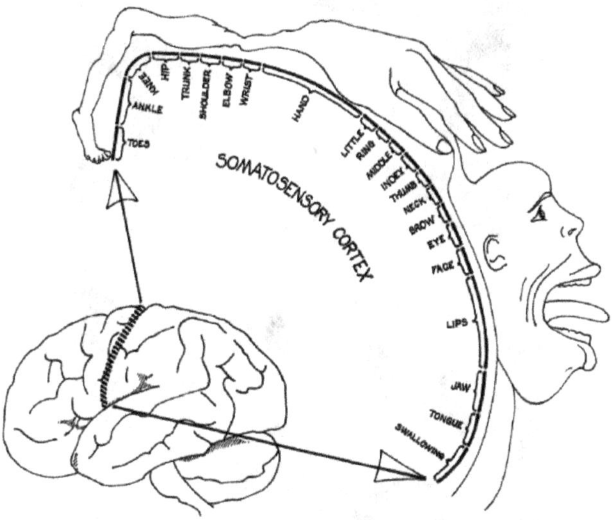

Figure 16.1. Diagram of two-dimensional homunculus.
Source: Walker *et al.* (2017). Reprinted with kind permission of Elsevier.

from the skin travel via the spinal cord, the brain stem, and the thalamus to the primary sensory cortex, which is in the post-central gyrus in the parietal cortex. Areas of the skin are projected in different parts of the primary sensory cortex in an orderly but unexpected way. Namely, the legs are represented in the top of the cerebral hemisphere, the genitals next to it, and the face is represented further down (see Figure 16.1).

As expected, the volume of the cortex dedicated to a particular area of the skin depends on the sensory importance of this area. Indeed, since the face and the hands are highly innervated, their representation is far greater than other parts of the body. The image, depicted in Figure 16.1, is called "sensory homunculus". This cartoon is adapted from the original sketched by Penfield in the 1950s. A three-dimensional version of the sensory homunculus, where each part is represented according to the volume it occupies in the sensory cortex, is shown in Figure 16.2. Penfield's somatotopic map has been confirmed and further delineated using fMRI.

Penfield succeeded in establishing a collaboration with his patients. He gently stimulated various parts of their brains and observed their reactions, while at the same time discussed their experiences with them. When Penfield stimulated the back of the cortex, i.e., the primary visual

Figure 16.2. Sculpture of a three-dimensional sensory homunculus.
Source: Wikimedia Commons.

cortex, patients saw lines, shadows, or crosses. When he stimulated the part of the cortex above the ears, i.e., the auditory cortex, the patients heard ringing, hissing, or thumbing. Stimulating the speech regions often made the patients sing against their will. Penfield began investigating the role of the temporal lobe by gently stimulating the brain of a female patient. She was immediately transported to the birth of her daughter 20 years earlier. Penfield was fascinated by this response and, for the remainder of his career, concentrated on the temporal lobe. He was so impressed by the multitude of reports generated via the stimulation of this part of the brain, which included psychiatric phenomena such as hallucinations, that he erroneously concluded that the temporal lobe provides the seat of the human consciousness.

Penfield also presented a detailed study linking the prefrontal cortex with cognition. Because of a specific brain tumour, his sister, Ruth, began suffering from headaches and seizures. Penfield cut out most of her right frontal lobe; this prolonged her life by a couple of years. After the tumour grew again and the seizures recurred, Penfield sent his sister to his friend Harvey Cushing, who operated on her. A few months later Ruth died from a stroke. The subtle changes experienced by Ruth after her surgery are reported in the case study written by Penfield and Joseph Evans. Her main complaints were that she felt "a little slow" and could not "think well enough". For example, she would feel overwhelmed if she had to organize a dinner involving several dishes. This was apparently due to a deficit in her working memory (Kean, 2014).

Penfield's work clearly established that, in the absence of any objective event,

the direct stimulation of neuronal circuits gives rise to consciousness.

A plethora of similar studies have verified this fact. For example, the stimulation of the fusiform gives rise to the objective experience of perceiving a face (Parvizi *et al.*, 2012). Similarly, stimulating the *insula*, which is a region beneath the frontal and temporal lobes, induces unpleasant feelings, including sensations of nausea, suffocation, and burning.

How do animals detect mechanical stimuli or force? This question was finally answered by the group of Ardem Patapoutian, who shared the Nobel Prize in Medicine in 2021 with David Julius, for their work related to the molecular receptors responsible for force and for temperature sensations. It turns out that a particular gene, the PIEZO gene, codes for a specific protein, which is a mechanosensitive ion channel.

In addition to direct brain stimulation, it is also possible to stimulate the brain non-invasively, as described in the following section.

BRAIN STIMULATION AND APPLICATIONS

In addition to yielding an important understanding of the function of different parts of the brain, the stimulation of the brain may also have therapeutic effects. In 50 AD, Scribonius Largus, who was the court physician of the famous Roman Emperor Claudius, used electrical torpedo fish to treat headache and gout. Millenia later, in 1774, Franz Anton Mesmer treated a woman who complained of pain and paralysis, by placing magnets on the woman's leg and stomach and then instructing her to drink a solution containing iron. Within an hour she reported "feeling a powerful force" and her symptoms disappeared. Following studies in medicine at the University of Vienna, Mesmer published his doctoral dissertation in 1766 with the Latin title *De planetarum influxu in corpus humanum* (*On the Influence of the Planets on the Human Body*). Building largely on Isaac Newton's investigation of tides and on positions of Richard Mead, an eminent English physician and Newton's friend, Mesmer claimed that certain illnesses might be caused by the movements of the sun and moon. These theories influenced the development of *hypnotism*, which at the time was also called mesmerism. Later, Mesmer established himself as a leading physician in Vienna and became a patron of the arts, supporting, among others, Wolfgang Amadeus Mozart, who included a comedic reference to Mesmer in his opera *Così fan tutte* (1790). Although a scientific commission conveyed by King Louis XVI, which included the famous chemist Antoine Lavoisier, concluded that there was no evidence of the existence of a "magnetic fluid" as claimed by Mesmer, the idea that such a fluid indeed exists continued to be considered. In analogy with the role of Newton's ether in the theory of gravity, the magnetic fluid was thought as the agent connecting the "world's souls".

In 1786, Luigi Galvani demonstrated that electricity could be conducted through the nerves of a frog's leg.[1] In 1801, Galvani's nephew, Giovanni Aldini, used direct current stimulation of the brain to improve the mood of melancholic patients (Parent, 2004).

The first established clinical application of brain electric stimulation is "electroconvulsive therapy", where electricity is used to induce a seizure with the goal of providing relief to patients suffering from certain mental disorders. This treatment, also known as "electroshock", is still used today in severe depression and in prolonged mania, in the cases that these mental disorders are resistant to pharmacological treatment. Moreover, it remains the first-line treatment for life-threatening *catatonia*, which is a severe form of depression (Rudorfer *et al.*, 2003). In 2018, the Food and Drug Administration (FDA) reclassified electroconvulsine therapy (ECT) from class III (high risk) to class III (moderate risk), stating that could

[1] A professional disagreement with Galvani, motivated Alessandro Volta to invent the "voltaic pile". This was an early electrical battery capable of producing a steady electric current.

be used in the treatment of severe major unipolar or bipolar depressive episodes or catatonia in persons 13 years of age or older, whose disorder is "treatment resistant or who require a rapid response due to the severity of their psychiatric or medical condition". ECT involves induction of brief general anaesthesia (lasting about 10 minutes), pharmacological muscle relaxation, and continuous monitoring of vital signs. An electrical charge is then delivered to the brain through scalp electrodes, which results in a generalized seizure typically lasting 20–60 seconds. Usually, patients receive 6–12 treatments spread over a period of 2–4 weeks. Refinements in technique have reduced, but not eliminated, side effects which result in approximately 2 deaths in 100,000 treatments. The most frequent complications are memory loss and acute cardiopulmonary events. ECT, despite its beneficial effect on severe psychiatric illnesses, including some types of schizophrenia, it remains underused, apparently due to stigma and lack of access to treatment (Espinoza & Kellner, 2022).

Electroshock is the precursor of transcranial magnetic stimulation (TMS), transcranial electrical stimulation (TES), and deep-brain stimulation (DBS). As discussed below, all these techniques involve neuro-stimulation.[2]

Transcranial magnetic stimulation

In TMS, brain activation is induced by the magnetic field generated via a strong electric current flowing in a coil placed atop of the head. The underlying mechanism of this non-invasive technique involves the generation of action potentials in the axons of the affected neurons. For example, TMS of the visual cortex of a subject placed in a completely dark room creates a *phosphene*, namely, the subject perceives a faint spot of light at a location that varies according to the site of the cortical stimulation. If the cortical stimulation is above the area V_5, the subject reports an impression of fleeting movement. Similarly, stimulating the area V_4 evokes sensations of seeing colour. Although the basic ideas of TMS were developed in the 1910s (Thomson, 1910), the first clinical device was not constructed until 1985 (Hallett, 2000).

[2] The stimulation of particular nerves can also have beneficial effects. For example, the stimulation of the vagus nerve has been approved for certain types of epilepsy.

TMS is useful for alleviating neuropathic pain, which is a chronic pain following an injury. Also, in 2008, the US Food and Drug Administration granted approval for the use of TMS to treat clinical depression.

An important application of TMS is the study of *functional connectivity* in both normal and pathological circumstances. In this regard, it is noted that TMS can be delivered to thc brain as a single-pulse or repetitive pulses or in the form of a specific pattern (such as a theta rhythm). The resulting activation may interfere with brain function creating a "virtual lesion". Alternatively, it may augment brain activity. By combining TMS with suitable functional imaging techniques, such as EEG, PET, or fMRI, it is then possible to isolate those neural networks that are activated in this process. This procedure can be aided by the use of a specific mathematical technique called graph theory (Sporns, *et al.*, 2004). This approach has been used for the evaluation of the brain's functional connectivity in several neurological conditions, including stroke, multiple sclerosis, amyotrophic lateral sclerosis, and various movement disorders (Groppa *et al.*, 2012).

TMS has also been used for the study of a plethora of mental functions (Hallett *et al.*, 2017). For example, by combining TMS with EEG it has been shown that neural oscillations in different parts of the brain have their own natural frequency. Also, by placing electrodes in specific parts of the brain during the process of deep brain stimulation therapy, direct connections between the basal ganglia and the cerebral cortex have been delineated. In addition, by combining TMS with PET, particular networks involved in dopaminergic neurons have been identified.

Transcranial electrical stimulation

In the 1960s, D. J. Albert established that current stimulation can change neuronal excitability and hence affect brain function. He also established that positive and negative stimulations have different effects on cortical excitability. Interest in TES was later renewed due to the success of TMS. In TES, current is applied via electrodes placed on the head. The current can be either a low direct current, called transcranial direct current stimulation, or alternating current called cranial electrotherapy stimulation. It is possible to deliver either positive stimulation, which increases neuronal excitability or negative stimulation that decreases spontaneous cell firing and thereby decreasing neuronal excitability. Regarding the basic underlying mechanism of TES, it is noted that direct current transcranial stimulation is insufficient to induce action potentials, as occurs in TMS. Instead, TES acts by modulating the spontaneous firing rates of the affected neurons. Although there are suggestions that TES is useful for treating depression and for alleviating neuropathic pain after spinal cord injury, the US Food and Drug Administration stated in 2012 that "there is no regulation for therapeutic TES". However, there are a plethora of applications of TES in elucidating the role of specific brain areas in specific cognitive tasks (Paulus, 2011; Harty *et al.*, 2014). Furthermore, it appears that TES provides a potential approach for interfering with a variety of cognitive functions.

Deep-brain stimulation

In contrast to TMS and TES, DBS is an invasive procedure. It involves implanting one or more electrodes attached to leads in specific regions of the brain. The electrodes are connected to a device which delivers electrical stimulation directly to the brain tissue. The clinician can change the DBS settings at bedside using a programming device. The main neurophysiological effect of DBS is the modulation of the electrical oscillations of neural circuits involved in the underlying neurological condition. There may also be local effects on neurochemistry, neuro-vascular structures, and neurogenesis (Okun, 2012). DBS has been used for the treatment of essential tremor, dystonia, epilepsy, obsessive-compulsive disorders, and most importantly, for the treatment of Parkinson's disease and major depression. These two prevalent disorders are briefly discussed in the next two subsections.

Essential tremor is a rather benign neurological disorder characterized by isolated bilateral upper-limb tremor, with or without tremor in other parts of the body, such as head, larynx (voice tremor), or lower limb tremor. In contrast to Parkinson's disease, there are no other neurological abnormalities. The actress Katharine Hepburn is a famous patient afflicted with this disorder.

Dystonia is also a movement disorder where abnormal muscle contractions cause twisting and repetitive movements or abnormal postures. For example, contractions of neck muscles or of muscles around the eyes cause, respectively, spasmodic torticollis (cervical dystonia) or blepharospasm (rapid blinking of the eyes).

Parkinson's disease

Parkinson's disease typically develops between the ages of 55 and 65 and occurs approximately in 1% of persons over the age of 60. It is a neurodegenerative disease caused by the cell death of dopaminergic neurons in the region of the brain called *substantia nigra*. Dopamine reaches most of the areas of the cortex via neurons originating from the *ventral tegmental area*, which is adjacent to the substantia nigra. Taking into consideration that dopamine affects many different parts of the brain, it is not surprising that Parkinson's patients present with a variety of symptoms. In particular, they present with characteristic motor symptoms, including tremor, small writing, rigidity of movement, shuffling with highly unstable gait, and expressionless face. Also, Parkinson's patients may have problems with the autonomic system that include sexual dysfunction, orthostatic hypotension, and gastrointestinal problems. Importantly, Parkinson's disease is associated with emotional and cognitive symptoms that include depression, apathy, anxiety, and cognitive difficulties.

Parkinson's disease was identified in 1817 by the British physician James Parkinson, who described six patients who presented with tremor, abnormal posture, and bradykinesia, i.e., slowness of movements. In 1912, the German-born American neurologist, Frederic Lewy, identified the occurrence of clumps of proteins inside certain neurons in the autopsy of patients with Parkinson's disease. In 1919, a Russian medical student, Konstantin Tretiakoff, observed the same clumps in neurons in substantia nigra, and called them Lewy bodies. Moreover, he noted that in these patients, this part of the brain was less dark, indicating neuronal loss. In 1958, the Swedish neuro-pharmacologist Arvid Carlsson (Nobel Prize in Medicine, 2000) discovered that there is a low concentration of dopamine in the brains of patients suffering from Parkinson's disease and speculated that this important neurotransmitter is of critical importance in this disease. Indeed, the Lewy bodies, play a key role in the death of dopaminergic neurons. In particular, a mutated version of the gene SNCA, which codes for the protein *alpha synuclein,* gives rise to misfolding, and this generates the Lewy bodies.

In 1967, the Greek-born neurologist, George Cotzias, introduced the medication levodopa, which was mentioned in Chapter 14. This treatment, as well as other medications, like amantadine, initially ameliorates the symptoms of patients with Parkinson's disease. However, after approximately 5 years of therapy, medication-related complications develop in the majority of patients. The most serious of such complications is the emergence of unpredictable deterioration of motor functions, called the "off-medication" state, and the presence of serious dyskinesias in the "on-medication" state.

Carefully screened patients with disabling on-off fluctuations are ideal candidates for DBS, which typically improves the "off" period by an average of 4–6 hours. The use of this procedure in Parkinson's disease has its origin in the pioneering work of the neurologist Mahlon DeLong. His seminal studies in the 1970s in primates provided a detailed description of the electrical activity occurring in the basal ganglia. In particular, he described the precise relationship between specific neurons in this area and the corresponding movement controlled by these neurons. Later, DeLong and collaborators (Alexander *et al.*, 1986) introduced the "segregated circuit hypothesis". This suggests that the basal ganglia, as well as associated areas of the cortex and the thalamus, can be divided into local neural networks "that are anatomically and functionally independent from each other". Moreover, these investigators claimed that many of the symptoms of several neurological and psychiatric diseases are the result of the dysfunction of some of these local circuits. This paved the way for the electrical modulation of the associated neural circuits. In 1986, a decisive step in this direction was taken by the neurosurgeon Alim-Louis Benabid. He placed a wire that could provide continuous current in the brain of an elderly patient suffering from tremor. Benabid established that low-frequency stimulation worsened the tremor, whereas faster pulses suppressed it.

DBS predominantly targets the *subthalamic nucleus* or the internal segment of *globus pallidus*, which although located in the cerebrum is considered part of the basal ganglia. The particular brain area chosen for stimulation depends on the characteristics of the individual patient. This invasive technique has ameliorated the symptoms of thousands of patients suffering from Parkinson's disease.

An alternative invasive treatment of Parkinson's disease is "ablation", namely, the cauterization of the subthalamic nucleus or of a segment of globus pallidus. This technique, which has similar benefits to DBS, also requires a craniotomy and has a risk of haemorrhage or stroke. It has the advantage of the absence of an implanted device but the disadvantage of the inability to modulate activity in the area under stimulation.

Recently, a new technique has been introduced, called "focus ultrasound". It has the major advantage of *not* requiring a craniotomy. This procedure was first approved for the treatment of essential tremor, where it produces lesions in the *ventral intermediate nucleus* of the thalamus (Elias *et al.*, 2016). Regarding the treatment of Parkinson's disease, a 2020 clinical study involving 27 patients (Martínez-Fernández *et al.*, 2020) established that a lesion caused by focus ultrasound in one hemisphere reduced the motor features of Parkinsonism in the opposite side by 50%. But, unfortunately, some patients had disturbing side effects which persisted after one year. Specifically, one patient had dysarthria, one had gait unsteadiness, and two had clumsiness; none of the patients who underwent a sham procedure had any side effects.

The results of a larger study based on ablation via focus ultrasound of the *globus pallidus internus* were published in February of 2023. It involved 94 patients who had at least mild motor fluctuations or dyskinesias (Krishna *et al.*, 2023). After 3 months, 45 of 65 patients (69%) who underwent this procedure showed a reduction in motor impairment in the off-medication state or a reduction in dyskinesias in the on-medication state. Interestingly, 7 of 22 (32%) of the patients who underwent a sham procedure had a similar improvement, confirming the importance of placebo in the treatment of Parkinson's disease. After 12 months the majority of the patients who had an initial response continued to show benefits. Adverse effects included gait disturbance, dysarthria, visual disturbance, loss of taste, and facial weakness, but they were generally mild and reversible.

Parkinson's disease, as well as Alzheimer's disease and frontotemporal dementia discussed in Chapter 14, are typical neurodegenerative diseases. Other diseases in this category include Huntington's disease, amyotrophic lateral sclerosis, and spinal muscular atrophy. There is emerging evidence that all these diseases involve aberrations in messenger RNA. This will be discussed in the second volume, as an example of the importance of unification in the search for common pathological mechanisms involved in different diseases.

Depression and bipolar disorder

The father of modern psychiatry, Emil Kraepelin, distinguished two major groups of psychotic disorders: disorders of thought and disorders of mood. He called the thought disorder, *dementia praecox*, i.e., dementia of the young, to distinguish it from the dementia of the old discovered by his colleague Alois Alzheimer. Dementia praecox is now called schizophrenia. Kraepelin called the second type of psychotic disorders manic-depressive illness. This type of illness, now called bipolar disorder, is characterized by periods of major depression and periods of mania. The majority of patients with depressive disorder are unipolar, namely they do not have manic episodes. Such patients are classified as suffering from major depression.

Although the distinction between disorders of thought and mood is highly simplistic — the ability to think is obviously affected by depression — there do exist fundamental differences between these two types of mental diseases. In particular, schizophrenia is characterized by continuous cognitive decline, whereas mood disorders are usually episodic; namely, despite the occurrence of psychotic episodes, patients, usually, do not deteriorate with time.

The concept of continuity suggests that pathological states are exaggerations of normal states. Indeed, depression can be considered as an exaggerated form of melancholia, starting with a loss of energy and ending up, in extreme cases, with the loss of the desire to live. The word melancholia means "black bile" in Greek, consistent with Hippocrates' position that depression is caused by an excess of black bile. We now know that, in contrast to the teachings of Hippocrates, diseases are not caused by the imbalance of the four "humours", blood, phlegm, yellow bile, and black bile. However, the underlying ideas of Hippocrates about the biological nature of diseases and of the importance of the imbalance of certain biochemical substances do have validity. Indeed, depression is characterized by an imbalance of various neurotransmitters, apparently depletion of serotonin and adrenalin. This results in anxiety, guilt, pain, despair, and cognitive difficulties. It was stated earlier in this volume that there exist several antidepressant medications that increase serotonin. In addition, there exists a different class of antidepressant medications, called MAO (monoamine oxidase) inhibitors, which inhibit the action of the enzyme MAO that breaks down noradrenalin and serotonin, and hence these medications also enhance the availability of these important neurotransmitters.

Imaging studies have identified several regions in the brain that are affected by depression, including the so-called "cortical area 25", which consists of the cingulate cortex and the right anterior insula (Mayberg, 2009). This area produces molecules that remove serotonin from the synaptic gap. Thus, the high activity of this area in depressed patients implies that the action of serotonin is indeed reduced. The right anterior insula is important for self-awareness, and perhaps this is related with symptoms of depersonalization experienced by depressed patients. The right anterior cingulate cortex is connected with many areas of the brain, including the prefrontal lobe. Thus, functional abnormalities in this part of the brain are related to the variety of emotional, cognitive, and behavioural symptoms observed in depressed patients. The hyperactivity observed in area 25 is accompanied by reduced activity in parts of the prefrontal cortex, consistent with the negative effect of depression on thinking (Kandel, 2018, Chapter 3). Both post-mortem and imaging studies have shown that patients suffering from long-term major depression have a decreased number of synapses in the prefrontal cortex and hippocampus, which is consistent with the flattening of emotions and impaired memory accompanying depression.

Currently, the standard treatment for patients suffering from depression is cognitive behavioural therapy with or without antidepressant medications. Enhancement of certain neurotransmitters, especially norepinephrine and serotonin, usually alleviates the symptoms of major depression. Typically, this pharmacological effect is seen after 4–6 weeks of treatment. The first generation of antidepressants, such as amitriptyline and imipramine, are thought to increase the brain activity of norepinephrine and serotonin, but they have not been uniformly effective. Second-generation medications, such as selective serotonin-reuptake inhibitors (SSRIs), have also not been effective for every patient. The lack of a deeper understanding of the causes of depression becomes evident by the following paradox: although serotonin levels increase immediately after the use of several antidepressant drugs like fluoxetine, clinical effects only appear after several weeks (interestingly, fluoxetine leads to the creation of new synapses and even new neurons in the hippocampus). Moreover, antidepressant medications can have different effects in different patients. For example, a patient may have a response to fluoxetine but not to escitalopram, despite the fact that these two medications belong to the same class of selective serotonin-reuptake inhibitors.

Somnolence, weight gain, and sexual dysfunction are common side effects of both the first and second generation of antidepressants. In particular, it has been reported in several studies using direct questioning that 40–50% of patients using SSRIs suffer from decreased libido. Patients who find such side effects intolerable may benefit from bupropion, which has a different mode of action: it inhibits the re-uptake of norepinephrine and dopamine neurotransmission without any significant direct effect on serotonin. It is a well-studied antidepressant with efficacy comparable to SSRIs and other antidepressants and, apparently, with the lowest incidence of sexual dysfunction, weight gain, and somnolence, among all commonly used antidepressants. Specifically, clinical trials showed that patients receiving bupropion did not have the weight gain commonly seen with other antidepressants, especially tricyclic antidepressants. Actually, compared to placebo, there was a weight loss. Of course, there is no medication without possible side effects; a rare such effect of bupropion is an epileptic attack, which is why an electroencephalogram is recommended before initiating treatment.

Incidentally, data indicate that bupropion may also be effective as a treatment of attention-deficit-hyperactivity disorder (ADHD) with efficacy similar to the standard treatment of methylphenidate. Also, it can help obese patients lose weight. However, the use of bupropion for these conditions is not currently approved by the FDA of the USA.

A recent study has presented a new approach to treating major depression: an oral neuro-steroid, called SAGE-217, which modulates GABA receptors, had a positive effect by the 14th day, whereas, as noted above, typical antidepressant medications usually have an effect after 6 weeks (Gunduz-Bruce *et al.*, 2019).

At least a third of patients do not respond after two or more trials of different SSRIs. The next step in treatment has generally been to augment the existing medication with either an antidepressant of a different class or an antipsychotic medication, such as olanzapine. For older patients, a recent study compared two different types of augmentation, bupropion and the atypical antipsychotic medication, aripiprazole. The percentage of patients who had remission was slightly higher in the aripiprazole group but overall the results were rather disappointing. Namely, only 29% of the patients in the augmented group had remission after treatment. Since the aripiprazole was associated with a lower frequency of falls in this group of older patients, the authors were in favour of augmenting an SSRI with aripiprazole instead of bupropion (Lenze *et al.*, 2023).

For patients resisting to SSRIs, the FDA has recently approved a derivative of ketamine, known as esketamine. Ketamine is an antagonist of the receptors of the excitatory neurotransmitter glutamate, which was approved in the 1970s as an anaesthetic. Subsequently, it gained notoriety as a drug of abuse ("special K"). Ketamine can be used as an antidepressant, with the unique advantage of rapid action (within a few hours). However, due to severe side effects, ketamine cannot be tolerated for long periods.

Another possible use of ketamine is that it provides an alternative to electro convulsion therapy (ECT). In a recent study, involving chronically ill men and women with severe depression in midlife, approximately half of the patients initially recommended for ECT were treated with ketamine infusion (Anand *et al.*, 2023). At the conclusion of the 3-week active-treatment period, 41% of the 170 patients treated with ECT and 55% of the 195 patients treated with ketamine reported at least a 50% reduction of their depressive symptoms. This is considered a moderate to excellent response to treatment. It should be noted that a longer duration of ketamine treatment increases the likelihood of the development of drug dependency and the appearance of serious side effects, including paranoia, dissociation (defined below), and other psychotic symptoms.

The use of ketamine as well as of ecstasy as recreational drugs is based on the fact that they have "psychedelic-like" properties. Namely, they induce "dissociative" experiences, where users feel disconnected from their bodies. These drugs should be distinguished from true psychedelics, like LSD and psilocybin, which induce "dreamlike" states. Incidentally, Albert Hofmann, the Swiss chemist who developed LSD in 1938 as a circulation booster, and who accidentally ingested some of the substance he was working on, in his book, *LSD: My Problem Child* (Hofmann 2013), described his experience as follows: "an uninterrupted stream of fantastic pictures, extraordinary shapes with intense, kaleidoscopic colours". He did not expect that LSD would be used as a recreational drug, since his experience was rather terrifying. However, he speculated that psychedelic drugs, combined with psychotherapy, could help patients "perceive their problems in their true significance".

Interestingly, this speculation of Hofmann is now evoked as one of the arguments in favour of treating depressed patients with psilocybin. This substance, which can be found in more than 200 species of mushroom, was isolated and then synthesized also by Hofmann. For example, a 2021 randomized trial considered two different groups: the psilocybin group was given two separate doses of 25 mg of psilocybin 3 weeks apart plus 6 weeks of daily placebo, whereas the escitalopram group was given two separate doses of 1 mg of psilocybin 3 weeks apart plus 6 weeks of daily oral escitalopram (a selective serotonin reuptake inhibitor). Interestingly, both groups had similar improvements in their antidepressant scores, but since a placebo group was not included, it is not possible to draw firm conclusions about the effectiveness of either treatment. Secondary outcomes generally favoured psilocybin over escitalopram but they did not reach a statistically significant level (Carhart-Harris *et al.*, 2021). In the most recent and most rigorous relevant clinical trial (Goodwin *et al.*, 2022), patients were treated with one dose of psilocybin of 25 mg or 10 mg or 1 mg (used as a control). The dose of 25 mg resulted in significantly lower levels of depressive symptoms after 3 weeks than the dose of 1 mg. However, the 37% incidence of response was lower than the large trial of conventional antidepressants. Because of its hallucinogen effect, patients had a special psychotherapy session of 6–8 hours duration, before administrating psilocybin. Although it is well known that psychedelics are tryptamine agonists, the particular mechanism responsible for their therapeutic effects is not known. There are ongoing efforts to create similar medications but without the hallucinogen effects.

There have been recent remarkable advances towards achieving the goal of choosing the proper treatment for a particular patient. In particular, it was shown by the University of Southern California neurologist, Helen Mayberg that patients with below-average electrical activity in their right anterior insula responded well to behavioural therapy but not to antidepressants. The situation is exactly the opposite with patients with above-average electrical activity (Mayberg, 2009). This pioneering study suggests that medical imaging is not only useful for following the effect of a given therapy but also for *choosing* the proper therapy.

For patients resistant to pharmacological treatment, TMS provides a viable option. In this treatment, high-frequency stimulation is applied on the left dorsolateral prefrontal cortex (Bersani *et al.*, 2013).

For patients resistant to the above treatments, the employment of DBS, despite having the disadvantage of being invasive, yields objective improvement: DBS in the right anterior insula in 25 patients whose depression was resistant to pharmacological treatment resulted in recovery and long-term stabilization (Kennedy *et al.*, 2011).

In his important book, *The Inflamed Mind* (Bullmore 2018), Cambridge neuroscientist, Edward Bullmore, presents compelling evidence of a direct connection between depression and inflammation. Taking into consideration the vital role of inflammation in a plethora of physiological and pathological circumstances, it would be indeed surprising if such a connection is not valid. This will be discussed in the second volume where inflammation will be presented within a unified framework.

One in four patients with major depression will experience an episode of mania, namely, an extreme form of elation and hyperactivity. Such episodes are usually precipitated by periods of high stress. Patients with manic episodes exhibit racing thoughts, decreased need for sleep, and high-risk behaviour including substance abuse, sexual promiscuity, excessive spending, and violence. Although aberrations in the hypothalamic-pituitary-adrenal axis play a major role in bipolar disorder (Carvalho *et al.*, 2020), the molecular understanding of this disorder is even less advanced than that of major depression. This is reflected in the lack of understanding of how to optimize the various existing treatment options.

Lithium remains the treatment of choice for manic patients. Since lithium enters neurons via the sodium ionic channels, its mode of action is different from the action of the usual antidepressant medications. The precise reason that it stabilizes mood swings remains unknown. Interestingly, it was first used by the Greek physician Soranus of Ephesus (1st/2nd century AD), who treated his manic patients with alkaline waters known to be high in lithium. The beneficial effect of lithium was rediscovered in 1948 by an Australian psychiatrist, John Cade, who observed that this substance made pigs lethargic. In addition to lithium, a variety of medications, including dilvaproex, carbamazepine, and lamotrigrine, can be employed as mood stabilizers. Moreover, several medications can be used to treat psychotic episodes, including quetiapine and chlorpromazine, as well as the extended release medications risperidone and zipprazidone.

Interestingly, the genetic component of bipolar disorder is the strongest among mental diseases: if one twin is bipolar, the other identical twin has a 70% chance of being a bipolar. The analogous numbers for depression and schizophrenia are 40% and 50%, respectively. In a recent study, the analysis of genetic material from nearly 10,000 bipolar patients identified 5 genetic regions that are associated with susceptibility to bipolar disorders (Collingwood, 2016).

Stress

In addition to precipitating manic episodes, stress is also one of the key factors leading to anxiety and depression. Overall, the major impact of stress on mental functions should not be underestimated. Stress is caused by a universal mechanism. Namely, in response to a perceived threat, the brain heightens its vigilance, and this is achieved with the release of cortisol and adrenalin. These powerful hormones have many different effects, including the *preparation* of the human body to respond to stressful situations. For example, they are essential for the body's "fight-or-flight" response. This physiological effect is clearly beneficial for short periods. However, long-term release of these hormones has a variety of detrimental effects, including changes in appetite, sleep, energy, and immunity. Increased levels of cortisol lead to increase in the blood sugar levels, dilating the blood arteries supplying the muscles, and importantly, suppressing the immune system (perhaps in order to save energy, which can be used for the flight-or-fight situation). This effect is related to the anti-inflammatory action of the medication cortisone, which constitutes a synthetic analogue of cortisol. Also, a high concentration of cortisol causes an increase in the neurotransmitter glutamate, which, in the long term, is poisonous to neurons.

The modern understanding of stress began in 1936 with the pioneering experiments of Hans Selye at McGill University. Selye, who was born in Vienna in 1907, established that rats exposed to stressful or harmful situations respond in a similar physiological manner independently of the precise stressor. It is now known that this is mainly due to the action of cortisol and adrenalin. The findings of Selye, presented in 33 books and 1,600 articles, have been supplemented by numerous modern studies. There is no doubt that the effect of cortisol is truly striking. As discussed by the leading investigator of stress, the Greek-born physician George Chrousos, cortisol affects the activity of 20% of our 23,000 genes (Chrousos, 2009). Excessive and persistent stress is a major contributor to morbidity and mortality (Sapolsky, 2004). The effect of stress on epigenetic mechanisms is discussed in the fourth volume.

In my opinion, there must exist robust evolutionary mechanisms ameliorating some of the deleterious effects of stress. These, so far largely unknown mechanisms, may involve the "healing" neurotransmitters enkephalin and endorphin. Perhaps, this provides an explanation for the longevity of highly successful individuals. If this is indeed the case, then the negative impact of the excessive levels of stress of such individuals is counteracted by the positive effects of their "healing" and "reward" systems.

Unfortunately, modern life is extremely stressful. I am particularly worried for young people. In addition to the high levels of stress caused by their continuous involvement with social media, they are increasingly aware of the unthinkable dangers of an environmental catastrophe, as well as of the inability of modern capitalism to provide a framework for the development of a just and humane society. In this sense, the alarmingly high rates of stress, anxiety, and depression among young people are not surprising, but of course are still extremely worrisome.

In this direction, attempts to "train the brain" should be seriously explored. For example, there are claims that "mindfulness", namely, the process of practising "non-judgmental, present-moment awareness", affects various parts of the brain. These include the hippocampus and the anterior cingulate cortex, which is critical for self-regulation, namely, suppressing inappropriate responses and directing attention and behaviour towards appropriate purposes (Congleton *et al.*, 2015). Mindfulness meditation, yoga, and other "mind-body practices", attempt to explore unconscious mechanisms for reducing stress. Several ancient cultures have incorporated mind–body practices, which provide further support for their value. Such practices are now becoming quite popular, supported by neurobiologically knowledgeable scholars, like Sam Harris. In my opinion, it is imperative to delineate the concrete benefits of such practices and to attempt to optimize their effectiveness by analyzing data from well-designed studies.

I consider the increasing gap between rich and poor, as well as the unprecedented high rates of stress, anxiety, and depression among the young, the most serious social problems facing modern developed societies. In my opinion, it is of vital importance that political leaders and the scientific community address these problems as a matter of emergency.

References

Alexander, G. E., DeLong, M. R., & Strick, P. L. 1986. Parallel organization of functionally segregated circuits linking basal ganglia and cortex. *Annual Review of Neuroscience* 9, 357–381.

Anand, A., Mathew, S. J., Sanacora, G., Murrough, J. W., Goes, F. S., Altinay, M., & Hu, B. 2023. Ketamine versus ECT for nonpsychotic treatment-resistant major depression. *New England Journal of Medicine* 388(25), 2315–2325.

Bersani, F. S., Minichino, A., Enticott, P. G., Mazzarini, L., Khan, N., Antonacci, G., Raccah, R. N. *et al.* 2013. Deep transcranial magnetic stimulation as a treatment for psychiatric disorders: A comprehensive review. *European Psychiatry* 28(1), 30–39.

Bullmore, E. 2018. *The Inflamed Mind.* Picador.

Carhart-Harris, R., Giribaldi, B., Watts, R., Baker-Jones, M., Murphy-Beiner, A., Murphy, R., Martell, J., Blemings, A., Erritzoe, D., & Nutt, D. J. 2021. Trial of psilocybin versus escitalopram for depression. *New England Journal of Medicine* 384(15), 1402–1411.

Carvalho, A. F., Firth, J., & Vieta, E. 2020. Bipolar disorder. *New England Journal of Medicine* 383(1), 58–66.

Chrousos, G. P. 2009. Stress and disorders of the stress system. *Nature Review of Endocrinology* 5, 374–381.

Collingwood, J. 2016. Bipolar disorder genes uncovered. *Psych Central.*

Congleton, C., Hölzel, B. K., & Lazar, S. W. 2015. Mindfulness can literally change your brain. *Harvard Business Review* 45(4), 1–3.

Elias, W. J., Lipsman, N., Ondo, W. G., Ghanouni, P., Kim, Y. G., Lee, W., Schwartz, M. *et al.* 2016. A randomized trial of focused ultrasound thalamotomy for essential tremor. *New England Journal of Medicine* 375(8), 730–739.

Espinoza, R. T., & Kellner, C. H. 2022. Electroconvusive therapy. *New England Journal of Medicine* 386(7), 667–672.

Goodwin, G. M., Aaronson, S. T., Alvarez, O., Arden, P. C., Baker, A., Bennett, J. C., Bird, C. *et al.* 2022. Single-dose psilocybin for a treatment-resistant episode of major depression. *New England Journal of Medicine* 387(18), 1637–1648.

Groppa, S., Oliviero, A., Eisen, A., Quartarone, A., Cohen, L. G., Mall, V., Kaelin-Lang, A. *et al.* 2012. A practical guide to diagnostic transcranial magnetic stimulation: Report of an IFCN committee. *Clinical Neurophysiology* 123(5), 858–882.

Gunduz-Bruce, H., Silber, C., Kaul, I., Rothschild, A. J., Riesenberg, R., Sankoh, A. J., Li, H. *et al.* 2019. Trial of SAGE-217 in patients with major depressive disorder. *New England Journal of Medicine* 381(10), 903–911.

Hallett, M. 2000. Transcranial magnetic stimulation and the human brain. *Nature* 406, 147–150.

Hallett, M., Di Iorio, R., Rossini, P. M., Park, J. E., Chen, R., Celnik, P., Strafella, A. P., Matsumoto, H., & Ugawa, Y. 2017. Contribution of transcranial magnetic stimulation to assessment of brain connectivity and networks. *Clinical Neurophysiology* 128(11), 2125–2139.

Harty, S., Robertson, I. H., Miniussi, C., Sheehy, O. C., Devine, C. A., McCreery, S., & O'Connell, R. G. 2014. Transcranial direct current stimulation over right dorsolateral prefrontal cortex enhances error awareness in older age. *Journal of Neuroscience* 34(10), 3646–3652.

Hofmann, A. 2013. *LSD: My Problem Child.* Oxford University Press.

Kandel, E. R. 2018. *The Disordered Mind.* Farrar, Straus & Giroux.

Kean, S. 2014. *The Tale of the Dueling Neurosurgeons.* Little, Brown and Company.

Kennedy, S. H., Giacobbe, P., Rizvi, S. J., Placenza, F. M., Nishikawa, Y., Mayberg, H. S., & Lozano, A. M. 2011 . Deep brain stimulation for treatment-resistant depression: Follow-up after 3 to 6 years. *American Journal of Psychiatry* 168(5), 502–510.

Krishna, V., Fishman, P. S., Eisenberg, H. M., Kaplitt, M., Baltuch, G., Chang, J. W., & Elias, W. J. 2023. Trial of globus pallidus focused ultrasound ablation in Parkinson's disease. *New England Journal of Medicine* 388(8), 683–693.

Lenze, E. J., Mulsant, B. H., Roose, S. P., Lavretsky, H., Reynolds III, Blumberger C. F., & Karp, J. F. 2023. Antidepressant augmentation versus switch in treatment-resistant geriatric depression. *New England Journal of Medicine* 388(12), 1067–1079.

Martínez-Fernández, R., Máñez-Miró, J. U., Rodríguez-Rojas, R., del Álamo, M., Shah, B. B., Hernández-Fernández, F., & Mata-Marín, D. 2020. Randomized trial of focused ultrasound subthalamotomy for Parkinson's disease. *New England Journal of Medicine* 383(26), 2501–2513.

Mayberg, H. S. 2009. Targeted electrode-based modulation of neural circuits for depression. *Journal of Clinical Investigation* 119(4), 717–725.

Okun, M. S. 2012. Deep-brain stimulation for Parkinson's disease. *New England Journal of Medicine* 367(16), 1529–1538.

Parent, A. 2004. Giovanni Aldini: From animal electricity to human brain stimulation. *Canadian Journal of Neurological Sciences* 31, 576–584.

Parvizi, J. *et al.* 2012. Electrical stimulation of human fusiform face-selective regions distorts face perception. *Journal of Neuroscience* 32, 14915–14920.

Paulus, W. 2011. Transcranial electrical stimulation (tES - tDCS; tRNS, tACS) methods. *Neuropsychological Rehabilitation* 21(5), 602–617.

Penfield, W. 1958. *The Excitable Cortex in Conscious Man.* Liverpool University Press.

Rudorfer, M. V., Henry, M. E., & Sackeim, H. A. 2003. Electroconvulsive therapy. In: Tasman, A., Kay, J., Lieberman, J. A., First, M. B., & Maj, M. *Psychiatry (2 Vol. Set), 1865–1901.* Wiley.

Sapolsky, R. 2004. *Why Zebras Do Not Develop Ulcer?* Holt Paperbacks.

Sporns, O., Chialvo, D. R., Kaiser, M., & Hilgetag, C. C. 2004. Organization, development and function of complex brain networks. *Trends in Cognitive Sciences* 8(9), 418–425.

Thomson, S. P. 1910. A physiological effect of an alternating magnetic field. *Proceedings of the Royal Society B* B82, 396–399.

Walker, S. C., Trotter, P. D., Woods, A., & McGlone, F. 2017. Vicarious ratings of social touch reflect the anatomical distribution & velocity tuning of C-tactile afferents: A hedonic homunculus? *Behavioural Brain Research* 320, 91–96.

THE INTERDISCIPLINARY NATURE OF MEDICINE

Medicine is a calling based on empathy and knowledge. This unique combination justifies the decision of great thinkers of the Enlightenment like Diderot, Voltaire, and Rousseau, to study medicine in order to expand their human understanding and compassion. The main focus of clinical medicine is to alleviate pain and disability, as well as to cure, or at least treat, diseases. This implies that Yuval Noah Harari's claim that "twenty-first century medicine is increasingly aiming to upgrade the healthy" (Harari, 2016: p. 405) is not entirely correct. Of course, only dedicated physicians who feel deep empathy for their patients can truly appreciate the calling of medicine. This feeling, like other transcendental experiences, cannot be communicated to outsiders.

Medicine provides a point of convergence of mathematics, many sciences, engineering, and technology. This convergence is indispensable to the discipline as it stands today, and in particular regarding its aim to become more "personalized". The goal of "precision medicine" is to approach prevention, diagnosis, and treatment in a highly personalized manner, taking into consideration each individual's genomics (the analysis of the genome), proteomics (the analysis of proteins), environment, health history, and medical records. This requires an interdisciplinary methodological framework, where physicians and clinical investigators interact closely with experts in many other fields in order to develop and apply specific approaches suitable for the treatment of each patient. Among these fields are biology, psychology, biostatistics, informatics, mathematics, physics, computer science, bioengineering, and material science.

In what follows, some striking examples illustrating the interdisciplinary nature of medicine are briefly discussed. This shows that medicine has been a great beneficiary of metarepresentations. Moreover, it illustrates yet another effect of the unlimited power of associations:

the stunning complexity of the brain's functional connectivity is reflected in the multitude of relations that exist among different disciplines.

MATHEMATICS AND COMPUTER SCIENCES

As discussed in Chapters 14 and 15 there are a variety of imaging techniques that are based on mathematics. An additional example of the importance of mathematics to medicine is provided by statistics, which impacts epidemiology, genetics, and genomics, as well as the design and interpretation of any clinical study. Statistical ideas and the mathematical discipline of probability theory emerged after the pioneering works of the French mathematicians Pierre de Fermat (1601–1665) and Blaise Pascal (1623–1662). Among the first applications of their ideas was the quantification of mortality statistics in 17th-century London at the time of the plague. Modern statistical reasoning and biostatistics were influenced, among others, by the works of Carl Friedrich Gauss (1777–1875), Thomas Bayes (1702–1761), Ronald Fisher (1890–1962), and Sir David Cox (1924–2022).

One of the earliest clinical trials was performed in 1747 and involved 12 scorbutic ship passengers. This trial compared the standard treatment of the time with a citrus-containing diet. The clear benefit of this diet resulted in the mandate of lime juice for all sailors, which led to the elimination of scurvy from the navy. Remarkably, this great success occurred almost 200 years before the discovery and production of vitamin C (ascorbic acid), which gave the final blow to scurvy. Incidentally, for the discovery of vitamin C, in 1937, Albert Szent-Györgyi and Walter Norman Haworth, respectively, were awarded the Nobel Prize for Medicine and Chemistry.

Landmark studies of quantitative observational trials as a tool for determining the causes of diseases were performed in 1950–1954 by Sir Richard Doll. These studies established smoking as a cause of lung cancer. In Dale's 1950 paper it is stated: "The risk of developing the disease [...] may be 50 times as great among those who smoke 25 or more cigarettes a day as among non-smokers" (Doll & Hill, 1950). Dale's 1954 study involving 40,000 physicians followed for 20 years, confirmed this risk, and led to the British government's warning that smoking and lung cancer rates are closely related.

Randomized clinical trials emerged in the UK in the 1950s and were adopted by the USA in the early 1960s. They led to an explosion of rigorous trials for the investigation of the treatment of heart diseases, cancer, diabetes, and a plethora of other pathological processes. Currently, every aspect of medicine is scrutinized by, the so-called, "double-blind studies". In order to clarify the meaning of such studies, I will look at a recent clinical trial involving the use of aspirin. Aspirin is a derivative of salicylate, which is a natural chemical found in plants that protects them against various diseases. Medications made from the willow plant, which is rich in salicylate, are mentioned in the *Ebers Papyrus* from ancient Egypt. Also, in around 400 BC, Hippocrates referred to the use of salicylic tea for reducing fever. The Greek physician and botanist, Dioscorides (40–90 AD), in his five-volume Greek encyclopaedia *De Materia Medica* (On Medical Material) classified 500 types of plants on the basis of their medicinal and perfumatory uses. This famous work, which included the willow plant, was updated by Leonhart Fuchs' *New Herbal*, published in 1543. John Vane (Nobel Prize in Medicine, 1982) established the main mechanism of the action of aspirin, which is the suppression of the production of *prostaglandin* and *thromboxane*. The hormone prostaglandin has a variety of actions, including facilitation of the transmission of pain information to the brain, modulation of the body's thermostat (that resides in the hypothalamus), and enhancement of inflammation. Thromboxane is produced by *platelets*, namely the cells in the blood responsible for the formation of a thrombus, i.e., a blood clot. Thromboxane increases platelet aggregation. The ability of aspirin to suppress the production of prostaglandin and thromboxane explains some of the benefits of this ancient medication, such as reduction of pain, fever, and inflammation, as well as protection against clot formation (anti-thrombotic action).

Taking into consideration that most heart attacks are due to blood clots, it is not surprising that rigorous clinical studies have shown that aspirin *reduces* the incidence of heart attacks in individuals who *already* had a heart attack or a certain type of stroke. However, the following question remained open for a long time: Should *healthy* individuals above 60 years of age be prescribed aspirin as a protective measure? The answer was finally provided in 2018 when the results of several clinical trials were announced. In one such typical double-blind study, more than 19,000 participants 70 years of age or older and free from cardiovascular disease, dementia, and any disability, were randomly assigned to receive 100 mg per day of enteric-coated aspirin or a *placebo*, i.e., a substance that has no medical effect. The essential requirement of a double-blind trial is that the patients and their physicians do *not* know if they have taken aspirin or placebo. Following these people for 5 years, it was shown that the use of aspirin conferred *no* benefits (McNeil *et al.*, 2018).

It should be noted that there is no medication without side effects. For example, aspirin, even in the protective enteric-coated form, can cause stomach irritation, which may even lead to an ulcer. Also, aspirin and other non-steroidal anti-inflammatory medications decrease prostaglandin synthesis. This shunt arachidonic acid, a predecessor of prostaglandin into the lipoxygenase pathway producing leukotrienes and other molecules that can cause intestinal inflammation and diarrhoea. More importantly, aspirin increases the diathesis for bleeding, making, for example, an emergency surgery very problematic. On the other hand, aspirin is a truly remarkable medication, in the sense that it has many additional, unexpected benefits. For example, its long-term use provides protection against colon and prostate cancers. A decision regarding the prescription of any medication involves a thoughtful risk-benefit analysis based on the specific characteristics of the particular patient.

Techniques developed in mathematics and computer sciences have had a tremendous impact on medical genomics and proteomics, including the technique of "sequence alignment". This is a way of arranging the sequences of DNA, RNA, or a protein, with the aim of identifying regions of similarity that may characterize evolutionary, structural, or functional relationships between different sequences. For example, the Smith–Waterman algorithm is the basis for many sequence-alignment programs (Smith and Waterman, 1981). An introduction to these remarkable developments in the thriving field of bioinformatics can be found in *Introduction to Computational Biology* (Waterman and Waterman, 1995).[1]

Using these mathematical techniques, it is possible to estimate the lifetime risk of many different conditions, including breast, colon, lung, prostrate, and stomach cancer, as well as melanoma; Graves' disease, lupus, psoriasis, rheumatoid arthritis, and several other autoimmune diseases; abdominal, aortic, and brain aneurisms; macular degeneration and glaucoma; heart attack and cardiac arrythmias; Crohn's disease and type 2 diabetes; Alzheimer's disease and multiple sclerosis. The existing genetic tests determining individual risk profiles are based on the fact that the great majority of DNA sequences are similar across the human population. Variations in DNA sequences take the form of Single Nucleotide Polymorphisms (SNPs), which means that in less than 0.5% of the population there is a substitution of a single nucleotide occurring at a specific position in the genome. By comparing a large number of people who have a particular medical condition with a large number of people who do not have it, specific SNPs associated with a specific disease can be identified. Moreover, the risk of developing this particular disease if one possesses these particular SNPs can be calculated. For example, single-base mutations in the *APOE* gene produce two variations of *apolipoprotein E*, a protein comprising 299 amino acids. $APOE_2$, where arginine in the position 158 is replaced with cystine, and $APOE_4$, where cysteine in the position 112 is replaced with arginine. $APOE_2$ reduces the risk of developing Alzheimer's disease, whereas $APOE_4$ increases it. The impact on medicine of Artificial Intelligence, especially of "deep learning," is discussed in the second volume.

PHYSICS

As noted in Chapters 14 and 15, several medical imaging techniques are based on the exploration of certain specific physical processes.

[1]This includes the formulation of the "Eulerian-De Bruijn sequence assembly", which provides the basis of several modern sequence developments.

Another example of the impact of physics on medicine is provided by radiotherapy, namely, the use of ionizing radiation to kill cells, utilized mostly to control the growth of cancer cells. This type of radiation can carry a sufficiently large amount of energy so that electrons can be detached from atoms or molecules. This causes DNA damage, resulting in cellular death. An important type of ionizing radiation is X-rays, which has been used for the treatment of cancer since 1896, one year after their discovery by Wilhelm Röntgen (Nobel Prize in Physics, 1901). The understanding of the effects of radiation was accelerated with the discovery of radium by Pierre and Marie Curie (they shared the Nobel Prize in Physics in 1903).[2] This will be further discussed in the second volume. A milestone was achieved in 1927 when it was shown that certain cancers of the head and neck can be *cured* via radiation treatment. Regarding other types of cancers, the modern era of radiotherapy began in 1950.

The most common form of delivery of ionizing radiation is *external beam radiotherapy*, where radiation is delivered via a linear accelerator. The development of CT, MRI, and PET have resulted in *image-guided radiation therapy*, where real-time imaging is combined with real-time adjustment of the therapeutic radiation beam.

CHEMISTRY

An example of the crucial impact of biochemistry on medicine is provided by the elucidation of the action of enzymes. In 1913, the development of quantitative physical chemistry by Antoine Lavoisier, Louis Pasteur, and other great pioneers, led Leonor Michaelis and Maud Menten to describe enzymatic reactions quantitatively via the use of mathematical equations. These equations, together with Amedeo Avogadro's basic law that 1 mol of every substance contains 6.02×10^{23} molecules, had a major impact in elucidating a variety of medical-biological mechanisms. In particular, they made it possible to achieve a quantitative understanding of the enzymatic oxidative reactions responsible for fuelling cells, as well as to quantify other

[2]Marie Curie was the first woman to win a Nobel Prize. In 1911, she was awarded a second Prize, the Nobel Prize in Chemistry. She remains the only female to have received two Nobel Prizes.

anabolic and catabolic processes. New techniques of chemical analysis allowed Hans Krebs (Nobel Prize in Medicine, 1953) and others to discover the Krebs cycle, the urea cycle, and several additional metabolic paths that are vital for life.

It is now appreciated that chemical reactions are facilitated by enzymes, whose catalytic activity is regulated by metabolites produced via a variety of different mechanisms. These processes, in addition to enzymes, are also affected by hormones, neurotransmitters, and other molecules, via a complex network of intercellular interactions.

The importance of interdisciplinarity is beautifully illustrated with the following specific example, which reveals unexpected relationships between chemistry, art, and medicine. It involves the chemist Michel Chevreul, the artist Georges Seurat, and the use of Seurat's Pointillism technique in immunology (Goldstein, 2019). Chevreul was one of the greatest chemists of the 19th century. Among his many contributions are his pioneering study of the chemistry of lipids (fats) and his discovery that body fats are composed of *triglycerides*. In 1812, while analyzing the chemical components of gallstones, he discovered and named the molecule of cholesterol. In addition, Chevreul's experiments on the chemical composition of dyes led him to construct a "colour wheel", mentioned in Chapter 13, which provided a new way to classify colours (Loske, 2019). As noted earlier this wheel contained 72 colours. Interestingly, virtually all of these colours appeared in Seurat's iconic painting *A Sunday Afternoon on the Island of La Grande Jatte* (1884), which is the first-ever painting in the style of *Pointillism* (Figure 17.1). This work, which contains about 220,000 dots of paint, depicts individuals of the Paris bourgeoise, in the mid-1880s, caught in a solemn and motionless manner (even the little girl in the orange dress, who attempts to move, is frozen in mid-air) (Burleigh, 2004).[3] Interestingly, in this work Seurat made extensive use of colour theory. Indeed, Seurat was impressed by Chevreul's "law of contrast of colour", which states that when two colours are juxtaposed to each other, each loses its own identity and the beholder perceives a new colour. This phenomenon is the result of "mixing"

[3]This vast canvas, nearly 7 by 11 feet, took the 25-year-old Seurat two years of intense work to complete it, involving 30 sketches and 3 preparatory canvases.

Figure 17.1. Georges Seurat, *A Sunday Afternoon on the Island of La Grande Jatte* (1884).

that takes place in the brain. By placing yellow and blue dots next to each other, Seurat was able to create new luminous and vibrant versions of these colours (instead of the pale green that would emerge if yellow and blue had been mixed).

Incidentally, Seurat's influenced many painters, including Bridget Riley. Her famous "Op Art" movement of the 1960s was motivated by the early black and white paintings of Seurat. This is exemplified in Riley's painting *Pause* (1964), which consists of black dots.

Seurat's pointillism motivated the development of the so-called "Seurat Plot", which provides a convenient way to represent clusters of different cell types. This is illustrated in the Figure 17.2, where each dot represents an individual cell and each cluster of dots of a given colour represents a specific type of cell of the immune system; details can be found in Butler *et al.* (2018).

BACTERIOLOGY–PHARMACOLOGY

It is impossible to separate medicine from bacteriology and pharmacology. Bacteriology was established as a science via the enormous achievements of Louis Pasteur (1822–1895), as well as Robert Koch (1843–1910) who was awarded the Nobel Prize in Medicine in 1905. Pasteur discovered

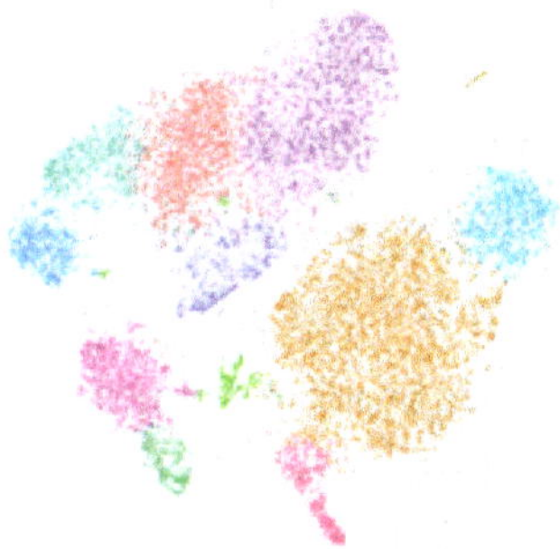

Figure 17.2. Alignment based on shared cell-type.
Source: Modified from Butler *et al.* (2018). Reproduced with kind permission of Springer Nature.

fermentation, whereby microorganisms take up glucose and metabolize it to simpler compounds. Importantly, he proposed the "germ theory", suggesting that microorganisms are often the cause of diseases.[4] Pasteur supported this theory by transmitting anthrax to a healthy sheep, via inoculating it with bacteria isolated from a diseased sheep.

Pasteur provided a powerful stimulus to the field of immunology by demonstrating that vaccination of sheep, with a vaccine constructed from the heat-attenuation of anthrax organisms, protected the sheep against death when injected with the same pathogen.

Actually, the first step towards vaccination was taken in England much earlier. In 1721, pus from a smallpox patient was rubbed in incisions made on the arms and legs of six convicts. The prisoners became ill for a couple of days but then they recovered.[5] Later, physician Edward Jenner observed that certain milkmaids never got smallpox, apparently because they were infected with cowpox from their cows. In 1776, he took pus from such a dairymaid and inoculated the 8-year-old son of the dairymaid's gardener. Although the boy was later given pus from a smallpox patient, he did

[4]Pasteur's germ theory had a significant impact on Art Nouveau, which employed the vocabulary of biomorphic forms derived from amoebas, protozoa, and cells. This is perfectly illustrated in the works of Odilon Redon. Later, Joan Miro adopted a vocabulary of cellular shapes under the influence of Hans Arp, who considered biomorphs as a universal language.

[5]These prisoners were pardoned by King George I. A few months later, the Prince and Princess of Wales had two of their daughters inoculated, after first testing the safety of the procedure by paying to have five orphan children inoculated.

not get sick. The term vaccine comes from the word *vacca*, meaning cow in Latin. Incidentally, in 1714, two Greek physicians, Iakovos Pylarinos and Emmanuel Timonis, published independently, in the famous British journal *Philosophical Transactions*, preliminary versions of the procedure used by Jenner.

The COVID-19 pandemic has brought again to the medical forefront the importance of vaccines against viruses. In this regard it is noted that, in 1885, Pasteur demonstrated that the spinal cord of rabbits that had been experimentally inoculated with rabies virus were no longer infectious after 15 days of desiccation (i.e., of drying out). Remarkably, using a series of inoculations with suspensions of desiccated rabbit spinal cords, Pasteur saved the life of a 9-year-old boy who was attacked by a rabid dog two days earlier. Following the decisive contribution of Pasteur to immunology, Max Theiler (Nobel Prize in Medicine, 1951) introduced in 1937 a systematic methodology for attenuating a virus in such a way that renders it less capable of causing disease but still capable of inducing protective immunity. This is achieved by inducing blind genetic alterations to the virus via its passage in certain non-human cells, such as mouse or chicken embryos. Using this technique, Albert Sabin, who had trained in Theiler's laboratory at the Rockefeller Foundation, New York, made a polio vaccine. Other such live viral attenuated vaccines include vaccines for preventing measles (1963), mumps (1967), rubella (1969), varicella (1905), and rotavirus (2008).[6]

The next breakthrough took place in 1980 at Stanford University, where the biochemist Richard Mulligan used the sophisticated technique of recombinant DNA (discussed in the last section of this chapter) to produce a vaccine containing a purified surface protein from the membrane of a bacterium that induces immunity against this bacterium. Vaccines containing purified proteins include those against the Hepatitis B virus (1986), human papillomavirus (2006), and influenza virus (2013).

[6]In addition to the development of live-virus attenuated vaccine, various investigators have developed killed-virus vaccines, including vaccines against polio by Jonas Salk in the mid-1950s, and against hepatitis A by Philip Provost and Maurice Hilleman in 1991.

The most remarkable development in vaccines is the recent production of the first vaccines used against COVID-19. These vaccines do not contain viral protein. Instead, either messenger RNA or DNA are introduced via viral vectors that provide instructions to cells for making appropriate proteins. Details of the effectiveness of these novel vaccines beyond COVID-19 and in particular against influenza will be presented in the fourth volume. There, a summary of mathematical models for modelling the dynamics of the COVID-19 pandemic will also be presented.

Returning to the founders of bacteriology, Pasteur also discovered the technique of *pasteurization*, namely the process of destroying microorganisms by heating a liquid below its boiling point. Moreover, he demonstrated that if this procedure is applied to milk, it protects against the transmission of several diseases. Koch, following in the steps of Pasteur, discovered the bacterium causing anthrax (*Bacillus anthracis*) and the bacteria causing several other diseases, including cholera (*Vibrio cholerae*) and tuberculosis (*Mycobacterium tuberculosis*). In addition, he proposed the famous "Koch's postulates". These are a series of steps that must be followed in order to prove that a given disease is indeed caused by a specific bacterium.

In 1910, the German bacteriologist Paul Ehrlich (1854–1915) (Nobel Prize in Medicine, 1908) discovered the substance salvarsan as a treatment for syphilis. This is also known as "606", because it was the 606th compound Ehrlich had tried. This discovery was the predecessor of the antibiotics era that began with a serendipitous observation made in 1928 by Sir Alexander Fleming, that a contaminant mould (*Penicillium notatum*) inhibited colonies of a particular bacterium (*Staphylococcus aureus*). At that time, penicillin was too unstable to be used in practise. Its clinical use began in 1942 following the studies of Sir Howard Florey and Sir Ernst Chain, which began at Oxford in 1939. This led to the purification of a sufficient amount of penicillin that could be used clinically (Fleming, Florey, and Chain shared the Nobel Prize in Medicine in 1945).

In 1943, Selman Waksman (Nobel Prize in Medicine, 1952) discovered the second clinically important antibiotic, streptomycin, which is effective against certain bacteria, called gram-negative, that are unaffected by penicillin. Streptomycin was the first antimicrobial drug used for the treatment of tuberculosis.

An important breakthrough in pharmacology took place in 1935 with the discovery of sulfonamide drugs, which have a wide range of actions, from their use as diuretics (such as hydrochlorothiazide) to their employment against a variety of conditions such as diabetes, seizures, and retroviruses. The first sulphonamide drug was named by its producer (the Bayer company) prontosil. This medication, later called sulphanilamide, inhibits an enzyme involved in the synthesis of folate, leading to the inhibition of the growth and multiplication of bacteria, but not to their death.[7]

Barry Marshall (Nobel Prize in Medicine, 2005) by infecting himself, discovered that the most common cause of gastritis (inflammation of the stomach) and of peptic ulcer, is an infection caused by the bacterium *Helicobacter pylori*. As a result of this unexpected discovery, these conditions have become curable.

The understanding of the intricate communication between cells led to the discovery of a variety of medications. Examples are mentioned throughout this volume. An important example is propranolol. This medication was developed at King's College, London, by Sir James Black (Nobel Prize in Medicine, 1988) following his understanding that the adrenergic effect on the heart is exerted via the so-called beta receptors, which are blocked by propranolol. Incidentally, propranolol, in addition to reducing heart rate and blood pressure, has many other beneficial effects, including substantially reducing the frequency of migraine episodes.

Deeper understanding of molecular mechanisms led to the discovery of a variety of drugs against many viruses. For example, several drugs are available against HIV (Human Immunodeficiency Virus), which causes AIDS (Acquired Immune Deficiency Syndrome). Another important development was the discovery of a variety of antiviral medications capable of curing chronic viral Hepatitis C, with a success rate of 90–98%.[8]

---

[7]Prontosil is actually a *pro-drug*, namely, it becomes a drug inside the body via the process of "bioavailability". It is metabolized into two pieces; a smaller, colourless, active compound called sulphanilamide, and an inactive dye portion.

[8]The first such medication, sofosbuvir, was developed in 2013. Later, additional effective medications were discovered, including simeprevir, ledipasvir, and daclatasvir.

MOLECULAR BIOLOGY

In a long paper published in 1928, the British bacteriologist Frederick Griffith detailed his work concerning *Streptococcus pneumoniae*, the bacterium that was the causative agent of secondary infections during the 1918–1919 pandemic, which may have caused more deaths than the influenza virus itself. At the time, it was known that this bacterium appeared in four different types. Griffith discovered that each of these types can occur in two different forms. The S, forming smooth colonies which is highly virulent, and the R, forming rough colonies which is mild. Remarkably, he was able to establish that the mild form of one type could be *transformed* into the virulent form of another type. This work raised the important question of identifying the component of the bacterium carrying this transforming capacity. This was answered in 1942–1944 by the pioneering molecular biologist Oswald Avery and his colleagues at the Rockefeller Institute Hospital (this institute was at that time the leading research centre in streptococcal pneumonia). *The transformation from a mild to a virulent form was caused by a change in DNA*. This result was highly surprising because, at the time, DNA was considered a "boring" molecule. Oswald privately speculated that DNA may be the substance of genes, but this hypothesis was not stated explicitly in his papers (Avery *et al*., 1944). In any case, this discovery was the first major step towards establishing that genetic information is *not* contained in proteins, as was widely assumed at the time, but in DNA. In addition to its crucial role in the discovery of genes, Griffiths' work provided the first experimental evidence of the fundamental mechanism of horizontal gene transfer, mentioned in Chapter 8.

The next important development was the discovery in 1949 by Erwin Chargaff of the four building bases of DNA, thymine (T), adenine (A), cytidine (C), and guanine (G), as well as of the rules of their pairing. Namely, the abundance of T equals the abundance of A, and the abundance of C equals the abundance of G. In 1952, Rosalind Franklin obtained an image of DNA using X-ray diffraction, which was communicated by Maurice Wilkins to James Watson and Francis Crick. Watson and Crick were at that time trying to decipher the structure of DNA using a model-building strategy, which was pioneered by the double Nobel Laureate

Linus Pauling. Franklin's data, combined with Chargaff's rules, allowed Watson and Crick to confirm the validity of one of their models, which was based on the *double-stranded structure of the DNA helix.* The two DNA strands are formed by the four nucleotides, T, A, C, G, as well as by a sugar (called deoxyribose), and a phosphate group.

In biology, the deciphering of the structure of DNA is considered the greatest discovery of all time. It was hugely important because, as Watson and Crick pointed out at the end of their short 1953 paper in *Nature*, the double-stranded structure of DNA made immediately clear how the genetic code was inherited across generations. This led to an unprecedented revolution in biology and medicine, which holds great promise to provide cures or at least ameliorate the symptoms of various human diseases, including cancer. Up until this historic development, such diseases were considered a tightly sealed black box. In 1962, Watson and Crick shared the Nobel Prize in Medicine with Wilkins.

In 1961, Marshall Nirenberg (Nobel Prize in Medicine, 1968) broke the genetic code and established the central dogma of biology: information is transmitted from DNA to RNA, dictating protein synthesis. In 1970, Frederick Sanger and Walter Gilbert (who shared the Nobel Prize in Chemistry in 1980) devised ways to determine the sequence of bases of DNA.[9] The same year, Hamilton Smith and Kent Wilcox (who shared the Nobel Prize in Medicine 1978) were able to cleave DNA at defined sequences called "restriction sites". This was achieved by employing certain enzymes that are used by bacteria to defend against *bacteriophages*, namely by viruses that infect them. This major discovery allowed researchers to manipulate the very large molecule of DNA and led to huge developments in biology and in biotechnology. It also paved the way for the sequencing of the full human genome, achieved between 2000 and 2003.

The largely unknown contributions of the crystallographer Arthur Lindo Patterson (1902–1966), who had degrees in mathematics and physics,

[9]This was the second Nobel Prize awarded to Sanger. The first, awarded in 1958, also in Chemistry, was for deciphering the structure of bovine insulin. Surprisingly, when he reached the age of 67, following the mandatory retirement from the University of Cambridge, he stopped doing any research and dedicated himself to gardening. Interestingly, he refused a Knighthood because he "did not want to be different from common people".

provide yet another illustration of the importance of interdisciplinarity. He developed a mathematical formula (the Patterson function, based on summing a particular Fourier series), which allows the interpretation of the X-ray diffraction patterns of complex molecules, such as DNA. In 1926, Patterson lived in Berlin, where he was fortunate to meet, among others, Nobel Laureates, Albert Einstein, Max Planck, Walther Nernst, Hans Bethe, Max von Laue, and Otto Hahn.

In summary, the human genome is composed of 23 pairs of distinct long DNA pieces, called chromosomes. Each of these consists of 50–250 million letters C, G, A, T (denoting the four nucleotides), whose sequence in the DNA defines the genetic code. In every normal cell, each of the 23 chromosomes is present in two copies, one inherited from each parent. The exception is in the sex chromosomes X and Y. In females there are two X chromosomes, whereas in males there is one X and one Y, with the Y chromosome always coming from the father. Additional genetic information is also contained in the cellular organelle *mitochondria*, which is the "energy factory of the cell". The genes of mitochondria are always inherited from the mother. Because of their bacterial origin, mitochondria contain a short, circular DNA, which encodes some, but not all the mitochondrial proteins. Incidentally, mutations in mitochondrial DNA give rise to devastating human diseases.

In order to use the instructions contained in their genome, cells construct the RNA molecule from the DNA template, via a process called "transcription". RNA has three of the same letters of DNA, but T is replaced with uracil (U). RNA transfers information from the nucleus (where the DNA is stored) to the outer region, the cytoplasm, where proteins are synthesized via the process of "translation". Namely, every three consecutive letters in RNA give rise to a specific amino acid.

Fundamental developments in molecular biology led to the birth of *genetic engineering* and *genome editing*. Genome editing, namely the process by which genes can be manipulated in almost any living organism with remarkable precision, ease, and speed, via the process of DNA insertions, deletions, or replacement, will be discussed in the second volume. Regarding genome engineering, it is noted that one of its early achievements was the production of insulin, which was marketed in 1982.

Another striking contribution was the development of the *recombinant human growth hormone*, approved by the Food and Drug Administration (FDA) of the USA in 1985. It was produced by the new technology of "recombinant DNA". In this technology, human genes are inserted into bacteria so that they can produce unlimited amounts of a desirable protein. The approval of the recombinant human growth hormone was accelerated after it was discovered in 1985 that four young adults who were treated in the 1960s with growth hormone extracted from cadavers developed the fatal Creutzfeldt–Jakob disease. The recombinant human hormone, which is produced by the company Genentech, is prescribed for the treatment of certain growth disorders in children, as well as for the treatment of deficiency of the growth hormone in adults. The growth hormone, also called somatotropin, is a 191-amino acid protein, synthesized in the anterior pituitary. It stimulates growth, cell reproduction, and cell regeneration; it also raises the concentration of glucose and fatty acids. Due to its anabolic effects, and since it cannot be detected in traditional urine analysis, human growth hormone was used illegally in sports since at least 1982. The ban on this hormone in competitions was not enforced until the early 2000s, when an effective blood test was developed. Some physicians prescribe it to the elderly in the hope of enhancing their vitality. However, neither its efficiency nor its safety has been tested in rigorous clinical trials.

An important achievement of genetic engineering was the production, in 1992 of the gene responsible for the "factor VIII". This gene, which was cloned eight years earlier, is used for the treatment of patients with Haemophilia A. This is a genetic bleeding disorder caused by a deficiency of the factor VIII, which is vital for blood coagulation. After replacement therapy with factor VIII administrated intravenously 3 or 4 times a week, patients' life expectancy became very close to that of unaffected males. The subsequent development of factor VIII with an extended plasma half-line, reduced the need for injections to only twice weekly. An alternative treatment is provided by the monoclonal antibody emicizumab, which *mimics* the activity of the factor VIII; this is administered subcutaneously every one or two weeks (Konkle *et al.*, 2020).

There are ongoing efforts to treat haemophilia A with gene therapy. This gene approach involves using an adeno-associated viral (AAV) vector directed to the liver with the aim of achieving sustained expression of factor VIII. This procedure was successful to 16 out of 18 participants (George *et al.*, 2021). It has already been demonstrated that a single infusion of the gene therapy is superior, with respect to the number of bleeding episodes, to prophylaxis using clotting-factors (Thornburg, 2022). Further progress will be presented in the second volume.

Severe Haemophilia A, corresponding to factor VIII activity level less than 1%, is characterized by repetitive bleeding in the joints beginning in early childhood. Also, there is a major risk for life-threatening haemorrhage. The early medical interventions mentioned earlier achieved levels of 1%–2%. However, it later became clear that this increase was insufficient to eliminate all joint bleeding. This led to the new goal of 12%, which, however, until very recently remained unattainable. In a breakthrough announced in the *New England Journal of Medicine*, January 2023, it was reported that patients receiving, once a week, efanesoctocog alfa, the first *recombinant* factor VIII, achieved levels of 15% and zero bleeding.

The recombinant production of factor XI, which is deficient in Haemophilia B, the second most prevalent disease of coagulation after Haemophilia A, was developed in 1997. Incidentally, the most famous carrier of the Haemophilia B gene was Queen Victoria (1819–1901). At least nine sons, grandsons, or great-grandsons of Queen Victoria have been identified as having Haemophilia B. The average age of death of these members of the royal families of United Kingdom, Spain, Germany and Russia was 24 years. In 2009, polymerase-chain-analysis of bones exhumed from the graves of Czar Nicholas II and his family identified this gene in his son, Alexei Romanov.

Efforts to employ gene therapy are also ongoing with respect to haemophilia B. In particular, in 2022, the FLT180a liver-directed AAV gene therapy, achieved normal range of the factor XI, but required immunosuppression (glucocorticoids with or without tacrolimus) (Chowdary *et al.*, 2022).

It is noted that genetic engineering also has had a significant impact in viral diseases. In particular, in 1986, the FDA approved a genetically engineered vaccine for the Hepatitis B virus.

In addition to the growth hormone, there exist a variety of *growth factors,* which are capable of acting on many different tissues. Among these are the *epidermal growth factor* that promotes cellular proliferation and differentiation, as well as the *fibroblast growth factor*, which is important in cell-signalling. Moreover, there exist many growth factors which are tissue specific and are capable of stimulating growth, proliferation, healing, and differentiation in specific cell-lines. The first such factor, the *nerve growth factor* (which stimulates nerve tissue), was discovered in 1952 by Rita Levi-Montalcini, who was awarded the Nobel Prize in medicine in 1986, jointly with Stanley Cohen, for their work on growth factors.

Growth factors, which are usually proteins or hormones, are naturally occurring substances. The name growth factor is sometimes used interchangeably with the term *cytokine.* This is due to the fact that early on, certain growth factors associated with blood cells and cells participating in the immune system were discovered, which were named cytokines. However, it was later realized that, while the term "growth factor" implies a positive effect on cell proliferation and maturation, there exist cytokines that have an inhibitory effect and some may even cause cell death, such as the Fas-ligand.[10] It is now possible to synthesize several growth factors via genetic engineering, as well as to produce similar molecules which either stimulate (agonists) or inhibit (antagonists) the receptors used by growth factors.

References

Avery, O. T., MacLeod, C. M., & McCarty, M. 1944. Studies on the chemical nature of the substance inducing transformation of pneumococcal types: Induction of transformation by a desoxyribonucleic acid fraction isolated from Pneumococcus type III. *The Journal of Experimental Medicine* 79(2), 137–158.

Burleigh, R. 2004. *Seurat and La Grande Jatte.* Harry N. Abrams Inc.

[10]Other specific growth factors are the *vascular endothelial growth* factor that stimulates angiogenesis (blood vessel formation) and various *hematopoietic growth* factors. The latter include *erythropoietin* (*EPO*), which stimulates red blood cell production and *granulocyte colony stimulating factor*, which stimulates white blood cell proliferation and maturation.

Butler, A., Hoffman, P., Smibert, P., Papalexi, E., & Satija, R. 2018. Integrating single-cell transcriptomic data across different conditions, technologies, and species. *Nature Biotechnology* 36(5), 411–420.

Chowdary, P., Shapiro, S., Makris, M., Evans, G., Boyce, S., Talks, K., Dolan, G. *et al.* 2022. Phase 1–2 trial of AAVS3 gene therapy in patients with hemophilia. *New England Journal of Medicine* 387(3), 237–247.

Doll, R. & Hill, A. B. 1950. Smoking and carcinoma of the lung. *British Medical Journal* 2, 739–748.

George, L. A., Monahan, P. E., Eyster, M. E., Sullivan, S. K., Ragni, M. V., Croteau, S. E., Rasko, J. E. *et al.* 2021. Multiyear factor VIII expression after AAV gene transfer for hemophilia A. *New England Journal of Medicine* 385(21), 1961–1973.

Goldstein, J. L. 2019. Seurat's dots: A shot heard'round the art world — fired by an artist, inspired by a scientist. *Cell* 179(1), 46–50.

Harari, Y. N. 2016. *Homo Deus.* Vintage.

Konkle, B. A., Shapiro, A. D., Quon, D. V., Staber, J. M., Kulkarni, R., Ragni, M. V., & Fruebis, J. 2020. BIVV001 fusion protein as factor VIII replacement therapy for hemophilia A. *New England Journal of Medicine* 383(11), 1018–1027.

Loske, A. 2019. *Colour: A Visual History from Newton to Modern Colour Matching Guides.* Smithsonian Books.

McNeil, J. J. *et al.* 2018. Effect of aspirin on all-cause mortality in healthy elderly. *New England Journal Medical* 379, 1519–1528.

Smith, T. F. & Waterman, M. S. 1981. Identification of common molecular subsequences. *Journal of Molecular Biology* 147(1), 195–197.

Thornburg, C. D. 2022. Prepare the way for hemophilia A gene therapy. *New England Journal of Medicine* 386(11), 1081–1082.

Waterman, M. R. & Waterman, M. S. 1995. *Introduction to Computational Biology: Maps, Sequences and Genomes.* Chapman and Hall.

THE QUEST FOR THE HIDDEN REALITY

The great ancient Greek philosopher Heraclitus, in addition to understanding that "Nature likes to hide", also appreciated the beauty of the hidden reality: "The hidden reality is more harmonious than the visible one".

The brain reaches hidden reality using associations:

Understanding is the process of elucidating relations. Associations provide the mechanism for constructing such relations. The highest achievement of the brain is the creation of an astounding web of associations, which, as explained in this part of the book, *generates the crucially important processes of unification and generalization.*

The best illustrations of these processes are, respectively, Physics and Mathematics. Physics will be discussed in the second volume. Mathematics is discussed in Chapter 18. In particular, the concrete example of the transition from Euclidean to Riemannian geometry illustrates the fact that the construction of new, complex mathematical structures is achieved via the vital process of generalizing.

Taking into consideration that the *essence of human thought is building associations and elucidating their mutual relations,* the indispensable role in our culture of mathematics becomes clear. Indeed, mathematics, with its rigorous and unambiguous language, provides an ideal tool for formalizing relations. In my opinion, *this is precisely the fundamental significance of mathematics, independent of applications.* In Chapter 18, an attempt is also made to explain the role of beauty in mathematical physics by appealing to the concept of symmetry in general and of *gauge symmetry* in particular.

The interaction between local and global neural processes expresses itself in a variety of forms. One of the most ubiquitous forms is the interplay between *simplicity* and *complexity*. For example, this interplay allows the brain to comprehend the infinitely complex physical world. Indeed, the *local* formulation of many of the fundamental physical laws makes these laws *simple*, and hence comprehensible. On the other hand, the *complexity* of the *global* solutions of the mathematical equations characterizing these laws captures the exceedingly complicated physical reality expressed by these laws.

The concepts of unification and local versus global processes are also important in art. For example, in the painting *Self-Portrait with Yellow Christ* (1890–1891), Paul Gauguin attempts to depict the unification of his dual identity, namely, the "divine" and "savage" elements of his

Figure PV.1. Paul Gauguin, *Self-Portrait with Yellow Christ* (1890–1891).

character. He portraits himself flanked by two of his earlier creations: The *Yellow Christ* (1889) and a ceramic pot depicting a grotesque head (Figure PV.1).[1]

In Chapter 19, an attempt is made to place into a proper perspective the question of "How the problem of consciousness *can* be solved?" In preparation for analyzing this question, the ideas of Sir Roger Penrose (Nobel Prize in Physics, 2020) regarding the ability of the brain to think non-algorithmically, and of the possible role of quantum gravity in the generation of consciousness, are discussed in detail.

Reference

Gauguin, P. 2005. *Noa Noa: The Tahiti Journal of Paul Gauguin (1894–1901).* Chronicle Books.

[1]The successful unification of the "civilized" and the "primitive" characterize many of Gauguin's creations. This is achieved by combining his "signature" use of colourful flat areas with a novel style of painting capable of creating the primitive forms needed for the depiction of an exotic framework. With the writing of the autobiographical tale of his first two years in Tahiti (Gauguin, 2005) *Noa, Noa* (1894–1901), Gauguin introduced another type of unification, namely, that of the "civilized" poet and the "primitive" painter. In writing *Noa, Noa*, Gauguin was perhaps influenced by his knowledge of the autobiographical notes of Eugène Delacroix, who in 1832 travelled to North Africa in search of a more primitive culture. Delacroix eventually produced over 100 works of scenes from or based on his experience of this new culture, which he found fascinating.

18 ABSTRACTION AND GENERALIZING ARE THE ESSENTIAL FEATURES OF MATHEMATICS

As will be discussed in detail in the second volume, the metaphorical Ariadne's thread that led Albert Einstein to his phenomenal contributions was his ability to *associate* seemingly disparate realms of nature. Associations generate specific *relationships* between different concepts. According to Einstein, establishing and using these relationships is the essence of thinking. In his attempt to answer the basic question "What, precisely, is 'thinking'"? Einstein wrote in 1946 that "sense impressions" give rise to "memory-pictures" (Einstein, 1946), which are referred to in this volume as mental images. He also noted that, if these mental images occur often and in different situations, then they become "concepts". Thinking is "operations with concepts [...] the creation and use of definite functional relations between concepts" (Miller, 1986).

The importance of "establishing relationships among concepts" was also appreciated by Henri Poincaré. In particular, elaborating on the notion of relativity, he wrote that the goal of science,

"[...] is not the things themselves, [...] it is the relations between things; outside of these relations there is no reality knowable" (Poincaré, 1902).

Motivated by the remarks of these pioneering thinkers, I propose the following definition:

Understanding is the process of elucidating relationships.

Mathematics decisively assists the process of establishing and analyzing relationships, and hence is vital for the deep understanding of a plethora of phenomena. Indeed,

the unambiguous and self-consistent language of mathematics provides the most efficient tool for a quantitative and rigorous approach to establishing, manipulating, and using relations.

David Hilbert's book, *The Foundations of Geometry* (Hilbert, 1902), is a great example of the importance of establishing relationships between different mathematical notions.[1] In this book, basic concepts are *not* defined at all, except through their *mutual relations*. For example, in Euclid's *Elements* (Euclid and Heath, 1956) it is written that "A point is that which has no parts". Hilbert omits this statement and only keeps the sentence expressing the relationship between two points: "Any two points lie on a line". The great mathematician emphasizes that it is not important to define points and lines. He writes, "Instead of calling these things points, lines, and planes, we could just as well call them tables, chairs and beer mugs".

By concentrating on relationships, the brain can overcome the impressions developed by sense perceptions, which can be *erroneous*. This was clearly shown by Einstein with the elucidation of the deep meaning of the concepts of time and gravity. The brain's ability to go beyond physical intuition is illustrated with the development of non-Euclidean geometries, discussed below. These geometries overcome the restriction of "flatness" imposed by our exposure to the "natural" geometry.

One of the axioms in Euclid's *Elements* is the "parallel axiom". This axiom states that, *if L denotes a straight line in a plane and if a point, P, on this plane is not located on this line, then there is only one line through P that does not intersect L.* In other words, there is only one line parallel

[1]In 1895, Hilbert took a position at the University of Gottingen, the Mecca of mathematics at that time, and became the successor of two of the greatest mathematicians of all time, Carl Friedrich Gauss and Bernhard Riemann.

to *L*. Although this axiom is certainly consistent with our geometrical intuition, it is not as obvious as the rest of Euclid's axioms. This is the reason that, for two millennia geometers tried, without success, to deduce it from the remaining Euclidean axioms.

What happens if we ignore our geometrical intuition and, instead of employing this axiom, we use an alternative one? In the 1820s, János Bolyai in Hungary and Nikolai Lobachevsky in Russia, independently, constructed a geometry where the parallel axiom is *replaced* by the requirement that there exist *many* lines through P that never intersect *L*. The resulting geometry, known as "hyperbolic geometry", is very different from our familiar Euclidean geometry. For example, the sum of the angles of a triangle in this geometry is always less than 180 degrees. However, it is logically as consistent as Euclidean geometry.

Later, Bernhard Riemann introduced "elliptic geometry", where the sum of the angles of a triangle is always greater than 180 degrees. The realization that there exist geometries that are not constrained by our spatial intuition led to further extensions. Indeed, following these developments, Riemann provided a unification of Euclidean, hyperbolic and elliptic geometries, as well as a vast generalization. He overcame the limitations of two and three dimensions and constructed a geometry in an arbitrary number of dimensions.

In summary,

associations generate relationships, and some of these relationships can be expressed in mathematical language. "Mathematizing" facilitates the process of abstraction, by allowing the process of dissociation of these relationships from their original realizations. The manipulation of abstract relationships gives rise to generalizations and to abstract mathematical constructions.

This statement suggests that the processes of abstraction and generalization are *intrinsic* to mathematics. Therefore, the question regarding the so-called "forward motion of mathematics", namely why mathematics continuously generates new concepts, raised by many mathematicians and philosophers, becomes mute. This question is eloquently expressed by George Steiner, who states:

"Is the unbounded, one may say 'fantastic', motion forward of mathematics from Pythagoras' triangle to elliptic functions, generated, energized from within itself, independent of either reality or applications? [...]. Mathematician themselves, philosophers have debated this issue across millennia. It remains unresolved" (Steiner, 2014: p. 20).

Noting that *mathematizing* is *generalizing*, it follows that the "forward motion" is an integral part of mathematics. I consider it one of its *defining* properties. Thus, the above question becomes, in my opinion, mute.

For example, in the case of geometry, starting from relationships that express spatial reality, the process of generalizing gives rise to abstract constructions which do not appear to reflect the reality of our world. However, occasionally, something truly surprising takes place: these abstract mathematical structures, unexpectedly, *do* represent hidden reality. Moreover, this reality can be revealed *only* through these abstract mathematical constructions!

Such a paradigm is provided by Riemannian geometry. Indeed, remarkably, Einstein's general theory of relativity, established that this non-intuitive geometry *is* the *physical geometry characterizing our universe and describing the effects of gravity on our universe.* Another example is provided by Boltzmann's mathematical formulation of a typical gas. Since the quantity, called a mole, contains 10^{23} number of molecules, it follows that an extremely large number of dimensions is needed for the mathematical formulation of this physical situation.

What is the explanation of the astonishing ability of mathematics to describe the physical world? It all starts from the great "gift" that the basic physical laws of our macrocosm are extremely simple. The reason for this remarkable fact remains, in my opinion, a mystery. The existence of such simple physical laws allowed physicists to find simple relationships satisfied by observational and experimental data. This led to *modelling*, namely the construction of simple mathematical equations consistent with these relations. For example, Newton's fundamental physical laws can be expressed by very simple mathematical relationships. Indeed, the basic law of motion is $F = ma$, stating that the force exerted on a particle of mass m moving with acceleration a, equals the product of m and a. Similarly,

Newton's gravitational law takes the very simple form, $F = GmM/r^2$, which states that the attractive force between two masses, m and M, equals the product of these masses, multiplied by a universal constant, G, divided by the square of their distance.

As soon as doubts are raised about the validity of a given physical law, either as a result of a logical inconsistency or because it disagrees with more accurate experimental data, the brain's immense cognitive capacity comes to the rescue. In particular, the unlimited ability of the brain for remote associations allows it to search for more elaborate mathematical relationships. Since these relations often express the existence of a non-intuitive and hidden reality, their formulation may require mathematical structures that are dissociated from our sensory intuition. In other words,

the mathematical modelling of forms of reality that are not perceived by our senses may require the use of abstract mathematical structures.

Since these structures *do* represent physical reality, albeit a hidden one, it follows that these abstract mathematical creations are indeed "physical". In other words, there exist abstract mathematical constructions which *do* express reality, despite the fact that these structures are generated via abstract thinking, instead of sensory perceptions. Hopefully, these remarks address, at least partly, the question about the "unreasonable effectiveness of Mathematics in the physical sciences" (Wigner, 1960), raised by Eugene Wigner (Nobel Prize in Physics, 1963).

This discussion illustrates the dual nature of mathematics: on the one hand, it involves the construction of abstract relations devoid of any empirical meaning. On the other hand, mathematizing leads to the derivation of useful structures modelling physical reality. This duality fascinated Ludwig Wittgenstein. The great philosopher, after attending, in 1928, a lecture in Vienna on the foundations of mathematics by the Dutch mathematician, L. E. J. Brouwer, devoted half of his writings in the next 20 years to mathematics. Wittgenstein finally concluded, erroneously in my opinion, that the essence of mathematics can be found *only* in applications. For example, according to him, the meaningless proposition $30 \times 30 = 900$, acquires meaning only if it is seen as a *tool* to be used in applications.

Wittgenstein stated that the "fundamental idea", which provided the foundation of his transformative book, *Tractatus Logico-Philosophicus*, was his realization that mathematical symbols have a purely *tautological* nature. Therefore, *logical propositions cannot reveal any objective truth about the empirical world.* According to Wittgenstein, the study of mathematics simply elucidates features that *already exist* in a latent form in the underlying mathematical axioms. Since these features are a logical consequence of the axioms, and since the axioms are by definition neither true nor false, these features have *no* implicit truth. Propositions 6.2–6.21 conclude that, "The propositions of mathematics are equations, and therefore pseudo-propositons" (Wittgenstein, 1922). In his earlier work, Wittgenstein had shown that pseudo-propositions are reducible to tautologies.[2]

Wittgenstein emphasized that arithmetic and Euclidean propositions are *systems of rules* concerning the number, magnitude, or other properties of *real things.* For Wittgenstein, it is a serious mistake to consider pure mathematics as descriptive of any *objective* domain beyond mathematics itself. According to him, mathematics becomes meaningful, or in other words *semantic as opposed to purely syntactic*, only in relationship to applications.

This preferential emphasis of Wittgenstein on applications, explains his attack on the foundational branch of mathematics, called "set theory". Another reason for attacking this theory was its broad generality. He writes: "nothing is more suspect than a generality which is too vast" (Waismann, 1979: p. 103). Incidentally, responding to the praise by Hilbert for Georg Cantor, the creator of set theory, that "No one will be able to drive us out of the paradise that Cantor created for us", Wittgenstein responded as follows: "I will try to show that it is not a paradise-so that you leave on your own accord" (Diamond, 1989: p. 102).

Incidentally, it is worth noting that Wittgenstein's views are clearly anti-Platonic. According to Plato, numbers, geometrical notions, and

[2]The above quote makes a subtle distinction between general logical pseudo-propositions that are reducible to tautologies, and mathematical pseudo-propositions that are reducible to equations. Wittgenstein's reluctance to state explicitly that mathematical equations are tautologies led to arguments with his Cambridge colleague, Frank Ramsey (Monk, 1990).

other universal “forms” or “ideas”, are “real objects of knowledge” (Marion, 1998).

I disagree with Wittgenstein’s view of a dichotomy within mathematics. In my opinion,

> the highest achievement of consciousness is the creation of an astounding web of associations. This is achieved with the crucial aid of metarepresentations, which include language and mathematics. Building such associations and elucidating their mutual relationships is indeed the essence of the human thought. Establishing connections between some of these associations with “reality” is of course important, but it is a secondary task.

In this context, it becomes clear that the dual nature of mathematics mentioned earlier is, indeed, its essential feature. Mathematics is based on a set of axioms and logical syllogisms. This means that, in the sense of Wittgenstein’s definitions, mathematics is indeed tautological. However, as stated earlier,

> the essence of understanding is elucidating relationships, and mathematics builds a strikingly complex apparatus for precisely this purpose. In this sense, mathematics is a great servant of consciousness, independent of its usefulness in applications.

The emphasis on applications is also problematic from an epistemological point of view. The notion of applications is based on the particular notion of “reality”, which, however, is often ephemeral. For example, Riemannian geometry is not consistent with human’s innate geometrical intuition. However, as shown by Einstein’s Theory of General Relativity, this non-intuitive geometry captures reality more accurately that the intuitive Euclidean geometry. If one follows Wittgenstein’s arguments, one reaches the obscure conclusion that Riemannian geometry was “senseless” before the brilliant insight of Einstein, and all of a sudden, after Einstein’s discovery, it acquired meaning! In analogy with Riemannian geometry, there exist many branches of mathematics that were considered areas of “pure mathematics” for many years, until it became clear that they are extremely useful in describing various forms of reality. Furthermore, just as with general relativity, this reality reveals itself only via the employment of these “pure mathematical constructions”.

Despite the emphasis on applications, it appears that Wittgenstein intuitively sensed that the impact of mathematics is much broader than its role in applications. In particular, he appreciated that mathematics is important in helping the brain *formulate proper questions*, which leads to deeper insight. For example, he writes, "it takes mathematics to define the *character* of what you are calling a *fact*". As a relevant example, he writes: "it is interesting to know *how many* vibrations this note has". Then, he elaborates that mathematics, via the complicated theory of waves, can indeed calculate the number of notes in a given type of vibration. Hence, he concludes that it took mathematics "to teach you this question; it taught you to see this kind of fact" (Wittgenstein, 1956: p. 381).

GENERALIZING AND THE INTERPLAY WITH PHYSICS

In my opinion, the astounding ability of the brain for generalizing reaches its apotheosis in mathematics. Actually,

the entire process of mathematizing can be considered synonymous to the process of generalizing.

One striking example of this process is the analysis of the formula for solving a quadratic algebraic equation. This formula contains the square root of an expression called the "determinant". A particular case of this formula was already known to the ancient Egyptians. It is contained in the Egyptian Berlin Papyrus, which is one of the main sources of ancient Egyptian mathematics, dating from the period of the Middle Kingdom of Egypt that lasted from around 2050 BC to around 1710 BC. Until the 15th century it was assumed that if the determinant is negative, then the given equation does *not* have a solution. This was the natural consequence of the understanding at the time that the square root of a negative number could *not* be defined. Indeed, it was thought that it was senseless to ask the question "What is the square root of −1?" However, at the beginning of the 15th century, Niccolo Tartaglia, and later Gerolamo Cardano, derived a complicated formula for the solution of a third order algebraic equation. This expression contains several square roots. It then became clear that there are cases where although some of these square roots involve negative numbers, the equation *does* have real solutions. This led to a paradox: on the one hand, it is apparently wrong to reject the square root of negative numbers, since the relevant expressions finally yield the correct answer; but on the other hand, such numbers were considered senseless. Cardano responded to this challenge in the *only* way that the brain can solve such paradoxes: *by generalizing*. He introduced the so-called *imaginary numbers*, which are usual numbers multiplied by the symbol i. This mathematical entity is *defined* via the following novel idea: the square root of -1 is *not* rejected, but is assigned the value of either i or $-i$.

The introduction of imaginary numbers led to the definition of *complex numbers*. These are entities of the form $a + ib$, where a and b are real numbers. The introduction of these numbers not only implies that every quadratic equation has two roots but also leads to the important result that any polynomial equation of degree n, where n is a positive integer, has n roots. Moreover, the introduction of the complex numbers gave rise to a further generalization. Namely, in the same way that a real quantity f can depend on a variable x, a mathematical quantity F can depend on the complex variable $x + iy$. Remarkably, mathematical functions which depend on *complex* variables are very important in describing *real* phenomena (Ablowitz and Fokas, 2003).[3] For example, the solution of the inverse problems arising in the medical imaging techniques described in Chapters 14 and 15 are based on the analysis of such functions!

Another example of generalizations can be traced to the dialogue *Timaeus*, where Plato hypothesized that the *classical elements* of his theory, which were earth, water, air, and fire, are made up of five solids. These entities, known as the *Platonic solids*, are the tetrahedron, cube, octahedron, dodecahedron, and icosahedron. These solids have 4, 6, 8, 12, and 20 faces, respectively. Geometers have studied the Platonic solids for thousands of years. The study of the *symmetry properties* of polyhedrons has given rise to the notion of a "mathematical group" (the mathematical notion of symmetry is discussed in the next section). This is the set of transformations that leave the given polyhedron unchanged.

---

[3]Functions depending on the complex variable $x + iy$ are called "analytic functions". Recently, functions which provide a generalization of analytic function, namely functions which depend on the *two* complex variables $x + iy$ and $x - iy$ have also appeared in applications (Fokas and Ablowitz, 1983).

The beautiful subject of "group theory" begins with the publications of Augustin-Louis Cauchy and Évariste Galois. Galois, in addition to laying the foundation of group theory also found a necessary and sufficient condition for a polynomial to be solvable by radicals, thereby solving a problem that had remained open for 350 years. Surprisingly, he proved that, in general, only polynomials of degree up to four can be solved via radicals. Incidentally, Galois, who was also a political activist, is the most tragic figure in the history of mathematics. He died at age 20 from wounds suffered in a duel, which may have been staged for political reasons. He stayed up the last night of his life writing letters to his Republican friends and composing a legendary letter to Auguste Chevalier, outlining some deep mathematical ideas. He also left to Chevalier some manuscripts, including a copy of a manuscript already submitted to the Academy, which was favourably reviewed later that year by Joseph Liouville.

The *classification* of groups has revealed a plethora of fascinating mathematical structures, including the "exceptional groups". Remarkably, the Platonic solids are related to these exceptional structures.[4]

Some of the most significant generalizations in mathematics are intimately related to corresponding developments in physics. Some important examples are catalogued below.

- It turns out that any geometry can be described *locally*, via a set of equations, called the Gauss–Codazzi equations. However, physics necessitated the need to understand these geometries *globally* and this motivated the discovery of "topology".
- The passage from classical mechanics to quantum mechanics made it imperative to study spaces of infinite dimensions, called "Hilbert spaces".
- Physical considerations motivated the introduction of mathematical entities that do *not* commute. This is in contrast to the usual cases, such as numbers, which are commutative; for example, 3×5 equals 5×3. Important non-commutative entities are the *quaternions*, introduced by William Hamilton in

[4]The tetrahedron is related to the exceptional group E_6, the cube to E_7, and the icosahedron to E_8. The octahedron and the dodecahedron are "dual", respectively, to the cube and the icosahedron. Thus, they share the same groups with their dual partners.

1843, the *octonions* discovered by Hamilton's friend John T. Graves soon after, and *matrices* developed by Arthur Cayley in 1858. The most important non-commutative algebra in physics is the one introduced by Heisenberg, which provides the basis for the algebraic formulation of quantum mechanics.

- Maxwell's equations of electromagnetism are *linear*. This means that the sum of two solutions of these equations is also a solution. However, Einstein's equations of general relativity are *nonlinear*. The mathematical study of nonlinear equations is far more complex than the study of linear equations. Nonlinear equations have very interesting properties, which cannot occur in linear equations. Moreover, some nonlinear equations express disordered behaviour that is studied by *chaos theory*. In contrast to this chaotic behaviour, there are certain distinguished nonlinear equations, called *integrable equations*, which exhibit well-organized behaviour and possess special solutions, called *solitons*. These solutions express physical entities which are so robust that they may be considered as mimicking the behaviour of individual particles.[5]

It is worth noting that there are numerous less well-known developments in mathematics that have been motivated by physics. Actually, the physical motivation is often kept a secret. For example, mathematician, Minhyong Kim, has recently revealed that his approach of using the so-called Selmer varieties for constructing rational solutions of Diophantine equations was motivated by electrodynamics (Hartnett, 2017). In particular, by the fact that among all possible paths that light rays can follow, they chose the unique path that minimizes the travelling time.

The role of beauty in mathematical physics

Beauty is apparently a necessary characteristic that an abstract mathematical structure must possess in order to be useful for the formulation of fundamental laws of nature. The deep relationship between abstraction and beauty was first emphasized by Plato. The "ideas" existing in the Platonic world are

[5]In will be noted in the next section that a particular type of non-geometrical symmetries, called gauge symmetries, are crucial for identifying physically significant theories. Similarly, particular types of non-geometrical symmetries can be used to identify integrable nonlinear equations. The identification of one of the first integrable equations via symmetry considerations was achieved in Fokas (1980).

abstract concepts that are, by definition, beautiful. A partial explanation for the importance of beauty in mathematical physics is provided by the concept of "symmetry". This concept is not only utilized in many aesthetic creations but it also plays a crucial role in the formulation of the laws of nature. An early argument for the existence of a relationship between symmetry and truth was made by Johannes Kepler. He argued that the elliptical orbits he had discovered for the motion of celestial bodies had a deeper symmetry than the circular ones, and thus they truly existed.

The mathematical definition of the intuitive concept of symmetry can be introduced using the example of the most symmetrical of all solids, namely, the sphere. The symmetrical nature of the sphere implies that it looks the same, independently of the particular angle from which one looks at it. Therefore, the equation defining a sphere cannot depend on angles. Hence, this equation must be highly symmetric, which is indeed the case. The equation specifying a sphere of radius r, with its centre at the origin, is given by the highly symmetrical expression $x^2 + y^2 + z^2 = r^2$, where (x, y, z) are the three coordinates of a point on this sphere. The above discussion implies that this equation should not change if the coordinates (x, y, z) are *rotated*. The technical statement expressing this intuitive fact is that, "this equation remains invariant under rotations". Rotations are a particular type of coordinate transformations. Other examples of such transformations are translations, such as adding to the coordinate x a constant, corresponding to shifting the body by this constant in the x direction. By generalizing, it is natural to pose the problem of determining all transformations that leave a given set of equations invariant. For several centuries, mathematicians have been studying this question, where the relevant coordinate transformations are specified in terms of a set of *constant* parameters. Such transformations are studied by the mathematical theory of "Lie Groups".

The highly original mathematician, Hermann Weyl (1885–1955), introduced a significant generalization of the geometrical transformations studied via Lie groups. Namely, he allowed the constant parameters specifying the changes of coordinates to depend on the spatial coordinates themselves. If an equation is invariant under such non-geometrical transformations, then it is said that "it possesses a local gauge symmetry". For example, the equations of general relativity possess the gauge symmetry of local translations. This means that they are invariant if x is replaced by $x + X(x, y, z)$, where X is an appropriately chosen function; similarly for y and z (Iliopoulos, 2017).

Remarkably, *the unified feature of all fundamental physical theories is that they are invariant under gauge symmetries.* In recent years, demanding invariance under a specific gauge symmetry has provided the most fruitful approach for deriving physically significant theories. In this sense,

modern physics constitutes the apotheosis of the concept of gauge symmetry. This concept provides a generalization of the aesthetic notion of symmetry.

The crucial role of beauty in discovering and expressing truth has been immortalized by John Keats' *Ode on a Grecian Urn*: "Beauty is truth, truth beauty, that is all Ye know on earth, and all ye need to know". Of course, many mathematicians and physicists have had a deep appreciation of the relationship between beauty and truth. This relationship is perfectly captured by Subrahmanyan Chandrasekhar (Nobel Prize in Physics, 1983): "Simplicity is the seal of truth and beauty is the splendour of truth".

IN SEARCH OF UNIFICATION

The genesis of new relationships among different concepts is facilitated by the brain's continuous pursuit of unification. What is the neuronal basis of this pursuit? Generalizing the mechanisms discussed in Chapter 4 with respect to visual perception, it is natural to speculate that *conceptual awareness* is achieved via the process of

associating different aspects of the given concept and integrating the unconscious structures that correspond to these different aspects.

Taking into consideration that the brain possesses neural architecture, as well as functional characteristics perfectly suited to *uniting* different aspects of the *same* concept, it is natural to expect that the brain is predisposed towards *uniting different* concepts. In other words, it is predisposed towards pursuing broad unification. This capacity is further facilitated by the redundancy of neuronal circuits. Actually, the brain is the organ par excellence both for integrating different ingredients of the same concept, which is the basis of consciousness, as well as for uniting different concepts, which provides the basis for the tendency for unification.

The rewarding feeling of *eudemonia* that accompanies the pursuit of unification encourages this endeavour. This *eudemonia*, was clearly felt by Einstein: when he realized, during his analysis of the Brownian motion, that he could unify phenomena both on small and large scales, he wrote to his friend Marcel Grossmann, that "It is a wonderful feeling to recognize the unity of complex phenomena which appear to sense observation as totally separate things" (Einstein, 1987).

It is surprising that the connection between modelling and unification was apparently not appreciated by Poincaré: according to this great mathematician, different models are *equally* valid, provided they match the observed or experimental data. However,

only encompassing models, consistent with a unified treatment of the laws of nature, survive the test of time.

For example, the study of the motion of the planet Mercury around the sun provided the first triumph of the general theory of relativity. The solution of Einstein's equations describing this motion, in contrast to the corresponding solution of Newton's equations, almost perfectly matched the relevant astronomical observations. At about the same time, physicist Ludwik Silberstein, who never accepted the theory of general relativity, proposed an *ad hoc* mathematical formula which also matched the above astronomical observations. This formula has the advantage of being much simpler than Einstein's equations (it involves only a slight modification of Newton's gravitational law) (Silberstein, 1917).[6] However, the limited applicability of Silberstein's model became clear a couple of years after its publication. Indeed, according to Einstein's theory, matter is capable of

[6]Silberstein simply multiplied Newton's gravitational formula, GmM/r^2, by the factor γ^n, where γ denoted the Lorenz factor, and showed that he could match the observations by the *ad hoc choice* of $n = 5$. It turns out that the success of Silberstein can be *a posteriori* explained. By using the fact that the ratio of the masses of the Sun and the Mercury is very large, Einstein's exact formula yields approximately a much simpler formula, which is precisely that of Silberstein (Fokas, 2020).

bending light. The first observation of such light deflection was reported by the famous Cambridge astronomer Arthur Eddington and his group, during the total solar eclipse on May 29, 1919.[7] This eclipse allowed the stars near the Sun to be observed. Simultaneous observations in Brazil and on the West coast of Africa demonstrated that light passing close to the sun was indeed slightly bent. The solution of Einstein's equations, in contrast to that of Silberstein, again almost perfectly matched these observations. This event made immediately Einstein world-famous, whereas Silberstein's model has been completely forgotten!

Unification in mathematics

The concept of generalization is intimately related to that of unification. For example, the generalization from real to complex numbers led to the unification of the most important functions used in applications. These functions are the exponential, the logarithm, and the trigonometric functions, sine and cosine. It turns out that the logarithm is the inverse of the exponential. Moreover, remarkably, the sine and the cosine can be expressed in terms of exponentials of purely imaginary numbers.[8] This means that with the use of complex numbers all these basic functions can be expressed in terms of the exponential function.

The tendency for unification is not as prominent in mathematics as in physics, but certainly mathematicians also strive for unification. For example, Felix Klein unified non-Euclidean geometries and the theory of symmetry groups. Currently, the far-reaching "Langlands program", considered "the grand unified theory of mathematics" attempts to unify geometry with number theory.

In what follows, I will discuss another example of unification, the *unified transform*, also known as the *Fokas method*, introduced in 1997 (Fokas, 1997, 2000; Wikipedia, n.d.).

---

[7]It may appear strange that light is affected by gravity, since light consists of photons which are *massless*. However, gravity not only pulls mass but also pulls energy, and obviously light has energy.

[8]Indeed, $\sin(x)=[\exp(ix)\text{-}\exp(-ix)]/2i$ and $\cos(x)=[\exp(ix)+\exp(-ix)]/2$.

There is an additional reason for including this example, beyond unification. It provides a clear illustration of the fact that progress in mathematics does not always follow the usual route of *increasing* complexity. Occasionally, it follows the opposite route, namely, from the complex to the relatively simple. Indeed, although as stated earlier the analysis of nonlinear equations is much harder than that of linear ones, the unified transform was first discovered for certain nonlinear equations, and it was then realized that it is also applicable to linear equations.

In order to place the Fokas method into a proper perspective, it is important to note that, since the middle of the 20th century, the range of applications of mathematics has been steadily expanding. It is now clear that mathematics is not only important in physics, chemistry, computer science, and technology but also in biology, medicine, economics, linguistics, and even in the arts and the social sciences. Many of the relevant phenomena can be described via *mathematical equations*. Among these equations the so-called "partial differential equations" are of particular importance. These equations describe phenomena that depend on two or more variables. For example, a vibrating string can be described by such an equation, known as the wave equation; it involves the time variable, t, and the space variable, x. If an equation involves only one variable, then it is called an "ordinary differential equation" and is much simpler to solve than a partial differential equation.

One of the approaches used by the brain to solve a new problem is to reduce it to a problem that already has been solved.

The implementation of this general concept to the solution of partial differential equations takes the form of the well-known method of "separation of variables". This method attempts to express the solution of a partial differential equation in terms of solutions of ordinary differential equations. Interestingly, as soon as Jean le Rond d'Alembert derived in 1746 the one-dimensional wave equation, which was the first partial differential equation ever formulated, d'Alembert, Daniel Bernoulli, and the great mathematician, Leonhard Euler, constructed particular solutions to this equation, via the use of separation of variables.

The famous French mathematician Joseph Fourier significantly extended the method of separation of variables by introducing the Fourier series and the Fourier transform (already mentioned in connection with the MRI). This breakthrough was needed in order to solve another important partial differential equation, the heat equation, which was derived by Fourier and which, among other applications, describes the propagation of heat. These mathematical constructions were later followed by the discovery of a variety of other transforms and series, giving rise to the powerful "method of transforms".[9]

Nonlinearity prevents the applicability of the method of separation of variables to nonlinear partial differential equations. However, the distinguished category of integrable nonlinear equations mentioned in the last section does admit a novel generalization of separation of variables, known as a Lax-pair formulation. Using this formulation, simple problems for integrable equations (known as initial-value problems) were solved in the early 1970s. But more complicated physically significant problems (known as boundary-value problems) remained open. In 1997, I presented a novel approach for solving such problems for these nonlinear equations. Later, I realized that this approach also provides a completely new method for solving the much simpler linear equations.

The distinguishing feature of the new method is that it employs the synthesis as opposed to the separation of variables. Until the discovery of the unified transform, the only other general technique for solving partial differential equations (PDEs), in addition to the method of transforms, was the so-called Green's functions, introduced by the University of Cambridge mathematicians George Green and Lord Kelvin. The new method provides a unification and generalization of the classical approaches of transforms and Green's functions. It also extends this unified approach to the important class of integrable nonlinear PDEs.

The Fokas method not only solves a variety of new problems that could not be solved with the earlier methods but it also constructs a more efficient representation of the solutions of problems that were solved earlier. This

[9]Such transforms include the sine, cosine, Laplace, Mellin, and Radon transforms, as well as corresponding infinite series, such as the sine and cosine series.

is exemplified by such classical problems as the heat equation (Fokas and Spence, 2012). For 200 years, since the pioneering contributions of Fourier, this equation was solved via the Fourier series. There is no mathematician who does not know the Fourier formalism. Remarkably, the Fokas method gives rise to a new representation for the solution of the heat equation, which has many analytical and computations advantages in comparison to the standard representation. In particular, it is straightforward to evaluate this expression numerically using any modern computer-based programming language. A pedagogical introduction of unified transform can be found in the book, *Modern Mathematical Methods for Scientists and Engineers* (Fokas and Kaxiras, 2022).

It is surprising that the great mathematicians of the 18th century, a time considered the gold epoch of mathematics, including Fourier, Pierre-Simon Laplace, and Augustin-Louis Cauchy did not discover this powerful new method. Perhaps this can be explained by the fact that the implementation of this method to linear equations was the by-product of the development of a methodology for solving the far more complicated integrable nonlinear equations, which were not analyzed by the classics.

References

Ablowitz, M. J. & Fokas, A. S. 2003. *Complex Variables: Introduction and Applications.* Cambridge University Press.

Diamond, C. 1989. *Wittgenstein's Lectures on the Foundations of Mathematics, Cambridge 1939.* University of Chicago Press.

Einstein, A. 1946. *Autobiographical Notes, in Albert Einstein: Philosopher-Scientist.* Open Court.

Einstein, A. 1987. *The Collected Papers of Albert Einstein: Volume 1, The Early Years, 1879–1902.*

Euclid & Heath, T.L. 1956. *The Thirteen Books of Euclid's Elements.* Dover Publications.

Fokas, A. S. 1980. A symmetry approach to exactly solvable evolution equations. *Journal of Mathematical Physics* 21(6), 1318–1325.

Fokas, A. S. 1997. A unified transform method for solving linear and certain nonlinear PDEs. *Proceedings of the Royal Society of London. Series A: Mathematical, Physical and Engineering Sciences* 453(1962), 1411–1443.

Fokas, A. S. 2000. On the integrability of linear and nonlinear partial differential equations. *Journal of Mathematical Physics* 41(6), 4188–4237.

Fokas, A. S. 2020. The justification of Silberstein's formula. *(preprint).*

Fokas, A. S. & Ablowitz, M. J. 1983. Method of solution for a class of multidimensional nonlinear evolution equations. *Physical Review Letters* 51(1), 7–10.

Fokas, A. S. & Kaxiras, E. 2022. *Modern Mathematical Methods for Scientists and Engineers: A Street-Smart Introduction.* World Scientific Publishing Company.

Fokas, A. S. & Spence, E. A. 2012. Synthesis, as opposed to separation, of variables. *Siam Review* 54(2), 291–324.

Hartnett, K. 2017. Secret link between pure math and physics. *Quanta Magazine.*

Hilbert, D. 1902. *The Foundations of Geometry.* The Open Court Publishing Company.

Iliopoulos, J. 2017. *The Origin of Mass.* Oxford University Press.

Marion, M. 1998. *Wittgenstein, Finitism, and the Foundations of Mathematics.* Clarendon Press.

Miller, I. 1986. *Imagery in Scientific Thought: Creating 20th-Century Physics.* MIT Press.

Monk, R. 1990. *Ludwig Wittgenstein: The Duty of Genius.* Free Press.

Poincaré, H. 1902. *La Science et l'hypothèse.* Flammarion.

Silberstein, L. 1917. The motion of the perihelion of Mercury deduced from the classical theory of relativity. *Monthly Notices of the Royal Astronomical Society* 77, 503–510.

Steiner, G. 2014. *The Poetry of Thought: From Hellenism to Celan.* New Directions.

Waismann, F. 1979. *Wittgenstein and the Vienna Circle.* Translated by J. Schute and B.F. McGuiness. Blackwell.

Wigner, E. P. 1960. The unreasonable effectiveness of Mathematics in the physical sciences. *Communications of Pure and Applied Mathematics* 13, 1–14.

Wikipedia. n.d. *Fokas Method.* Accessed February 2019. https://en.wikipedia.org/wiki/Fokas_method.

Wittgenstein, L. 1922. *Tractatus Logico-Philosophicus.* Routledge & Kegan Paul.

Wittgenstein, L. 1956. *Remarks on the Foundations of Mathematics.* MIT Press.

19 HOW CAN THE PROBLEM OF CONSCIOUSNESS BE SOLVED?

Is mathematics of vital importance for understanding consciousness? According to many distinguished neuroscientists, as already mentioned in several parts of this volume, the answer is yes. For example, Stanislas Dehaene states:

"Only mathematical theory can explain how the mental reduces to the neural. Neuroscience needs a series of bridging laws, analogous to the Maxwell–Boltzmann theory of gases, that connect one domain with the other" (Dehaene, 2014: p. 163).

Before concentrating on the possible value of mathematics in solving the mystery of consciousness, it is useful to discuss the role of mathematics in physics versus its role in biology.

THE ROLE OF MATHEMATICS IN PHYSICS VERSUS BIOLOGY

Israel M. Gelfand is recognized as one of the greatest mathematicians of the 20th century. It remains mostly unknown in the mathematics community that Gelfand also made important contributions to biology, in particular to the study of the cerebellum (Arshavsky *et al.*, 1986), as well as to the problem of protein folding.

Since a given protein is uniquely identified by the sequence of its amino acids, it should be possible to predict the three-dimensional structure of a protein purely from the knowledge of its amino acid sequence. However, no such predictive algorithm is available, despite the involvement of many leading biologists. The problem of designing an algorithm for predicting the topology of a protein, i.e., understanding how a protein "folds", is known as "protein folding". It remains one of the most important problems in the history of biology. This problem was essentially solved in 2020, via the AlphaFold programme developed by the group of Demis Hassabis (Senior *et al.*, 2020).[1] This represents an astounding achievement of Artificial Intelligence.

Solving the problem of protein folding will have a significant impact in clinical medicine. In particular, as mentioned already in this volume, several degenerative neurological diseases are, apparently, caused by the misfolding of specific proteins. For example, an important feature of Parkinson's disease is the occurrence of Lewy bodies, which, as noted in Chapter 16, is caused by the misfolding of the protein α-synuclein. Interestingly, the importance of this process was first understood by studying a familiar type of Parkinson's disease caused by the mutation of the gene encoding this protein. Although patients with the typical form of the disease, the sporadic form, do not possess this mutated gene, they still have the defect of a misfolded α-synuclein. Remarkably, mutations in the gene encoding DJ-1, which apparently assists the process of normal protein folding, also cause a familiar form of Parkinson's disease. Thus, protein misfolding is a vital component of Parkinson's pathophysiology.

Incidentally, insight gained from rare familiar forms of Parkinson's disease has been quite helpful for understanding the prevalent sporadic form. For example, the discovery that a familiar form of Parkinson's disease involves a mutation in the PTEN-1 gene that encodes a particular mitochondrial protein kinase, led to the understanding that mitochondrial pathology contributes to Parkinson's disease. Similarly, the discovery that the mutation in two distinct genes (the genes encoding the proteins parkin and ubiquitin

[1]Demis Hassabis co-founded the start-up company DeepMind in 2010. It was bought by Google for 400 million pounds in 2014, but it is independently operated.

C-terminal hydrolase L1), which are involved in the mechanism of protein degradation cause familiar Parkinson's disease, suggests that abnormalities in the clearance of proteins are important in the disease (Feany, 2004).

In what follows, I will briefly discuss my collaboration with Gelfand on the analysis of an important class of proteins known as "sandwich proteins". Antibodies are examples of such proteins. Taking into consideration that antibodies are crucial for therapeutic purposes and also are important for identifying or locating intracellular and extracellular proteins, the significance of understanding "how these protein fold" becomes apparent. The author, along with Gelfand and collaborators (Fokas *et al.*, 2004, 2005; Kister *et al.*, 2006) discovered a set of rules that can be used for predicting the topological structure of sandwich proteins. More precisely, these rules make it possible to predict the possible arrangements of strands in the two β-sheets forming a given sandwich protein. As a consequence of the validity of these unexpected rules, the number of permissible arrangements is dramatically reduced. By combining our rules with the powerful mathematical technique of "optimization", the chemical engineer Christodoulos Floudas and his group at Princeton University, were able to construct an algorithm predicting the three-dimensional structure of sandwich proteins with a high rate of success. Floudas wrote: "We are now able to predict the 3D structure of beta proteins with a success rate of 80%. This is based on the seminal works of Fokas and Gelfand who discovered unexpected topological properties of these proteins, which we have incorporated in our optimization mathematical model" (Floudas, 2015).

I was very surprised that Gelfand, unfortunately, refused to follow my suggestion of combining our protein-folding rules with the mathematical language of optimization, which, as shown later by Floudas and his collaborators, was indeed needed for making further progress. Gelfand's refusal was consistent with the fact that he lived two *parallel* academic lives; one in mathematics and one in biology. He dogmatically refused to use mathematics in biology, believing that "the existing mathematics is not the correct language for biology". This claim is contradicted by the plethora of examples where mathematics led to important insight in both biology and medicine. However, in my opinion, Gelfand's position does contain an element of truth. Indeed, it appears that, *although the role of mathematics in physics is decisive, its role in biology appears to be only supportive.*

In this regard, it is useful to recall that *physics is invariant and predictive.* This is the result of the *existence of fundamental physical laws.* As discussed in Chapter 18, mathematics is crucial both for formulating and analyzing these laws. Moreover, in the case of quantum mechanics, mathematical notions provide the *only way of introducing certain basic physical entities.* Specifically, the fundamental physical entity needed for the characterization of a quantum mechanical state is not defined via physical considerations but via the solution of a particular mathematical equation, the Schrodinger equation (this will be discussed in detail in the second volume). In this way, mathematics reveals the "secrets of nature" that would have remained forever hidden without its astounding powers.

In contrast to physical theories, the theory of biological evolution, while providing a useful framework for understanding biology, is non-predictive. *The modus operandi of biological evolution appears to be "trial and error". Moreover, new solutions are usually sought within the constraints of what already has been constructed.* An example of the latter tendency is provided by the development in humans of phonation and articulation. At some point in evolution, the larynx dropped deeper, it took a 90-degree turn, and divided into two tubes of approximately equal length. This was crucial for the emergence of language, but the close proximity of the systems of breathing and talking increases the risk of choking.[2] This development illustrates that evolution tinkers what already exists instead of creating a completely new system (Marcus, 2009: p. 107). Evolution's trial and error approach explains the abundance in biology of *redundancy*, namely, the fact that the same task can be accomplished with varying degrees of efficiency, via completely different mechanisms. Evolution invents new improvements, which are used within the existing framework, without necessarily abandoning the use of older inventions. *This gives rise to overlapping systems of high redundancy and immense "illogical" complexity.* An example of this process is provided by a plethora of completely different mechanisms used for neuronal plasticity and employed for learning, as

[2]The epiglottis, which is in the pharynx, protects the *trachea* (windpipe) during swallowing, directing food to the *oesophagus*.

discussed in Chapters 9 and 10. This crucial feature of biological evolution, in addition to arguing against the position that organisms are the result of a "grand design", presumably eliminates the possibility that fundamental biological processes follow *global fundamental laws*. Since mathematics reveals its power precisely in relation to such laws, these arguments suggest that perhaps mathematics cannot have the pivotal impact on biology that it has had on physics.

Incidentally, in contrast to biology, physics, apparently, does *not* employ redundancy. Physical phenomena are described by *unique* laws. Actually, the controversy generated by the "dual" nature of light, namely, the fact that light is both particles and waves (which, as discussed in Chapter 13, was finally resolved by Einstein), emphasizes precisely the belief of physicists that the nature of every physical phenomenon can be explained via a *single* mechanism.

As stated in Chapter 4, this trial and error *opus operandi* of biology is consistent with the "bag-of-tricks hypothesis", which claims that biological organisms have evolved a *specialized set of optimal algorithms for* solving *particular problems*. Anthony Zador has claimed that this *highly specialized* nature of biological algorithms explains their great efficiency in comparison to algorithms of artificial intelligence, which, at least until recently, are designed to solve a *wide range* of problems (Zador, 2016). Another difference with physics is that biology does not follow the model of the "scientific revolutions", described by Thomas Kuhn. This has been emphasized by the leading evolutionary biologist Ernst Mayr (1904–2005) in *The Growth of Biological Thought* (Mayr, 1982).

Of course, it should be remembered that *mathematics is the language of complexity*. Therefore, the modelling of any sufficiently complex process certainly benefits from mathematics.

In summary, *the lack of global fundamental laws in biology, suggests that, apparently, mathematics cannot have in biology the crucial impact it has had in physics. On the other hand, the modelling of a variety of complicated biological processes benefits decisively from the use of mathematics.*

The diverse applicability of mathematics in biological processes becomes evident, for example, in the works of Joseph Keller, who is recognized as the preeminent applied mathematician of his time. Keller's attitude was the opposite to that of Gelfand. He was one of the strongest exponents of the broad importance of mathematics in applications in general, and in biology in particular. I was fortunate to collaborate with Keller and Bayard Clarkson, a physician at Memorial Sloan Kettering Cancer Center, USA, and one of the world experts on chronic myelogenous leukaemia (CML). More precisely, in the late 1980s, first, we formulated a mathematical model for *granulopoiesis*, i.e., a model for the formation of granulocytes (*poiesis* comes from the Greek word *poiēsis* meaning creation). A granulocyte is one of the two subtypes of a white blood cell (the other subtype being a monocyte).[3] Then, by altering the parameters of this model, we showed that it is possible to predict the dramatic increase in granulocyte production observed in CML (Fokas *et al.*, 1991). The creation of granulocytes is a dynamic, very complicated process. Therefore, it is not surprising that the mathematical modelling of normal, as well as abnormal granulopoiesis, yields useful biological insight.

In general, the usefulness of a mathematical model is measured by the insight gained as a result of the use of mathematics. For example, regarding our model for granulopoiesis, it is noted that the progenitors of granulocytes are myeloblasts, promyelocytes, myelocytes, and metamyelocytes. A promyelocyte is formed from a myeloblast via the process of division. Prior to our work, it was thought that similar considerations were valid for the other cells involved in the creation of a granulocyte. However, our mathematical model could match the experimental data *only* if metamyelocytes could be created from myelocytes via *two* different mechanisms: either via division or via the process of maturation, without division. Observations during the filming of the process of granulopoiesis confirmed this mathematical prediction. In addition to this insight into the mode of genesis of metamyelocytes, this model has provided the basis for several studies regarding the dynamics of CML.[4]

---

[3]Similar models can be developed for the other two main categories of blood cells (in addition to the white cells), namely, red blood cells (erythrocytes) and platelets (thrombocytes).
[4]For example, it has been used for the dynamic characterization of particular subpopulations of cells derived from the bone marrows of CML patients, which do not express the cell markers of mature myelocytes. In addition, it has been used for the elucidation of the role of the so-called *human c-kit* ligand, which is an important molecule affecting normal and abnormal granulopoiesis.

Incidentally, thinking of the decisive role of mathematics in physics brings into mind Galileo Galilei (1564–1642), the "father of modern physics". The following statement is considered by many scholars the first clear proclamation of the importance of mathematics for understanding the physical world:

"Philosophy is written in this grand book — the universe I say — that is widely open in front of our eyes. But the book cannot be understood unless we first learn to understand the language, and know the characters, in which it is written. It is written in the language of mathematics, and its characters are triangles, circles, and other geometric figures, without which it is humanly impossible to understand a single word of it; without these, it is like wandering in vain in an obscure labyrinth".

However, this insight is actually much older. Indeed, Bertrand Russell, considered Pythagoras one of the greatest intellectuals of all time precisely because Pythagoras claimed that "the universe could be understood mathematically". According to this ancient Greek philosopher, "everything is made of numbers". Later, the neo-Platonic philosopher Iamblichus (295–345 AD) wrote:

"The shapes created [by the orbits of the planets] and the forces which exist between them, as well as the illuminations of the moon and the order of the spheres, and the distance between them, and the centers of the circles on which they move, everything is expressed in numbers".

In addition, the exceptionally prolific philosopher Proclus (412–484 AD), who wrote extensively on Plato, Aristotle, and Euclid, wondered: "How is the sensible world organized? According to what principles? What principles was it born from, if not from mathematical ones?"

Mathematical modelling of aspects of consciousness

The apparent lack of underlying fundamental laws in biology does not mean that certain aspects of neuronal activation cannot be modelled. For example, in 1986, psychologist Bernard Baars proposed the "global workspace theory". This is an architectural model of cognition based on the following simplistic abstraction of the brain's functions: the brain is modelled via a plethora of local processors, each capable of a specific operation. At any given moment, the workspace selects a subset of these processors, extracts a representation of the information they contain, and then disseminates this information to all the processor modelling the brain (Baars, 1988). Dehaene and Jean-Pierre Changeux presented a generalization and a specific implementation of Baars' model, called the "global neuronal workspace theory". Specifically, they modelled processors by neural networks organized in cortical columns. Moreover, they modelled additional anatomical and functional aspects of the brain, including multiple connections at the thalamocortical level. Numerical simulations of the Dehaene–Changeux model have produced neuronal activation patterns similar to the patterns that are known to be associated with consciousness (Dehaene, 2014: pp. 180–186). There exist several other modelling approaches to consciousness including the approach of University College's British neuroscientist Karl Friston, which is based on the "free-energy principle" (Friston, 2010). Also, Christos Papadimitriou is in the process of constructive a comprehensive model. Despite the success of such models to capture aspects of the function of the brain, they could not answer the fundamental question, "What is the specific neuronal mechanism giving rise to consciousness?"

Incidentally, some authors, like Antonio Damasio and Friston, regard self-consciousness a necessary ingredient of consciousness (Friston, 2018). In particular, Damasio states that consciousness is "the self in the act of knowing". Damasio also writes: "I doubt that the neural basis for the conscious mind can be comprehensively elucidated without first accounting for the self-as-object-the material-me and for the self-as-knower" (Damasio, 2012). On the other hand, Gerald Edelman requires self-consciousness only for the occurrence of "an advanced form of consciousness" that he refers to as "higher-order conscious" ("to be conscious of being conscious"). In addition, Edelman associates higher-order conscious with the "internal ability to deal with tokens of symbols", placing special emphasis on language (Edelman, 2005: p. 9).[5] As already

[5]Ravi Valluri endorses the importance of language, stating that "It seems likely that fully modern language and higher-order consciousness were, as Edelman argues, linked; it is impossible to have one without the other" (Valluri, 2016: p. 189).

noted in Chapter 4, I am in agreement with several other authors who do not consider self-consciousness a requirement for consciousness.

Sir Roger Penrose's approach to deciphering consciousness

I believe that,

a measure of the significance of a specific notion is its relationship with functions of the brain. The deeper the relationship, the higher the importance.

Within this framework, the immense importance of *algorithms* becomes evident: many of the metarepresentations constructed by the brain take the form of the triptych of mathematics-computing-technology, which is driven by algorithms. As a mathematician, I have a huge appreciation for the rational; thus, it was difficult for me to look beyond the certainty of what is strictly logical and algorithmic. Actually, in my opinion, *this bias is a* characteristic feature of most of the current approaches to consciousness. Today's Western thinking tends to focus only on the importance of the algorithmic.

Having overcome the above bias, I am now convinced of *the existence and huge importance of a non-algorithmic mode of thinking.* The capacity of the brain to "think" non-algorithmically was proposed several years ago by the versatile and deep thinker, Sir Roger Penrose (Penrose, 1989). Understanding the importance of the issue of whether the brain "thinks" algorithmically, Harari in his *Homo Deus* (Harari, 2015) states: "At present we have no idea how or why data flows could produce consciousness and subjective experience. Maybe we will have a good explanation in twenty years. But maybe we will discover that organisms aren't algorithms at all".

Regarding Penrose's proposal, there is a need, in my opinion, to clarify the relationship between *our thinking process* on the one hand and *the operations of the brain*, on the other. In what follows, I will refer to our thinking based on logical considerations, including mathematics and computations, as *algorithmic thinking.* There is no doubt that the *algorithmic thinking* of humans is *not* capable of fully capturing several aspects of reality. Regarding language this was first appreciated by Ludwig Wittgenstein. Then, as discussed in Chapter 7, this fact was rigorously demonstrated for mathematics by Gödel, and for computing by Turing. However, humans *can* decide intuitively about the validity of some of the statements that are mathematically undecidable. This implies that some of the brain's operations do *not give rise* to algorithmic thinking. Actually, more generally, such operations cannot be *reflected* in literal terms. I will call such operations *transcendental. These operations are the result of processes in the mysterious world of the unconscious and beyond.*

Penrose, aware of Gödel's result as well as of the ability of humans to decide intuitively about the validity of some of the propositions that mathematics cannot provide answers to, inferred that the brain has the ability to "think" non-algorithmically. In other words, from the fact that *we* can think non-algorithmically, he concluded that the *operations* which I call transcendental are non-algorithmic. However, *since no one has established a one-to-one correspondence between the way we think and the way that the brain operates, it does not follow that our ability to think non-algorithmically implies that the brain's transcendental operations are also non-algorithmic.*

Penrose made a serious attempt to "locate" the part of the brain where transcendental operations actually occur. In this direction, Sir Roger made a further questionable assumption. Namely, motivated by the existence of fundamental laws in physics, he assumed that brain's operations can also be described in terms of fundamental laws. However, as stated earlier, due to the "trial and error" approach of evolution, it is doubtful that this assumption is valid. The fact that the brain's operations are based on physical and chemical processes does *not* imply that these operations can be adequately described via the concrete formulation of well-defined laws. I have always been attracted to the rationally aesthetic appeal of *materialistic reductionism*, namely, the dogma which claims that insight can be reduced to a well-defined set of principles and appropriate syllogisms. However, as already noted earlier, the "trial and error" nature of biological evolution, suggests the need for a broader perspective.

Penrose, consistent with his assumption that the non-algorithmic way of our thinking can be described by neuronal mechanisms which obey non-algorithmic laws, attempted to find a physical realm which *cannot* be described by algorithmic physical laws. Does such a physical realm exist?

According to Penrose, it is the "fine grain space-time" of the quantum world. Interestingly, when Susan Blackmore asked him about his choice of this particular world, his answer was quite revealing. Following the deductive argument style reminiscent of Sherlock Holmes' that, "if all other possibilities are eliminated then even the most improbable must be the correct one", Penrose stated:

"Where do you see non-computability in physics? You do not see it anywhere else [other than the quantum world], so this, therefore presumably is where it is" (Blackmore, 2005: p. 181).

Sir Roger knows that quantum mechanics *is* dictated by algorithmic laws, albeit probabilistic ones, so he had to go beyond the known world of quantum mechanics, and indeed he did. His co-author, the anaesthesiologist Stuart Hameroff, stated this position very clearly:

"this non-algorithmic, non-computable-factor which distinguishes our choice from those of computers, is due to the brain's operations taking place in certain tiny structures called *microtubules*. There, 'non-computability' is due to the fine grain of space-time geometry, superimposed to the deterministic quantum mechanics" (Blackmore, 2005: p. 120).

What are the microtubules? They are hollow cylindrical polymers made up of an individual protein called *tubulin*. This protein can exist as a superposition of two slightly different shapes. According to Penrose and Hameroff, the microtubules are ideal for the occurrence of two fundamental processes taking place in the quantum world, namely of "superposition", as well as of the "collapse" of Schrödinger's wavefunction.[6]

The arguments presented so far, suggest that Sir Roger made two questionable assumptions: first that our non-algorithmic thinking is

[6]In this connection, it is important to note that according to Sir Roger, this collapse is *not* the result of measurements (as commonly assumed) but the effect of gravity: if a particle is in a superposition of being simultaneously in two different space-time geometries, the curvature of the surrounding space-time will be affected by where the mass is more likely to be. The incongruities of these geometries will finally reach a critical state, and this will cause collapse. Then, according to Penrose, consciousness will emerge. Sir Roger acknowledges that he cannot provide more details of this process since the development of a theory of quantum gravity remains elusive.

due to non-algorithmic operations of the brain that are dictated by non-algorithmic laws, and second that the brain's functions can be comprehensively characterized via global fundamental laws.

In addition to these assumptions, there are further problems with Penrose's arguments. First, the "fine-grain of space-time geometry", usually referred to as "quantum gravity", remains *terra incognita*. However, it is expected that even in this physical reality there *do* exist mathematical laws. Discovering these laws is part of physics' holy grail of developing a mathematical formalism that will unify all physical interactions (this was named by my colleague in the Academy of Athens, Dimitris Nanopoulos as the "Theory of Everything"). Second, there are specific estimates suggesting that the relevant time and space scales valid in the microtubules are inconsistent with the scales needed for Penrose's model.

In my opinion,

deciphering the brain's operations is essentially equivalent to the elucidation of the fundamental mechanisms that give rise to the continuum of unconscious–conscious processes.

Therefore, not only does the question of understanding the brain's transcendental operations remain a mystery but identifying the subset of the brain's operations that generate the algorithmic part of our thoughts also remains open.

Overall, regarding the claims that quantum gravity is at the heart of consciousness, I am reminded of the answer of the physicist, Eugene Wigner, who when asked whether Physics could explain consciousness, replied: "Physics cannot even explain physics", let alone consciousness.

Incidentally, it seems ironic that the great thinkers Francis Crick and Gerald Edelman were scornful of Penrose's theory. Penrose's intuition that the brain's operations *cannot* be described by laws that take a mathematical form is certainly closer in spirit to the evolutionary approach to the function of the brain embraced by the above great thinkers than the approach advocated by their closest disciples, Christof Koch and Giulio Tononi, respectively. As stated in Chapter 7, these leading neuroscientists now claim that the brain's operations can be adequately described by the fundamental law of "integrated information theory", which takes a simple mathematical form!

Interestingly, Gerald Edelman also claimed that the brain uses two types of operations, namely, "logic" and "pattern recognition". He stated: "Two main modes of thought, logic and pattern recognition [...]. For example, how to choose axioms is via intuition" (Edelman, 2005: p. 147). Also, he noted that "Brain's *prima facie* mode of operation is not by logic but rather via pattern recognition". However, to my knowledge, no attempt was made by Edelman to explain the different neurophysiological mechanisms characterizing these two modes.

THE NECESSITY OF THE PROPER DEFINITION OF CONSCIOUSNESS

How can the problem of consciousness be solved? From the way that this question is posed, it is clear that I have *assumed* that this problem *can* be solved. I believe that this is indeed the case, provided that a vitally important clarification is achieved, namely, the difference between the *defining property* of consciousness on the one hand, and its astonishingly rich *manifestations* on the other, is fully elucidated. My position is antithetical to the claim of some scholars, known as *mysterians*, who, following Noam Chomsky, believe that consciousness belongs to the class of problems which are "mysteries", i.e., it cannot be solved. Several arguments against this unproven claim of the mysterians are presented by the leading philosopher of consciousness Daniel Dennett (Dennett, 2017).

Incidentally, the more elaborate the conscious functions, the more global, in space and time, the corresponding activation of the brain. This implies that statements claiming that consciousness is associated with only particular areas of the cortex become increasingly unfounded. Actually, "virtually all the brain's regions can participate in both conscious and unconscious thought" (Dehaene, 2014: p. 53).[7]

---

[7]Francis Crick proposed that a particular area of the brain, the *clustrum*, is vital for consciousness. Although electrical stimulation of this area reversibly disrupts consciousness (Koubeissi *et al.*, 2014), it is now accepted that virtually any part of the cortex can support consciousness.

It is clear that conscious processes have evolved. In particular, the interaction of metarepresentations with basic neural mechanisms has enormously enriched the scope of consciousness. This makes it imperative to identify the *defining property* of consciousness. Indeed, without first defining the *essence* of consciousness, terms like "self-consciousness" and "higher-order consciousness" cannot be properly understood. Furthermore, only after the formulation of such a strict definition does the important question of "which is the most primitive organism possessing consciousness?" become well-posed.

As a result of the lack of an accepted definition of consciousness there is currently a wide range of answers to this question. For example, according to Gerald Edelman, lobsters do not possess consciousness, whereas Baroness Susan Greenfield claims that "anything with any brain has consciousness" (Blackmore, 2005: p. 97). Interestingly, since 1986, in the United Kingdom, octopi (*Octopus vulgaris*) have been classified as "honourable vertebrate" in the sense that they have the same legal protection as mammals, birds, and reptiles.

The evolution of consciousness has introduced the need for assigning more and more characterizations to conscious processes in order to describe its emerging capabilities. For example, consciousness has been characterized as "an arbiter" needed to resolve conflicts. This is indeed what happens in the case of binocular rivalry. Also, the role of consciousness in elaborate forms of planning (typically associated with humans) is captured by the following statement: "The ability to integrate information over time and space by employing different modalities, to retain it for a sufficiently long time and to recall it in the future, is a crucial component of consciousness" (Dehaene, 2014: p. 101). I believe that such statements are useful for characterizing the remarkable enrichment of conscious processes, but they do *not* describe the essence of consciousness.

As Benjamin Libet emphasized, the basic property of consciousness is *awareness*. Defining consciousness as awareness, the problem of consciousness can be solved, in my opinion, following three steps:

First, *isolate objective "signatures" of awareness*. Second, *identify the earliest organism possessing this defining property of consciousness*. Third, *decipher the specific neural mechanisms giving rise to awareness*.

The implementation of the first two steps is within reach. Indeed, several neuroscientists, and in particular Stanislas Dehaene and his collaborators, have introduced ingenious techniques for distinguishing conscious from unconscious processes (Dehaene's discussion of "signatures' of consciousness" is outlined in Chapters 4 and 6 of his book (Dehaene, 2014)). The difficulty of distinguishing conscious from unconscious processes becomes evident with the following observation: until recently it was assumed that activation in the frontal lobe, expressed in the form of neural synchronization, is a characteristic of consciousness. However, it is now known that unconscious processes can also be accompanied by increased synchronization in the frontal lobe (Greenfield, 2016).

Identifying appropriate signatures of consciousness should allow investigators to search systematically for the earliest organism that had the historic privilege to experience, for the first time, the miracle of awareness. Undoubtedly, this organism is quite primitive. Identifying differences between this organism and its evolutionary predecessor should subsequently facilitate the search to decipher the neurological basis of awareness. *I expect that this breakthrough will finally be achieved as a result of painstaking analysis of concrete neural circuits, as opposed to the use of "general principles" and fancy mathematical formalisms such as "integrated information theory" or "quantum gravity".*

The first two steps needed for solving the mystery of consciousness should be considered within an evolutionary framework, perhaps starting with the sponges. As Aristotle first realized, sponges constitute the simplest multicellular "animals". These organisms lack nerve and muscle cells. Nevertheless, in some sponges, coordinated signals can propagate over the whole cellular net. Nerve cells appeared in the next level of sophistication, namely in *coelenterates* and *ctenophores*, which include jellyfish and sea anemone. These phyla were followed, among others, by the *platyhelminths*, including flatworms and nematodes. Then, 570 million years ago, the Precambrian era came to an end and our closer relatives appeared including *molluscs* (clams, snails, squid) and *arthropods* (insects, spider). The nervous system of these more advanced organisms can support learning, communication, and social behaviour.

A vital role in such an evolutionary framework will be played by genetic studies. For example, the gene ASPM, which occurs in many organisms, from humans to nematode worms, apparently regulates the number of neuronal stem-cells divisions that occur at the very early stage of the genesis of an organism (Ridley, 2003: p. 110).[8]

Incidentally, taking into consideration the vital importance of homeostasis, it is natural to expect that, as noted in Chapter 5, the first instance of awareness is associated with a particular homeostatic mechanism. Presumably pain, temperature, or thirst.

After the basic three steps for identifying the nature of awareness are implemented, there will be a need to model some of the *global manifestations* of consciousness. Such situations are characterized by high complexity, and therefore, since as stated earlier, mathematics is the language of complexity, mathematical formalisms, including integrated information theory, should be useful. In particular, employing the mathematical techniques of "chaos theory" will be beneficial. Indeed, concepts and tools of this theory have already been quite useful in studies of global properties of the brain. For example, the concept of "an emergent property" is borrowed from the mathematical theory of chaos. Similarly, "synchronization" is an example of global network dynamics, which results from many dynamic network interactions.

Incidentally, the mention of "chaos" generates an interesting association between two disparate areas: chaotic dynamical systems and different phenotypes of a given biological organism. Indeed, the defining characteristic of chaotic systems is "their exquisite sensitivity on initial data". This means that a slight change in initial conditions gives rise to a completely different outcome. Taking into consideration that a variety of human physiological mechanisms can be modelled through chaotic dynamical systems, it is not surprising that humans exhibit such broad diversity of phenotypes.

---

[8]The important role of mitochondria in evolution has been elucidated by Nick Lane (Lane, 2005).

It is worth noting that the definition of consciousness as awareness also covers the so-called "altered states of consciousness", including the following:

- Dreaming, in which the capacity of the brain for creating unconscious structures and mental images remains intact.
- The generation of internal imagery as a result of sensory deprivation or of meditative techniques.
- Near-death experiences.
- Hallucinations, for example occurring in schizophrenia and temporal lobe epilepsy.
- The geometrical objects seen during the prodromal state of classical migraines.
- The imagery created by some psychotropic substances.

It is noted that there have been several reports concluding that normal subjects placed in a sound-proof, dark room, start, after a short time, to report hallucinations (Daniel & Mason, 2015). Similarly, "self-imposed" sensory deprivation is used in many Eastern meditative techniques, where practitioners are able to shut out environmental stimuli and concentrate only on a repeated mantra or visual symbol.

Regarding hallucinations, it is recalled that memory involves the *reactivation* of particular neural circuits. This process is characterized by ambiguity and lack of precision, prompting Gerald Edelman to state that, "Every act of remembering is an act of imagination" (Edelman, 2006). In mental disorders, this 'act of imagination' is exacerbated, reaching a pathological form.

With respect to psychotropic substances, it is noted that *yaje* (commonly known as ayahuasca) is a psychedelic vine used extensively by the Tukano people. This vine causes "grid patterns, zigzag lines and undulating lines alternating with eye-shaped motifs, many-coloured concentric circles or endless chains of brilliant dots" (Valluri, 2016: p. 131). Interestingly, some of these geometrical objects are similar to those seen before the headache that occurs in classical migraine attacks.

Concluding this chapter, it should be emphasized that, with the emergence of awareness, evolution began a truly transformative process. Namely, nature was able to supplement "competence" with "comprehension". Indeed, up to that point, organisms could survive and reproduce, but they could *not* comprehend. Of course, the *motto* of "competence without comprehension" continues to dominate many functions of the organisms that possess consciousness. Indeed, for such organisms, Nature makes extensive use of the "Need to Know" principle. This implies that the "comprehension" of any organism, including humans, is highly incomplete. For example, as emphasized throughout this volume, we are unaware of a plethora of complicated neural mechanisms keeping us alive, as well as solving (on our behalf) many difficult cognitive problems.

The emergence of comprehension not only allows us to scrutinize the external world but also to penetrate deeper and deeper into the essence of our humanity. Among the most important metarepresentations enabling us to achieve comprehension are those of computing and technology. Interestingly, artificial intelligence (AI), which provides the most advanced synthesis of these two powerful metarepresentations, provides an example of the "negation" of the advanced state of "competence *with* comprehension". Indeed, AI can be considered a remarkably efficient and powerful creation of "competence *without* comprehension". The relationship between artificial and human intelligence will be discussed in the second volume.

References

Arshavsky, Y., Gelfand, I., & Orlovsky, G. 1986. *Cerebellum and Rhythmical Movements (Studies of Brain Function).* Springer-Verlag.

Baars, B. 1988. *A Cognitive Theory of Consciousness.* Cambridge University Press.

Blackmore, S. 2005. *Conversations on Consciousness.* Oxford University Press.

Damasio, A. 2012. *Self Comes to Mind.* Vintage Books.

Daniel, C. & Mason, O. 2015. Predicting psychotic-like experiences during sensory deprivation. *BioMed Research International* 2015(1), 439379.

Dehaene, S. 2014. *Consciousness and the Brain: Deciphering How the Brain Codes Our Thoughts.* Viking.

Dennett, D. 2017. *From Bacteria to Bach and Back.* Penguin Press.

Edelman, G. 2005. *Wider than the Sky: A Revolutionary View of Consciousness.* Penguin.

Edelman, G. 2006. *Second Nature.* Yale University Press.

Feany, M. 2004. New genetic insights into Parkinson's disease. *New England Journal of Medicine* 351(19), 1937–1940.

Floudas, C. 2015. Optimization in medicine and energy. *Proceedings of the Academy of Athens* 90A(138), 129–144.

Fokas, A., Gelfand, I., & Kister, A. 2004. Prediction of the structural motifs of sandwich proteins. *PNAS* 101, 16780–16783.

Fokas, A., Keller, J., & Clarkson, B. 1991. Mathematical model of granulocytopoiesis and chronic myelogenous leukemia. *Cancer Research* 51(8), 2084–2091.

Fokas, A., Papatheodorou, T., Kister, A., & Gelfand, I. 2005. A geometric construction determines all permissible strand arrangements of sandwich proteins. *Proceedings of the National Academy of Sciences* 102(44), 15851–15853.

Friston, K. 2010. The free-energy principle: A unified brain theory? *Nature Reviews Neuroscience* 11, 127–138.

Friston, K. 2018. Am I Self-Conscious? *Frontiers in Psychology* 9, 579.

Greenfield, S. 2016. *A Day in the Life of the Brain.* Allen Lane.

Harari, Y. N. 2015. *Homo Deus: A Brief History of Tomorrow.* Harvill Secker.

Kister, A., Fokas, A., Papatheodorou, T., & Gelfand, I. 2006. Strict rules determine arrangements of strands in sandwich proteins. *Proceedings of the National Academy of Sciences* 103(11), 4107–4110.

Koubeissi, M., Bartolomei, F., Beltagy, A., & Picard, F. 2014. Electrical stimulation of a small brain area reversibly disrupts consciousness. *Epilepsy & Behavior* 37, 32–35.

Lane, N. 2005. *Power, Sex, Suicide: Mitochondria and the Meaning of Life.* Oxford University Press.

Marcus, G. 2009. *Kluge.* Faber and Faber.

Mayr, E. 1982. *The Growth of Biological Thought.* Belknap Press.

Penrose, R. 1989. *The Emperor's New Mind.* Oxford University Press.

Ridley, M. 2003. *Nature via Nurture.* Harper Collins.

Senior, A. W., Evans, R., Jumper, J., Kirkpatrick, J., Sifre, L., Green, T., & Hassabis, D. 2020. Improved protein structure prediction using potentials from deep learning. *Nature* 577(7792), 706–710.

Valluri, R. 2016. *The Matter of Mind.* The Write Place.

Zador, A. 2016. The connectome as a DNA sequencing problem. In: Marcus, G., & Freeman, J. (eds.), *The Future of the Brain.* Princeton University Press.

EPILOGUE

As stated repeatedly in this volume, consciousness works towards presenting a *complete* and *unified* picture of reality. This tendency is clearly reflected both in the sciences and humanities. Examples include, respectively, the theory of evolution and various ideologies, such as Marxism, Existentialism, and several other "isms". Although I understand and admire attempts towards unification, I believe that formulating such "complete" theories and ideologies is at best utopic and at worst detrimental to society. Indeed, taking into consideration the enormous complexity of physical and biological processes, it follows that even comprehensive theories such as the "theory of everything", which aims to present a unified formalism of the laws dictating physical reality, cannot be "complete". There will always exist "local" fundamental physical mechanisms that remain elusive. As a result of the dynamic character of biological processes, the situation with biological theories is even more complicated. Regarding the *impossibility* of any ideology to fully capture reality, it is sufficient to note that a comprehensive ideology, in addition to providing a complete framework for the understanding of physical and biological processes, should also decipher the laws dictating social interactions.

In what follows, I will provide support for my position that every theory is incomplete, and every philosophy highly limited, by criticizing representative examples of well-known theories and influential philosophies. Namely, the theory of evolution, and Platonism, respectively.

According to Darwin and Wallace, biological evolution takes place within a framework of hereditary stability, where only small random variations occur over long stretches of time, dictated by natural selection. However, in response to their environmental changes, bacteria acquire new genetic traits not only via mutations and the modification of gene function within their cell but also by employing horizontal gene transfer. Actually, this mechanism, which results in the acquisition of new genes from other bacteria is the *preferred* mode of response in cases of extraordinary evolutionary pressure. For example, bacteria can survive the effects of antibiotics by employing horizontal gene transfer. In general, this powerful mechanism, completely unsuspected by Darwin and Wallace, provides a faster way of creating heritable variations in the genome of all forms of life than the traditional mechanisms of mutation and vertical inheritance. Horizontal gene transfer is of crucial importance for archaea, bacteria, and eukaryotes. For example, it has been shown that 18% of the genes of *E. coli* entered the genome of this bacterium via horizontal gene transfer. Furthermore, it was later understood that complex eukaryote organisms are also significantly affected by this mechanism. For example, it was shown in 2008 that 22 genes in the genome of bdelloid rotifers have bacterial, fungal, and even plant origin (Gladyshev *et al.*, 2008). Similar considerations are valid for the human genome. This implies, as clearly stated in an influential article by Ford Doolittle, published in *Science* in 1999, that the universal tree of life is actually a "reticulate tree" (Doolittle, 1999). In other words,

> we are connected with myriad evolutionary predecessors, in far more complicated ways than envisioned by Darwin, Wallace, and their followers.

In this sense, "the web of life" provides a better metaphor than "the tree of life" (Quammen, 2019: p. 287).

In addition to creating completely new genetic possibilities, horizontal gene transfer also allows organisms to adapt quickly to new ecological niches without the danger of going through several phases of slow adaptation with potentially dangerous implications. The tremendous impact of this mechanism, which was widely accepted only in the early 2000s, is evident by the fact that there exist many genes whose "history" cannot be explained in the traditional way of following the evolution of related species.

For example, after analyzing 66 fundamental proteins, James Brown and Ford Doolittle concluded in 1997, that "each gene has its own history" (Brown and Doolittle, 1997). Following this new understanding, Nigel Goldenfeld and Carl Woese claimed, in an essay published in *Nature* (Goldenfeld and Woese, 2007) that,

"The emerging picture of microbes as gene-swapping collectives, demands a revision of such concepts as organism, species and evolution itself".

These authors further noted that, since microbes absorb and discard genes as needed depending on their environment, the concept of "species" is useless, and the notion of "an organism" difficult to define. In the same paper, they also wrote: "Thus, we regard as regrettable the conventional concatenation of Darwin's name with evolution because other modalities must be considered".[1]

These developments clearly show that the transformative ideas of Darwin and Wallace need to be evolved and that their theories must be replaced by a new, more dynamic framework. This is of course consistent with how science progresses.

Can we reach a "complete" evolution theory? In my opinion,

the importance of any discovery is not only measured by the significance of the questions it answers but also by the multitude of the questions it raises.

This implies that even this new dynamic framework that has replaced the Darwinian ideas remains, and will always remain, incomplete. For example, the discovery of endosymbiosis certainly elucidated the origin of two fundamental organelles of eukaryotes, namely, mitochondria and chloroplasts. However, this fundamental discovery raises key questions: What were the first recipient cells of mitochondria and chloroplasts,

---

[1]Unfortunately, around the time that this essay was written, Woese embraced the dubious ideas expressed in Roy Davies' book titled, *The Darwin Conspiracy*, where Darwin is accused of plagiarism from, among others, Jean-Baptiste Lamarck, the English zoologist Edward Blyth, his own grandfather Erasmus Darwin, and especially Alfred Wallace. There is no doubt that several scholars, starting with Aristotle, had expressed some of the ideas later developed by Darwin. Wallace, after four years of fieldwork in the Amazon and four more in Malay Archipelago, formulated the principle of natural selection independently of Darwin. However, the cordial correspondence between Wallace and Darwin provides, in my opinion, strong evidence for the fallacy of Davies' accusations.

and what was the impact of these organelles on the genesis of a nucleus in the recipient cells? In this connection, it is noted that phylogenomic analyses suggest that archaea, and in particular the *Asgard superphylum*, are the closest prokaryotic relatives of eukaryotes (Eme *et al.*, 2017). These developments are truly remarkable, taking into consideration that archaea were only discovered in 1977. In the paper (Eme *et al.*, 2017) the following statement is made, which is consistent with my position regarding the limitless nature of deep research activities:

"Fully understanding the process of eukaryogenesis requires finding answers to several challenging and intertwined questions. Although we have seemingly answered some of these questions, others remain fiercely debated, and new questions continue to arise".

Let us now turn to the limitations, as well as potential harm, caused by "complete" philosophical theories and ideologies. It is straightforward to argue that as a result of the brain's capacity for plasticity, analogical thinking, unification, and generalization, *any* philosophical theory or ideology,

will finally give rise to a highly restrictive cognitive framework. Indeed, the brain becomes trained to interpret any phenomenon within this rigid structure. This leads to highly biased and distorted views, and worst, occasionally to inhumane positions and actions.

As an example of the limitations of any philosophical theory, it is noted that Platonism is based on the apotheosis of *reason* and *permanence* (time invariance). These properties characterize mathematical rules, which explains Plato's admiration for mathematics. The brain's ability for analogical thinking and generalization, led Plato to the erroneous conclusion that *everything important can be expressed via infallible rules*. Hence, an ideal society should be governed according to a set of strict statutes. Indeed, specific mandates were presented in the Platonic dialogue *Laws*, which, if implemented, would have given rise to a joyless, authoritarian state. Incidentally, the autocratic character of this state is in contradiction with the lack of dogmatism characterizing the early Platonic dialogues. In particular, the effectiveness of the method of *elenchus* was based on the ability of Socrates to *persuade* his interlocutors, as opposed to imposing his opinion on them. Furthermore, Socrates was constantly expressing his willingness to review his arguments and re-examine his points of view.

The banning in Plato's ideal society of poetry and tragedies, from Homer to Aeschylus, is in my opinion consistent with the cognitive restrictions imposed on Plato's brain by his own ideology. Indeed, first, according to Plato the "sensible reality" is nothing but an *inferior* representation of the "true reality" contained in his Forms. Hence, since poetry constructs "fake representations of the sensible reality", poetry is doubly *inauthentic*. Second, for Plato, idealizations (the *mimetic eidola*) and fake representations of the sensible reality created by poets, not only lack authenticity but also have the ability to arouse emotions, empathies, and terrors. This ill effect, together with the claim that poetry is best suited for depicting inferior subjects, led Plato to accuse poetry of corrupting the souls of even "the best among us". So, the "narcotics of phantasms" had to be banned! Interestingly, in book III of the *Republic*, Plato does accept the educational value of poetry for the future guardians of the city. However, after reviewing in books IIX and IX the need to protect the republic from a variety of dangers, Plato announces in book X that all poets have been expelled.[2]

As discussed in Chapter 7, artistic creations are metarepresentations expressing mental images and unconscious structures. Thus, high-quality art provides examples *par excellence* of *originality*. Furthermore, the genesis of metaphors and the employment of ambiguity illustrate the astounding capacity of the brain for creativity. The restrictions imposed on the brain of Plato by his philosophical framework, and in particular by his idealization of the notion of precision, did not allow him to accept that,

> the beauty of poetry is precisely its ability to express ambivalence and metaphors in a highly economical manner.

Instead, Plato concentrated on the fact that metaphors, by definition, lack precision. Thus, he criticized poetry for presenting a distorted "representation of reality": "[…] all poets from Homer on give [only]

[2]In the *Republic*, Plato attempts to prove "the mimetic, fake character of arts in general" by using the example of painting. However, he does not ban the painters from his ideal city, because he considers their creations "merely play (*paidia*) and not anything serious (*spoude*)". Thus, in contrast to poetry, painting cannot corrupt the souls.

representations [...] art is something that has no serious value". Incidentally, in his youth, Plato wrote tragedies. This explains how painful it was for him, as he confesses in the *Republic*, to free himself from the enchantment of poetry. He wrote that, "We shall behave like lovers who see their passion is disastrous, and violently force themselves away from the object of their love".

It is worth noting that there are similarities between some of Plato's general ideas and the Christian doctrine. This is related to the fact that some of the early influential Christian theologians were Platonists. A clear illustration of the influence of Platonism on Christianity is Plato's theory of souls, summarized in Chapter 7. In his fascinating book *Hidden Harmony* (2021), the priest and theologian Nikolaos Loudovikos, argues that the *Christian orthodox dogma provides the fulfilment of deep existential questions raised by Plato in particular, and the Ancient Greek philosophy in general.* I will return to this most interesting issue in the third volume. Here, it is only noted that the relationship between ancient Greek philosophy and the Christian dogma provides yet another illustration of the notion of continuity, which often is expressed in a nonlinear way.

Incidentally, Plato's adherence to strict rules and the priority of the intellect over the flesh provides the theoretical foundation of asceticism. Christianity, for different reasons, also arrived at asceticism, emphasizing the priority of the soul over flesh. In this way, ascetism provides another similarity between Platonism and Christianity. Further connections of Platonism with religious movements are established by Iain McGilchrist in his highly erudite book, *The Master and the Emissary* (McGilchrist, 2010). For example, the iconoclastic movement was due to the inability of its followers to appreciate the metaphoric significance of icons as mere *symbols* of the divine.

The catastrophic effects of various ideologies, including those inspired by religious beliefs, are of course well documented. At an individual level, an illustrative example is provided by the "scientific racism" advocated by the polymath Ernst Haeckel, who was mentioned in Chapter 8. Haeckel's apotheosis of Darwin's dogma of "the survival of the fittest", together with the brain's capacity for analogical thinking and generalization, compelled him to propose that this dogma is also applicable in societies, thereby advocating for the killing of 200,000 mentally and congenitally ill people.

THE INSIGHT OF ALBERT CAMUS

Perhaps no one understood deeper the contradiction between the need of the consciousness to paint a "complete picture" of reality, and the impossibility of this endeavour, than the writer Albert Camus (Nobel Prize in Literature, 1957). Actually, the understanding of this contradiction provided the foundation of Camus' philosophy of *Absurdism*, which is elaborated on in his book *The Myth of Sisyphus* (Camus, 1955). In the same way that the mythological hero was condemned to a meaningless life, carrying out an "absurd task" of having to push a rock up to the top of a mountain, where upon reaching the top the rock would roll down again, forcing Sisyphus to start over, humans also face the following "absurd condition": how to resolve the conflict between the innate human need "for the absolute and for unity" and "the impossibility of reducing this world to a rational and reasonable principle". For the Nobel Laureate, this absurdity was so overwhelming that it led him to pose, what he considered to be, the *only* basic question in philosophy: "There is but a serious philosophical problem, that of suicide". It should be noted that Camus answered this question in favour of life: "Even if one does not believe in God, suicide is not legitimate". Sisyphus' desire to live, despite his tormented life, is another reason for Camus' affinity towards the mythological hero. After all, Sisyphus was punished by the gods precisely because he attempted to defeat death.

Several other writers and philosophers also noted this contradiction. Some of these philosophers were severely criticized by Camus, including Martin Heidegger, Karl Jaspers, Søren Kierkegaard, and Edmund Husserl. According to Camus, these deep thinkers committed "philosophical suicide" because they reached conclusions that contradicted their starting point of the "absurd position". In particular, Kierkegaard abandoned rationality and embraced God, while Husserl erred in the opposite direction, namely, following Plato, he based his arguments on the apotheosis of rationality.

In my opinion, none of the philosophers resolved satisfactorily the fundamental question posed by Camus. Perhaps this was the result of a limited understanding of science, and especially of neuroscience. For Camus, the main origin of the absurdity of life, in addition to the highly depressing fact that every moment of life brings us closer to our only invincible enemy, death, is his belief that "true knowledge is impossible and that science, which is based on rationality, cannot explain the world". Ironically, this part of Camus' analysis was biased by the tendency of consciousness for "the absolute and the unity", for which, he correctly, was so critical. It is, of course, true that *ultimate* knowledge and *full* understanding *cannot* be achieved. However,

the human brain can approach these goals via a limiting process, tending ever closer to an unattainable limit. Furthermore, during this journey, humans experience inexhaustible *eudemonia*, which more than compensates for the knowledge that the ultimate limit will never be reached.

Is there a framework capable of filling the huge intellectual gap arising from the apparent inadequacy of any *ideological approach to answer fundamental questions arising in nature and society?*

The material presented in this volume suggests an affirmative answer:

Let us be guided by the processes used by the brain.

Specifically, first, the analysis of any phenomenon should be based on the hypothesis that,

unconscious and conscious processes form a continuum.

Second, taking into consideration that the unconscious is less biased by the presence of pre-conceptions and the need for "completion" than consciousness, it follows that unconscious processes are more capable of reaching the essence of things than consciousness. This suggests that the best way to avoid the cognitive constraints imposed by any ideology is for us to

search for ways to submerge in our unconscious.

Third,

let us try to employ the variety of concrete notions that reflect fundamental brain mechanisms.

Some of these concepts, including *associations, continuity, local versus global processes, simplicity versus complexity, reduction versus unification, generalization, analogical thinking, plasticity, and interconnectedness*, have been articulated and employed throughout this volume. Others remain to be discovered.

Fourth,

in our search for truth, let us be guided by scientific knowledge and at the same time let us enjoy the irreplaceable *eudemonia* that accompanies deep insight. During this journey, we should appreciate that reaching the truth is a limiting process, where the limit can be approached but can never be reached.

It is my hope that this novel approach can be instrumental in eliminating various misconceptions, dangerous preconceptions, and inhumane attitudes. In what follows the usefulness of the new approach is illustrated by arguing that the concept of *continuity* strongly suggests that homosexuality is a normal gender variation. In this connection it is noted that there exist several anatomical differences between the brains of men and women. For example, in men the right hemisphere is slightly larger than the left, whereas in women the two hemispheres are more symmetrical. Also, the corpus callosum is proportionally larger in women.[3] In addition, there are several subtle differences. For example, a particular nucleus of hypothalamus, called *nucleus* $INAH_3$, which affects the regulation of sexual behaviour, is on average 2.5 times larger in men than in women. A study published in 1991, using data obtained from autopsies of homosexual men who died of AIDS, claimed that the *nucleus* $INAH_3$ is smaller in homosexual than heterosexual men (LeVay, 1991). Later, it was argued that this is due to the fact that homosexual men have higher neuronal packing density, i.e.,

[3]The corpus callosum of women is actually the same size of that of men. However, since men have larger brains than women (reflecting their larger bodies) men would be expected to have larger corpus callosum.

the number of neurons per cubic millimetre is higher in homosexual men (Byne *et al.*, 2001). In addition, an MRI study of 25 heterosexual men, 25 heterosexual women, 20 homosexual men, and 20 homosexual women, suggested that heterosexual men and homosexual women had a rightward cerebral asymmetry, whereas the cerebral hemispheres of homosexual men and heterosexual women were symmetrical (Savic and Lindsrom, 2008). Although these studies were later criticized, it appears that *certain brain anatomical structures of homosexual men are in size somewhere between heterosexual men and women.* For example, MRI studies have shown that the corpus callosum of homosexual men is larger than that of heterosexual men, particularly in the *isthmus* area, which is the posterior part of the corpus callosum, connecting the parietotemporal cortical regions.

The neuroscientist Dick Swaab, whose early studies regarding the biological basis of homosexuality were viciously attacked by various prejudiced groups, discusses homosexuality extensively in his book *We Are Our Brains* (Swaab, 2014). There, he states that studies in identical twins show that homosexuality is 50% genetically determined.[4] These and other similar studies strongly suggest that homosexuality has a biological basis. Importantly, it should be noted that,

> the notion of evolutionary continuity implies that Nature usually explores the full range of values in a given interval. The fact that the values of certain anatomical structures for homosexuals are between the values of heterosexual men and women, provides strong evidence that homosexuality is a normal gender variation.

In the same way that the notion of continuity can be used to argue against homophobia, different notions elucidated in the new approach can be used against other discriminations. For example, the notion of *interconnectedness* is useful for formulating arguments against racism. Indeed, as noted in Chapter 8, the work of the Nobel Laureate Svante Pääbo and his team suggests that cultural evolution allowed Neanderthals and our European ancestors to intermingle. The historic consequence of

[4]In the same book, several behavioural differences between boys and girls are discussed. For example, girls prefer to draw human figures in red, orange, and yellow, whereas boys prefer mechanical objects in blue.

this interconnectedness is that humans originating in Europe carry 1%–3% of Neanderthals' genes. Hopefully,

this unexpected discovery can be used as the scientific basis of a manifesto promoting the final and complete rejection of poisonous racism and the embrace of the unified nature of humanity.

Incidentally, the impossibility of defining a "pure" organism mentioned earlier in Chapter 8 and in this Epilogue, as well as the crucial role of *symbiosis* that will be illustrated in a future volume via the discussion of microbiota, imply that,

interconnectedness is not a vague philosophical notion, but the only mode that life can exist.

Having visited, during this long journey, 19 harbours, the reader is ready for deeper reflections. If studying this volume has induced eudaemonic pleasure, then certainly this will motivate the reader's quest for further contemplation. In this case, the purpose of this book will have been fully achieved.

References

Brown, J. R. & Doolittle, W. F. 1997. Archaea and the prokaryote-to-eukaryote transition. *Microbiology and Molecular Biology* 61.

Byne, W., Tobet, S., Mattiace, L. A., Lasco, M. S., Kemether, E., Edgar, M. A., Morgello, S., Buchsbaum, M. S., & Jones, L.B. 2001. The interstitial nuclei of the human anterior hypothalamus: An investigation of variation with sex, sexual orientation, and HIV status. *Hormones and Behavior* 40(2), 86–92.

Camus, A. 1955. *The Myth of Sisyphus and Other Essays.* A. A. Knopf.

Doolittle, W. F. 1999. Phylogenetic classification and the universal tree. *Science* 284(5423), 2124–2128.

Eme, L., Spang, A., Lombard, J., Stairs, C. W., & Ettema, T. J. 2017. Archaea and the origin of eukaryotes. *Nature Reviews Microbiology* 15(12), 711–723.

Gladyshev, E. A., Meselson, M., & Arkhipova, I. R. 2008. Massive horizontal gene transfer in bdelloid rotifers. *Science* 320(5880), 1210–1213.

Goldenfeld, N. & Woese, C. 2007. Biology's next revolution. *Nature* 445(7126), 369.

Huffington, A. 2016. *The Sleep Revolution: Transforming Your Life, One Night at a Time*. Virgin Digital.

Kok, B. E. 2013. How positive emotions build physical health: Perceived positive social connections account for the upward spiral between positive emotions and vagal tone. *Psychological Science* 24(7), 1123–1132.

LeVay, S. 1991. A difference in hypothalamic structure between heterosexual and homosexual men. *Science* 253, 1034–1037.

Loudovikos, N. 2021. *Αφανής αρμονία*. Αρμός.

McGilchrist, I. 2010. *The Master and the Emissary*. Yale University Press. 2008.

Quammen, D. 2019. *The Tangled Tree*. William Collins.

Savic, I. & Lindsrom, P. 2008. PET and MRI show differences in cerebral asymmetry and functional connectivity between homo- and heterosexual subjects. *PNAS* 105, 9403–9408.

Swaab, F. D. 2014. *We Are Our Brains: A Neurobiography of the Brain, from the Womb to Alzheimer's*. Random House.

Trivers, R. 2013. *Deceit and Self-Deception: Fooling Yourself the Better to Fool Others* (Original publication: 2011). Penguin Press.

Walker, M. 2017. *Why We Sleep*. Scribner.

INDEX

F

G

H

www.ingramcontent.com/pod-product-compliance
Lightning Source LLC
Chambersburg PA
CBHW071956280226
40402CB00004B/21
9781800615137